Fighting Infection in the 21st Century

Fighting Infection in the 21st Century

EDITED BY

Professor P.W. Andrew
Professor of Microbial Pathogenesis
Department of Microbiology and Immunology
University of Leicester
Leicester
UK

Dr P. Oyston
Microbiology
CBD Porton Down
Salisbury
UK

Professor G.L. Smith
Sir William Dunn School of Pathology
University of Oxford
Oxford
UK

Professor D.E. Stewart-Tull
Division of Infection and Immunity
Faculty of Biomedical and Life Sciences
University of Glasgow
Glasgow
UK

Blackwell
Science

© 2000 the Society for Applied Microbiology and
Society for General Microbiology and published for
them by Blackwell Science Ltd
Editorial Offices:
Osney Mead, Oxford OX2 0EL
25 John Street, London WC1N 2BS
23 Ainslie Place, Edinburgh EH3 6AJ
350 Main Street, Malden
    MA 02148-5018, USA
54 University Street, Carlton
    Victoria 3053, Australia
10, rue Casimir Delavigne
    75006 Paris, France

Other Editorial Offices:
Blackwell Wissenschafts- Verlag GmbH
Kurfürstendamm 57
10707 Berlin, Germany

Blackwell Science KK
MG Kodenmacho Building
7–10 Kodenmacho Nihombashi
Chuo-ku, Tokyo 104, Japan

The right of the Authors to be
identified as the Authors of this Work
has been asserted in accordance
with the Copyright, Designs and
Patents Act 1988.

First published 2000

Set by Best-set Typesetters Ltd., Hong Kong
Printed and bound in Great Britain

DISTRIBUTORS

Marston Book Services Ltd
PO Box 269
Abingdon, Oxon OX14 4YN
(*Orders*:   Tel: 01235 465500
            Fax: 01235 465555)

USA
  Blackwell Science, Inc.
  Commerce Place
  350 Main Street
  Malden, MA 02148-5018
  (*Orders*:   Tel: 800 759 6102
              781 388 8250
              Fax: 781 388 8255)

Canada
  Login Brothers Book Company
  324 Saulteaux Crescent
  Winnipeg, Manitoba R3J 3T2
  (*Orders*:   Tel: 204 837 2987)

Australia
  Blackwell Science Pty Ltd
  54 University Street
  Carlton, Victoria 3053
  (*Orders*:   Tel: 3 9347 0300
              Fax: 3 9347 5001)

A catalogue record for this title
is available from the British Library

ISBN 0-632-0581-7-X

Library of Congress
Cataloging-in-publication Data
Fighting infection in the 21st century/
edited by P.W. Andrew . . . [et al.].
    p. ; cm.
    Includes bibliographical references.
    ISBN 0-632-05817-X
    1. Communicable diseases. 2. Anti-infective agents.
    3. Vaccines. I. Andrew, P.W.
    [DNLM: 1. Communicable Disease Control—
    methods. WA 110 F471 2000]
    RC111 .F55 2000
    616.9—dc21

                                    00-057979

For further information on
Blackwell Science, visit our website:
www.blackwell-science.com

The Blackwell Science logo is a
trade mark of Blackwell Science Ltd,
registered at the United Kingdom
Trade Marks Registry

Contents

Contributors

Eileen Barry
Center for Vaccine Development, University of Maryland School of Medicine, Baltimore, MD 21201, USA

Jamie Bartram
Water, Sanitation and Health Programme, World Health Organization, Geneva, Switzerland

Stig Bengmark
Ideon Research Park, Scheelevägen 18, SE-223 70, Lund, Sweden

Lore Brade
Research Center Borstel, Center for Medicine and Biosciences, Department for Immunochemistry and Biochemical Microbiology, Parkallee 22, D-23845 Borstel, Germany

Gilbert Domingue
PHLS Food Microbiology Research Unit, Church Lane, Heavitree, Exeter EX2 5AD, UK

James E. Galen
Center for Vaccine Development, University of Maryland School of Medicine, Baltimore, MD 21201, USA

Oscar Gomez-Duarte
Center for Vaccine Development, University of Maryland School of Medicine, Baltimore, MD 21201, USA

Coenraad F.M. Hendriksen
National Institute of Public Health and the Environment (RIVM), PO Box 1, 3720 BA Bilthoven, The Netherlands

Didier Heumann
Division of Infectious Diseases, CHUV, CH-1011 Lausanne, Switzerland

Adrian V.S. Hill
Molecular Immunology Group, Institute of Molecular Medicine, University of Oxford, John Radcliffe Hospital, Oxford OX3 9DU, UK

José Hueb
Water, Sanitation and Health Programme, World Health Organization, Geneva, Switzerland

Thomas J. Humphrey
PHLS Food Microbiology Research Unit, Church Lane, Heavitree, Exeter EX2 5AD, UK

Contributors

Karen Kotloff
Center for Vaccine Development, University of Maryland School of Medicine, Baltimore, MD 21201, USA

Myron M. Levine
Center for Vaccine Development, University of Maryland School of Medicine, Baltimore, MD 21201, USA

Ling Lissolo
Aventis Pasteur, Campus Mérieux, 1541 avenue Marcel Mérieux, 69280 Marcy L'Etoile, France

Brian W.J. Mahy
National Center for Infectious Diseases, Centers for Disease Control and Prevention, Atlanta, GA 30333, USA

Martin C.J. Maiden
The Wellcome Trust Centre for the Epidemiology of Infectious Disease, Department of Zoology, University of Oxford, South Parks Road, Oxford OX1 3FY, UK

Sven Müller-Loennies
Research Center Borstel, Center for Medicine and Biosciences, Department for Immunochemistry and Biochemical Microbiology, Parkallee 22, D-23845 Borstel, Germany

Frank C. Odds
Department of Molecular and Cell Biology, Institute of Medical Sciences, University of Aberdeen, Foresterhill, Aberdeen AB25 2ZD, UK

Franco di Padova
Preclinical Research Novartis, CH-4002 Basel, Switzerland

Marcela Pasetti
Center for Vaccine Development, University of Maryland School of Medicine, Baltimore, MD 21201, USA

Thames Pickett
Center for Vaccine Development, University of Maryland School of Medicine, Baltimore, MD 21201, USA

Marie-José Quentin-Millet
Aventis Pasteur, Campus Mérieux, 1541 avenue Marcel Mérieux, 69280 Marcy L'Etoile, France

Ernst T. Rietschel
Research Center Borstel, Center for Medicine and Biosciences, Department for Immunochemistry and Biochemical Microbiology, Parkallee 22, D-23845 Borstel, Germany

Peter H. Stephens
Oxoid Ltd, Wade Road, Basingstoke RG24 8PW, UK

Marcelo Sztein
Center for Vaccine Development, University of Maryland School of Medicine, Baltimore, MD 21201, USA

Contributors

Carol Tacket
Center for Vaccine Development, University of Maryland School of Medicine, Baltimore, MD 21201, USA

Paul Taylor
Director, CBD Porton Down, Salisbury SP4 0JQ, UK

Alexander Tomasz
The Rockefeller University, 1230 York Avenue, New York, NY 10021, USA

Garry C. Whitelam
Department of Biology, University of Leicester, University Road, Leicester LE1 7RH, UK

Preface

Major microbiological advances were made in many fields during the last century, but there are many real and possibly severe challenges to face in the new century. The two major British microbiological societies, the Society for General Microbiology and the Society for Applied Microbiology, agreed that the millennium presented a golden opportunity to join forces and celebrate with a joint meeting with the highlight a symposium on *Fighting Infection in the 21st Century*. The meeting was organized and edited by a team from both societies led by Pat Goodwin, and our thanks must go to Melanie Scourfield, who helped enormously as our Desk Editor. The speakers were set a hard task as the team posed specific questions for them to address, but their endeavours were rewarded by the crowded lecture theatre throughout the 2-day symposium. On behalf of our two societies we thank all the speakers for all their efforts in making this a very successful joint meeting.

The two societies hope that the printed versions of the lectures in this book will appeal to a wide audience. There are many themes, from emerging and re-emerging diseases to our battles to contain them, from the benefits of producing clean water for people from all countries to the misuse of our science as a means of killing innocent people through biological warfare. We hope that the younger generations of microbiologists may consider pursuing the more beneficial of these avenues during their research careers and help to achieve the various objectives.

Peter Andrew Martin Easter Pat Goodwin
Petra Oyston Geoff Smith
Duncan Stewart-Tull

The global threat of emerging infectious diseases

Brian W.J. Mahy

National Center for Infectious Diseases, Centers for Disease Control and Prevention, Atlanta, GA 30333, USA

Introduction

Emerging infectious diseases remain a continuing threat to the health of people worldwide, despite the development of some complacency around the middle of the 20th century, when it was widely believed that infectious diseases could be conquered by the application of antibiotics combined with prophylactic immunization. For example, in 1962 Sir Macfarlane Burnet wrote: 'In many ways one can think of the middle of the twentieth century as the end of one of the most important social revolutions in history, the virtual elimination of infectious disease as a significant factor in social life'. Certainly, remarkable successes have been achieved. During the 20th century in the US, for example, infectious disease mortality declined during the first eight decades from 797 deaths per 100 000 population in 1900 to 36 deaths per 100 000 in 1980; this decline was interrupted by a sharp spike in mortality caused by the 1918 influenza pandemic (Fig. 1) (Armstrong, Conn & Pinner, 1999). However, with the advent of human immunodeficiency virus (HIV) and the recognition of human acquired immunodeficiency syndrome (AIDS) cases beginning around 1980, infectious disease mortality has increased steadily. This upward trend has resulted not just from HIV infection itself, but from opportunistic infections in persons immunocompromised not only with AIDS but also as a result of cancer chemotherapy or immunosuppressive therapy associated with organ transplantation. The emergence of unexpected new disease threats, such as hantavirus pulmonary syndrome, new variant Creutzfeldt–Jakob disease (nvCJD) and drug-resistant forms of bacterial pneumonias, tuberculosis and malaria, has finally altered the complacency of the 1950s into real and immediate concern as we enter the new millennium.

In 1992, the Institute of Medicine published a report prepared by a Committee on Emerging Microbial Threats to Health which defined emerging infectious diseases as those diseases whose incidence in humans has increased within the past two decades or threatens to increase in the future (Lederberg, Shope & Oaks, 1992). This definition could be widened to include diseases of farm livestock, of animals

1

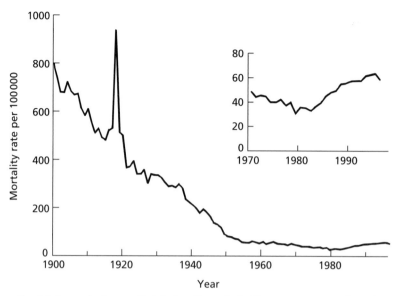

Fig. 1. Trends in human deaths caused by infectious diseases in the US, 1900–1996. Figure adapted from that published by Armstrong, Conn & Pinner (1999).

in general, and of plants, but this chapter will concentrate on a selected sample of infectious diseases that affect the human population.

Major factors in the emergence of infectious diseases

Many factors contribute to the transmission and spread of infectious diseases, but in considering why disease emergence an increasing problem in recent years, modern demographic and environmental conditions are of prime importance. These factors include microbial adaptation and evolution, population growth and movements, human behaviour, ecological and environmental changes, and technologies for food and water processing.

Microbial adaptation and evolution

It is now generally agreed that microbial adaptation and evolution may give rise to new or drug-resistant pathogens at any time. This phenomenon is particularly true of RNA viruses, which mutate with high frequency in the absence of proofreading enzymes and so give rise to diverse quasi-species populations during replication (Domingo & Holland, 1997). For example, the development of the antiretroviral drug azidothymidine (AZT) was an important breakthrough in the treatment of HIV infection (De Clercq, 1995); however, HIV replication generates RNA quasi-species populations that contain mutants resistant to AZT that multiply in the presence of the drug since they have a selective replication advantage. To combat HIV, therefore, it has been necessary to develop more complex, triple-drug therapies active against several different targets (highly active antiretroviral therapy, HAART) which are more difficult to administer and very expensive, limiting their use in developing countries.

It is likely that the opportunity for new RNA viruses to emerge is greater than for DNA viruses or other more complex pathogenic microbes, and indeed most known RNA virus infections of humans appear to have entered the human population relatively recently. However, the development and widespread application of antimicrobial drugs has itself stimulated the generation of mutant microbes able to multiply in their presence. Although this problem was recognized many years ago, it has escalated during the last 10 years to the point where there is alarm that effective antibiotics might soon become unavailable. For example, *Staphylococcus aureus* is a common nosocomial infection, and more than 200 000 infections are estimated to occur in hospitalized patients every year in the US alone.

Initially, fluoroquinolones were effective in treating methicillin-resistant forms of *S. aureus* (MRSA), but with increasing use these bacteria have become resistant to quinolones as well, leaving vancomycin as the only drug available to treat methicillin-resistant staphylococci (Jacoby & Archer, 1991). Recently, strains of MRSA that are resistant to vancomycin have been detected in the US, Japan and Europe. Similarly, vancomycin tolerance, a precursor to vancomycin resistance, has emerged in the commonest organism that causes bacterial pneumonia, *Streptococcus pneumoniae* (Novak *et al.*, 1999). Multiple drug resistance is now common, and *Salmonella* strains resistant to up to as many as five major drugs have been reported. Factors contributing to this situation include the widespread unnecessary use of antibiotics in human medicine (e.g. when prescribed for colds, and other viral infections), and the use of common antibiotics as growth-promoting substances in animal feed. A more detailed discussion of antibiotic resistance can be found in the chapter by A. Tomasz in this volume (pp. 198–216).

Population growth and movements

The world population will approach 7 billion persons in 2000, but of most concern to infectious disease spread is that more than half of this population will be urbanized (Dentler, 1977), with almost 500 cities worldwide having more than a million citizens. Each year the world population will continue to grow by about 70 million persons. Large urban conurbations in many parts of the world are associated with a lack or breakdown of public health infrastructure, poor hygiene, insufficient supply of clean water and inadequate sanitation, all of which can lead to a flourishing of microbial pathogens.

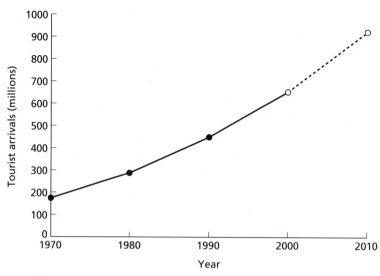

Fig. 2. World Tourist Organization (WTO) estimates of the number of arrivals internationally, 1970–2010.

Population movements due to wars, famines and natural or man-made disasters can also favour the spread of infectious diseases. The availability of rapid, relatively inexpensive global air travel has resulted in massive increases in the movement of individuals over long distances, so that a new infectious disease in one part of the world can rapidly be spread to another. For example, the number of airline passengers within the US is expected to increase over 56% from 630 million passengers in 1997 to 987 million in 2009 and world travel is expected to increase to almost 1000 million arrivals per year by 2010 (Fig. 2).

Human behaviour

Despite widespread public health information and education programmes, people in many parts of the world still place themselves at risk from blood-borne infectious diseases by risky sexual behaviour and intravenous drug injection practices. Although AIDS can usually be prevented by behavioural changes, such as appropriate condom use, an estimated 16 000 new HIV infections occur every day, adding to the more than 33·4 million HIV-infected persons still alive in the world. This large population of immunosuppressed persons provides a breeding ground for numerous other infectious diseases caused by bacteria, fungi, rickettsiae or viruses (Kaplan *et al.*, 1998).

Even more common than HIV on a worldwide basis are blood-borne hepatitis viruses, such as hepatitis B and hepatitis C viruses. Considerable advances are being made in the control of hepatitis B by universal infant immunization in many countries, but prospects for the control of hepatitis C, which causes a persistent infection that can lead to serious liver disease, are poor, and no vaccine is likely in the next several years. The best available data suggest that 4 million persons are infected with hepatitis C virus in the US alone, and probably more than 170 million worldwide. There is an urgent need to better understand and control the major routes of transmission of hepatitis C in the population. Most cases of hepatitis C virus infection were acquired through transfusion of blood before 1992, the year in which adequate screening of the blood supply became possible through the application of improved antibody tests. Nevertheless, most infected persons in developed countries are unaware of their infection, and in some countries, notably Egypt, up to 20% of the adult population became infected before adequate screening procedures were introduced. Few data are available on the prevalence of hepatitis C virus infection in underdeveloped countries, but it is certain that hepatitis C is a worldwide problem. Bacterial contamination of blood and blood products can also be a problem (Wagner, Friedman & Dodd, 1994).

Ecological and environmental changes

Alterations in the ecology brought about by natural or man-made events can lead to conditions that favour the emergence and spread of infectious diseases. For example, climatic changes occurring in association with the El Niño Southern Oscillation (ENSO) can cause alterations in wildlife or vegetation that can influence infectious disease vector species, such as rodents. In 1998, the increased rainfall over the southwestern US caused by ENSO nourished the vegetation and set the stage for an explosive growth in the deer mice population (*Peromyscus maniculatus*), which may carry Sin Nombre virus, which causes hantavirus pulmonary syndrome (Hughes *et al.*, 1993).

Global environmental changes are also constantly under way as a result of human intervention into new ecosystems (Vitousek *et al.*, 1998). The building of dams, in particular, has been cited as a contributing factor in the expansion of mosquito populations which vector the spread of Rift Valley fever (RVF) virus. Formerly a problem only in sub-Saharan Africa, a major epidemic of human disease involving almost 600 deaths occurred in 1977 in Egypt and was linked to the construction of the Aswan Dam, completed in 1970 (Lederberg, Shope & Oaks, 1992). Similarly, the construction of the Diama Dam in the Senegal River Basin probably spawned a large RVF outbreak in West Africa (Jouan *et al.*, 1988). Human activities such as deforestation may also increase the opportunities for contact with disease vector species such as mosquitoes or other arthropods, as well as rodents. Clearing a forest region in South America and replacing it with human habitation can greatly increase the chances of contact with sigmodontine rodents, which may enter buildings for shelter and carry dangerous haemorrhagic fever viruses, such as

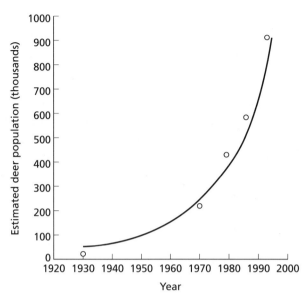

Fig. 3. White-tailed deer population, Virginia, USA, 1930–1995. Data from Knox (1997).

Junin, Machupo or Guanarito virus (Childs & Peters, 1993). On the other hand, reforestation may also be associated with disease emergence, as in the northeastern and mid-Atlantic US, where it has been accompanied by a dramatic rise in the deer population (Fig. 3). As a consequence, there is now greatly increased contact between people and deer ticks of *Ixodes* spp., which are vectors of *Borrelia burgdorferi*, the spirochaete that transmits Lyme disease. This ecological change resulted in the emergence of Lyme disease as one of the most common vector-borne diseases in the US. Other tick-borne diseases, such as human granulocytic ehrlichiosis (HGE), caused by a rickettsia that was first recognized in 1992, have also emerged (Walker, 1998), and indeed coinfections with Lyme disease and HGE have been reported (Duffy *et al.*, 1997). These pathogens are vectored to humans by the deer tick from their natural reservoir in white-footed mice that also inhabit the forest.

Technologies for food and water processing

In recent years, globalization of the food supply has brought about a redistribution of known food-borne pathogens from one country to another, as well as the recognition of new disease agents. In addition, the emergence of large centralized food-processing plants carries with it the risk of widespread food-borne disease outbreaks involving large numbers of people. For example, *Cyclospora cayetanensis*, first associated with human disease in 1993, emerged as a serious infectious disease problem when widespread outbreaks of cyclosporiasis occurred in the US and Canada in 1996 in association with the ingestion of contaminated raspberries from Guatemala (Herwaldt & Ackers, 1997). The outbreak was clearly associated with an increase in raspberry importations from Guatemala to North America.

A new strain of *Escherichia coli* that produces *Shigella*-like toxins (*E. coli* O157 : H7) was first recognized in 1982, but became the cause of serious concern in the US in 1993, following a large outbreak of diarrhoeal disease and the deaths of four children from haemolytic uraemic syndrome (Griffin & Boyce, 1998). Most cases of *E. coli* O157 : H7 infection have originated in ground meat that became contaminated by organisms present in the intestines of food animals (usually sheep or cattle) during slaughter and meat processing. The increase in large centralized meat-processing plants that turn beef into hamburgers, which are then frozen and distributed over a wide geographic area, carries considerable risk in infectious disease contamination occurring at any time during the process. In 1997, the Colorado Department of Public Health and Environment detected a cluster of human cases of *E. coli* O157 : H7 infection and linked them to one brand of frozen hamburger patty. As a result, 25 million pounds of ground beef were recalled, averting a potential national outbreak of disease. *E. coli* O157 : H7 is now recognized as the cause of disease outbreaks in several other countries in addition to the US, including Canada, Japan, Africa and the UK, associated not only with meat, but also with milk, cider, lettuce and contaminated water.

Also in 1997, outbreaks of hepatitis A virus infection in school children in several parts of the US were linked to consumption of strawberries distributed through a Department of Agriculture-sponsored school lunch programme. Nucleotide sequence analysis of the virus present in patient stools confirmed the link to strawberries of Mexican origin packed by a company in California, which subsequently recalled the implicated product lots. The production of eggs involving large-scale housing and management of laying hens has resulted in an increased risk of widespread human infection with *Salmonella* serotype Enteritidis. A large egg-production facility in the US today may comprise several linked houses, each containing 100 000 hens (Bell, 1995). *Salmonella* serotype Enteritidis can be transmitted easily from breeding flocks to egg-laying hens and is a difficult organism to eliminate from such large facilities without considerable economic loss. In consequence, outbreaks of *Salmonella* serotype Enteritidis disease have increased in number and extent in the past 15 years (Tauxe, 1997).

Technology-based recognition of infectious agents

As we enter the 21st century, techniques are becoming available for the detection of infectious agents without the need for traditional methods of characterization, such as *in vitro* growth or culture. A now classic example of this technological progress was the identification of hepatitis C virus by Bradley and coworkers in 1989 (Choo *et al.*, 1989), who cloned the genome of hepatitis C virus from the blood of an experimentally infected chimpanzee. RNA extracted from the chimpanzee's blood was reverse-transcribed to make cDNAs by using random primers. The cDNAs were cloned into bacteriophage expression vectors, and the resultant bacterial colonies were screened by using sera from patients with antibodies to non-A non-B hepatitis. Thousands of clones were screened before one was found to react positively to the antisera. DNA from this clone was then used to search for other positive clones by hybridization, until eventually the full-length RNA sequence of hepatitis C virus was determined. This extremely important discovery has enabled screening of the blood supply worldwide to reduce the risk of hepatitis C virus transmission, even though it is still not possible to grow hepatitis C virus in cell culture.

Another molecular approach, known as representational difference analysis, can be used to compare DNA preparations from different tissues and was instrumental in the discovery of human herpesvirus 8, the aetiological agent of Kaposi's sarcoma (Chang *et al.*, 1994). Once the virus was discovered and its genome nucleotide sequence determined, it was possible to develop specific PCR primers to search for the presence of the virus in other malignancies, such as multicentric Castleman's disease (Spira & Jaffe, 1998).

Detection of 'orphan' infections

In the early days of virus recognition by specific cytopathic effects in cell culture, several viruses were identified that had no apparent association with any disease, and so were termed 'orphan' viruses. Enteric cytopathic human orphan (ECHO) and respiratory enteric orphan (REO) viruses are examples of this type. Now the application of new technologies for virus genome detection has led to the emergence of some apparently widespread infections that have not so far been associated with a specific disease. A new flavivirus, currently known as GB virus C/hepatitis G virus, was discovered independently by two groups searching for new causative agents of transfusion-transmitted hepatitis not caused by hepatitis viruses B or C (Simons *et al.*, 1995; Linnen *et al.*, 1996). However, although distributed widely in human populations, GB virus C/hepatitis G virus appears to cause a persistent infection that has not yet been associated with any overt disease. It has been suggested that this RNA virus may be co-evolving worldwide in human and some primate populations without causing harm to the host (Simmonds & Smith, 1999).

Another virus that was discovered in the search for the cause of some cases of post-transfusion hepatitis is called TT, after the initials of the original Japanese patient in whom it was found (Nishizawa *et al.*, 1997). Despite a large number of studies, an association of TT virus with hepatitis or elevated serum transaminase levels has not been confirmed (Watanabe *et al.*, 1999), but clearly the virus is widespread in human populations worldwide. Although infection is detected as a viraemia by PCR, the early age of acquisition of infection in young children suggests that transmission may involve the faecal–oral route, as with hepatitis A virus (Davidson *et al.*, 1999). TT virus has an unusual genome, consisting of a small, circular, negatively stranded DNA 3852 nucleotides in length, and on the basis of its novel sequence it has been proposed as the first recognized member of a new virus family, the *Circinoviridae* (Mushahwar *et al.*, 1999). Further investigations will be needed to determine whether GB virus C/hepatitis G virus and TT virus are indeed examples of harmless persistent virus infections in the human population, as currently seems possible. Nevertheless, it seems certain that other presently unrecognized viruses will be discovered in the future, not on the basis of disease, but through molecular studies. So far, few groups have searched deliberately for new infectious agents in healthy populations. An example of what might be found is illustrated by the work of Herniou *et al.* (1998), who searched for evidence of retrovirus infection in 18 orders of vertebrates and discovered more than 30 previously unrecognized retroviruses by using a purely molecular approach.

Vector-borne and zoonotic diseases

In recent years, there have been numerous examples of the global threat of human infection by diseases that have their origin in non-human species. Recent evidence indicates that HIV was passed into the human population from a non-human primate reservoir (Gao *et al.*, 1999). Many of the most serious infectious diseases, such as rabies, plague, Ebola, haemorrhagic fever and yellow fever, have their origin in zoonotic reservoirs. Excellent vaccines are available to control rabies and yellow fever infections in the human population, but there is a continued concern regarding the emergence of other diseases for which there are presently no adequate public health prevention measures.

Dengue

Dengue is caused by a flavivirus that exists in four major serotypes that are distributed in tropical regions throughout the world. Because cross-protection between the four serotypes lasts only for a few months, as many as four dengue infections can occur over a period of years in a single individual. The virus is transmitted by mosquito bites, the principal vector being *Aedes aegypti*. Infection can result in mild to severe dengue fever, but the most serious form, dengue haemorrhagic fever (DHF), emerged in epidemic form in the Philippines in 1954 and is believed to result from sequential infection with different dengue virus serotypes (Halstead,

1988). DHF has become an increasingly important problem and is now one of the most rapidly emerging diseases, with thousands of cases reported each year. The Americas, in particular, have experienced increasing numbers of epidemics, largely due to expansion of the mosquito populations in urban habitats as a result of failed mosquito control measures (Gubler & Meltzer, 1999). Currently, no vaccines are available to prevent dengue virus infection, and with increasing mosquito density not only in developing countries but also in many developed countries, the potential for future large dengue epidemics is considerable.

Marburg and Ebola

Marburg and Ebola haemorrhagic fevers are caused by filoviruses that are reasonably well characterized in terms of their molecular biological properties and clinical manifestations, but their believed zoonotic origin remains unknown (Feldmann, Sanchez & Klenk, 1998). One species of Marburg virus and four species of Ebola virus have been identified. Both viruses cause sudden febrile illness accompanied by haemorrhagic manifestations and high mortality rates, ranging from 22% with Marburg virus up to 88% with Ebola Zaire virus. One species of Ebola virus, Ebola Reston virus, has caused disease only in non-primate species and may even be apathogenic in humans. Outbreaks of Ebola Reston virus infection in non-human primates have all been associated with monkeys originating in the Philippines (Miranda *et al.*, 1996), but all other filovirus outbreaks can be related to East and Central Africa. The species in which natural filovirus infection occurs remains unknown, despite extensive investigation following recent outbreaks of both Ebola and Marburg haemorrhagic fever, and it is this uncertainty that has caused the notoriety associated with these relatively rare infections. Global spread of human filovirus disease has never been reported apart from one recent case in a physician infected in Gabon, who travelled to South Africa to receive hospital care and infected a nurse, who died as a result (Georges *et al.*, 1999; World Health Organization, 1996). Although a great deal of fear has been generated regarding the filoviruses, rightfully so because of their high disease potential, which requires that they may be studied safely only in biosafety level four laboratories, they do not constitute a major global emerging disease threat.

Hendra and Nipah

In 1994, a horse trainer in the Hendra suburb of Brisbane, Queensland, Australia, died of severe respiratory disease caused by a previously unknown paramyxovirus that had infected 20 horses in his stables, killing 13 of them. A stable hand was also infected and became severely ill, but survived. Initially, the virus was named equine morbillivirus (Murray *et al.*, 1995), but was subsequently renamed Hendra virus. In another episode that came to light a year later, a farmer became infected in

August 1994 whilst performing an autopsy on two dead horses, suffered a mild febrile infection, and recovered, but died a year later of fatal encephalitis. One of the horses was found to be infected with Hendra virus. A search for the reservoir of Hendra virus showed that it caused frequent infections of *Pteropus* species, commonly known as fruit bats or flying foxes, as demonstrated by the presence of antibodies in 20–30% of all four known *Pteropus* species in Australia (Young *et al.*, 1996). Serological evidence for Hendra virus infection was sought in thousands of horses in Australia and elsewhere, but none was found. However, in January 1999, a horse that died in Cairns, in northeastern Australia, was found to be infected with Hendra virus.

Beginning in October 1998, an epidemic of severe febrile encephalitis with up to 40% fatality occurred in peninsular Malaysia in persons who had close contact with pigs. Disease was also seen in pigs on farms associated with human cases, although the fatality rate in pigs was less than 5%, and most pigs exhibited respiratory rather than encephalitic symptoms. Overall, more than 100 people died in Malaysia, and there was one death among 13 cases that occurred in Singapore associated with an abattoir processing Malaysian pigs. A previously unknown paramyxovirus, Nipah virus, was isolated from a patient who lived in Sungai Nipah, in the Negeri Sembilan district of Malaysia, and this proved to be the cause of the encephalitis outbreak in humans as well as the respiratory disease in pigs. The virus was first identified because it cross-reacted with antisera made against Hendra virus, but subsequent molecular characterization showed that the two viruses, now considered to represent a hitherto unknown genus within the family *Paramyxoviridae*, differed by some 30% at the nucleotide sequence level across the genome (Chua *et al.*, 2000).

An unusual feature of both these viruses is their ability to infect and cause severe disease in a wide range of species. For example, Nipah virus infection causes fatalities in cats and dogs as well as pigs and humans. The reservoir of natural infection for both viruses appears to be fruit bats, although a great deal more work will be needed to understand the natural history and origin of these unusual virus infections. Although there is some evidence in the case of Nipah virus that infection was occurring in pig farms as early as 1997, it is remarkable that such a virulent infection was previously unknown, and there is an urgent need to understand the nature of its emergence into the pig and human populations. The slaughter of more than a million pigs in Malaysia was necessary to control the outbreak (Fig. 4).

Influenza

Although not always considered as such, human pandemic influenza, caused by influenza A virus, is a zoonotic disease that has its origins in influenza infections of birds and pigs. Once it enters the human population, influenza virus is very effi-

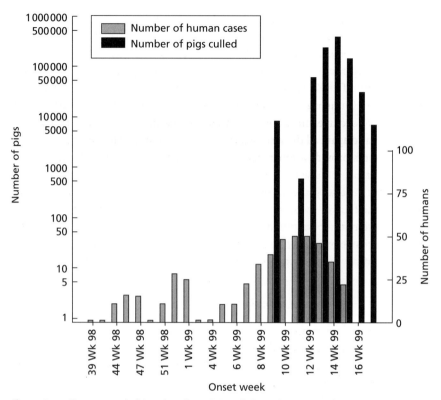

Fig. 4. Cases of human encephalitis and numbers of pigs culled in Malaysia, September 1998–May 1999.

ciently transmitted from person to person if it displays a novel surface haemagglutinin antigen. For all recent pandemics, this antigen was acquired by genetic reassortment of a virus of human origin with an avian influenza virus (Webster & Kawaoka, 1994). In the worst pandemic ever recorded, in 1918, more than 25 million persons died worldwide, and ever since the global threat of influenza has outweighed all other emerging infectious disease threats. Since 1918, two other pandemics have occurred, in 1957 and 1968, each associated with severe mortality and morbidity. More than 30 years have elapsed since the last pandemic, and many public health observers believe that the world is overdue for a new influenza pandemic.

A characteristic feature of the 1957 and 1968 pandemics was the introduction of a new surface antigen derived from an avian influenza virus onto a background of human influenza virus genes. It was very surprising, therefore, when a virus containing only genes derived from avian influenza virus caused 18 human infections with six deaths in Hong Kong in 1997. This virus contained a surface haemagglutinin antigen (subtype H5) that previously had been observed only in birds and was able to infect humans without the need for gene reassortment with a human virus (Subbarao *et al.*, 1998). Further cases of human infection were probably prevented by slaughter of most of the chickens in Hong Kong, and the outbreak died out at the end of 1997. However, in 1999, two human influenza cases caused by another avian influenza virus, this time with a different surface haemagglutinin (subtype H9), occurred in young children in Hong Kong (World Health Organization, 1999), and again all the virus genes were found to be avian in origin.

These events provide a sober reminder that even though our understanding of infectious disease emergence and epidemiology is increasing, we must always be prepared for the unexpected. Until 1997, it was widely believed that only influenza viruses of the HI, H2 or H3 subtypes were capable of causing disease in humans. There are presently 15 known haemagglutinin subtypes in avian species; thus the recent experience in Hong Kong would suggest that the potential threat of a new pandemic influenza strain may be even greater than was previously believed.

Conclusions and prospects

Infectious diseases are emerging globally due to many factors, including changing human demographics and behaviour, microbial adaptation and evolution, ecological and environmental changes, and increasing numbers of immunodeficient people worldwide. Multiple drug-resistant micro-organisms causing old diseases such as tuberculosis are emerging at a time when there is little prospect for the development of new antibiotics to control them. The application of newly developed molecular technologies is beginning to reveal new viruses that were previously unsuspected and are still not linked to any disease. Finally, new technologies for food production and processing have led to the emergence of deadly strains of *E. coli* and devastating new infectious agents, such as the prion causing bovine spongiform encephalopathy that has crossed the species barrier and now threatens to become the human plague of this millennium in the UK.

To combat these microbial adversaries, it will be necessary in the coming century to bring together individuals, communities and nations in a common effort to improve surveillance for infectious diseases worldwide so that appropriate responses can be made and control measures taken. The CDC has developed a plan for the new millennium entitled 'Preventing Emerging Infectious Diseases: A Strategy for the 21st Century'. The plan outlines four interdependent goals: Surveillance and Response, Applied Research, Infrastructure and Training, and Prevention and Control. Implementation of the plan will require not just public agencies but also the participation of academic research institutions and private sector pharmaceutical companies. If the world is to take decisive preventive action against emerging infectious diseases, there is a need in the 21st century to apply the kind of global commitment that was required in the 20th century to eradicate smallpox. Time will tell whether we are up to the challenge.

Acknowledgements

I thank Martin Cetron, Christopher Paddock, Robert Pinner and Sherif Zaki for their help in preparing the figures, and Veronica Brown, Sue Driesner and John O'Connor for assisting in preparing the manuscript.

References

Armstrong, G. L., Conn, L. A. & Pinner, R. W. (1999). Trends in infectious disease mortality in the United States during the 20th century. *Journal of the American Medical Association* **281**, 61–66.

Bell, D. (1995). Forces that have helped shape the U.S. egg industry: the last 100 years. *Poultry Tribune* (September 1995), 30–43.

Burnet, M. (1962). *Natural History of Infectious Disease*, 3rd edn, p. 3. Cambridge: Cambridge University Press.

Chang, Y., Ceasarman, E., Pessin, M. S., Lee, F., Culpper, J. J., Knowles, D. M. & Moore, P. S. (1994). Identification of herpesvirus-like DNA sequences in AIDS-associated Kaposi's sarcoma. *Science* **266**, 1865–1869.

Childs, J. E. & Peters, C. J. (1993). Ecology and epidemiology of arenaviruses and their hosts. In *The Arenaviridae*, pp. 331–384. Edited by M. S. Salvato. New York: Plenum.

Choo, Q.-L., Kuo, G., Weiner, A. J., Overby, L. R., Bradley, D. W. & Houghton, M. (1989). Isolation of a cDNA clone derived from a blood-borne non-A, non-B viral hepatitis genome. *Science* **244**, 359–362.

Chua, K. B., Bellini, W. J., Rota, P. A. & 19 other authors (2000). Nipah virus: a recently emergent deadly Paramyxovirus. *Science* **288**, 1432–1435.

Davidson, F., MacDonald, D., Mokili, J. L. K., Prescott, L. E., Graham, S. & Simmonds, P. (1999). Early acquisition of TT virus (TTV) in an area endemic for TTV infection. *Journal of Infectious Diseases* **179**, 1070–1076.

De Clercq, E. (1995). Antiviral chemotherapy: where do we stand and what can we expect? *International Antiviral News* **3**, 52–54.

Dentler, R. A. (1977). *Urban Problems.* Chicago: Rand McNally.

Domingo, E. & Holland, J. J. (1997). RNA virus mutations and fitness for survival. *Annual Review of Microbiology* **51**, 151–178.

Duffy, J., Pittlekow, M. R., Kolbert, C. P., Rutledge, B. J. & Persing, D. H. (1997). Coinfection with *Borrelia burgdorferi* and the agent of human granulocytic ehrlichiosis. *Lancet* **349**, 399.

Feldmann, H., Sanchez, A. & Klenk, H.-D. (1998). Filoviruses. In *Virology*, volume I of *Topley & Wilson's Microbiology and Microbial Infections*, pp. 651–664. Edited by B. W. J. Mahy & L. Collier. London: Edward Arnold.

Gao, F., Bailes, E., Robertson, D. L. & 9 other authors (1999). Origin of HIV-1 in the chimpanzee *Pan troglodytes troglodytes*. *Nature* **397**, 436–441.

Georges, A. J., Leroy, E. M., Renaut, A. A. & 13 other authors (1999). Ebola haemorrhagic fever outbreaks in Gabon, 1994–1997: epidemiologic and health control issues. *Journal of Infectious Diseases* **179** (Suppl. 1), S65–S75.

Griffin, P. M. & Boyce, T. G. (1998). *Escherichia coli* O157 : H7. In *Emerging Infections* 1, pp. 137–145. Edited by W. M. Scheld, D. Armstrong & J. M. Hughes. Washington, DC: American Society for Microbiology.

Gubler, D. J. & Meltzer, M. (1999). Impact of dengue/dengue hemorrhagic fever on the developing world. *Advances in Virus Research* **53**, 35–70.

Halstead, S. B. (1988). Pathogenesis of dengue:

challenges to molecular biology. *Science* 239, 476–481.

Herniou, E., Martin, J., Miller, K., Cook, J., Wilkinson, M. & Tristem, M. (1998). Retroviral diversity and distribution in vertebrates. *Journal of Virology* 72, 5955–5966.

Herwaldt, B. L. & Ackers, M. L. (1997). An outbreak in 1996 of cyclosporiasis associated with imported raspberries. *New England Journal of Medicine* 336, 1548–1556.

Hughes, J. M., Peters, C. J., Cohen, M. L. & Mahy, B. W. J. (1993). Hantavirus pulmonary syndrome: an emerging infectious disease. *Science* 262, 850–851.

Jacoby, G. A. & Archer, G. L. (1991). New mechanisms of bacterial resistance to antimicrobial agents. *New England Journal of Medicine* 324, 601–612.

Jouan, A., LeGuenno, B., Digoutte, J. P., Philippe, B., Riou, O. & Adam, F. (1988). A RVF epidemic in southern Mauritania. *Annales de l'Institut Pasteur Virology* 139, 307–308.

Kaplan, J. E., Hanson, D. J., Jones, J. L., Beard, C. B., Juranek, D. D. & Dykewicz, C. A. (1998). Opportunistic infections (OIs) as emerging infectious diseases: challenges posed by OIs in the 1990s and beyond. In *Emerging Infections* 2, pp. 257–272. Edited by W. M. Scheld, W. A. Craig & J. M. Hughes. Washington, DC: American Society for Microbiology.

Knox, W. M. (1997). Historical changes in the abundance and distribution of deer in Virginia. In *The Science of Overabundance*, pp. 27–36. Edited by W. J. McShea, H. B. Underwood & J. H. Rappole. Washington & London: Smithsonian Institution Press.

Lederberg, J., Shope, R. E. & Oaks, S. C., Jr (editors) (1992). *Emerging Infections: Microbial Threats to Health in the United States*. Institute of Medicine. Washington, DC: National Academy Press.

Linnen, J., Wages, J., Zhang-Keck, Z. Y. & 27 other authors (1996). Molecular cloning and disease association of hepatitis G virus: a transfusion-transmissible agent. *Science* 271, 505–508.

Miranda, M. E., Ksiazek, T. G., Retuya, T. J. & 9 other authors (1996). Epidemiology of Ebola (subtype Reston) virus in the Philippines. *Journal of Infectious Diseases* 179 (Suppl. 1), S115–S119.

Murray, K., Rogers, R., Selvey, L., Selleck, P.,

Hyatt, A., Gould, A., Gleason, L., Hooper, P. & Westbury, H. (1995). A morbillivirus that caused fatal disease in horses and humans. *Science* 268, 94–96.

Mushahwar, I. K., Erker, J. C., Muerhoff, A. S., Leary, T. P., Simons, J. N., Birkenmeyer, L. G., Chalmers, M. L., Pilot-Matias, T. J. & Desai, S. M. (1999). Molecular and biophysical characterization of TT virus: evidence for a new virus family infecting humans. *Proceedings of the National Academy of Sciences, USA* 96, 3177–3182.

Nishizawa, T., Okamoto, H., Konishi, K., Yoshizawa, H., Miyakawa, Y. & Mayumi, M. (1997). A novel DNA virus (TTV) associated with elevated transaminase levels in posttransfusion hepatitis of unknown etiology. *Biochemical and Biophysical Research Communications* 241, 92–97.

Novak, R., Henriques, B., Charpentier, E., Normark, E. & Tuomanen, E. (1999). Emergence of vanamycin tolerance in *Streptococcus pneumoniae*. *Nature* 399, 590–593.

Simmonds, P. & Smith, D. B. (1999). Structural constraints on RNA virus evolution. *Journal of Virology* 73, 5787–5794.

Simons, J. N., Leary, T. P., Dawson, G. J., Pilot-Matias, T. J., Muerhoff, A. S., Schlauder, G. G., Desai, S. M. & Mushahwar, I. K. (1995). Isolation of novel virus-like sequences associated with human hepatitis. *Nature Medicine* 1, 564–569.

Spira, T. J. & Jaffe, H. W. (1998). Human herpesvirus 8 and Kaposi's sarcoma. In *Emerging Infections* 2, pp. 81–104. Edited by W. M. Scheld, W. A. Craig & J. M. Hughes. Washington, DC: American Society for Microbiology.

Subbarao, K., Klimov, A., Katz, J. & 13 other authors (1998). Characterization of an avian influenza A (H5NI) virus isolated from a child with a fatal respiratory illness. *Science* 279, 393–396.

Tauxe, R. V. (1997). Emerging food-borne diseases: an evolving public health challenge. *Emerging Infectious Diseases* 3, 425–434.

Vitousek, P. M., D'Antonio, C. M., Loope, L. L. & Westbrooks, R. (1998). Biological invasions as global environmental change. *American Scientist* 84, 468–478.

Wagner, S. J., Friedman, L. I. & Dodd, R. U. (1994). Transfusion associated bacterial species. *Clinical Microbiology Reviews* 7, 290–302.

Walker, D. H. (1998). Emerging human ehrlichioses: recently recognized, widely distributed, life-threatening tick-borne disease. In *Emerging Infections* 1, pp. 81–91. Edited by W. M. Scheld, D. Armstrong & J. M. Hughes. Washington, DC: American Society for Microbiology.

Watanabe, H., Shinzawa, H., Shao, L. & Takahashi, T. (1999). Relationship of TT virus infection with prevalence of hepatitis C virus infection and transaminase levels. *Journal of Medical Virology* 58, 235–238.

Webster, R. G. & Kawaoka, Y. (1994). Influenza — an emerging and reemerging disease. *Seminars in Virology* 5, 103–111.

World Health Organization (1996). Ebola haemorrhagic fever — South Africa. *Weekly Epidemiology Record* 71, 359.

World Health Organization (1999). Influenza. *Weekly Epidemiology Record* 14, 111.

Young, P. L., Halpin, K., Selleck, P. W., Field, H., Gravel, J. L., Kelly, M. A. & Mackenzie, J. S. (1996). Serologic evidence for the presence in *Pteropus* bats of a paramyxovirus related to equine morbillivirus. *Emerging Infectious Diseases* 2, 239–240.

Can molecular techniques be used in the prevention of contamination of processed food by pathogens?

Thomas J. Humphrey,[1] Gilbert Domingue[1] and Peter H. Stephens[2]

[1]*PHLS Food Microbiology Research Unit, Church Lane, Heavitree, Exeter EX2 5AD, UK*
[2]*Oxoid Ltd, Wade Road, Basingstoke RG24 8PW, UK*

Introduction

Food-borne disease continues to be a major international public health problem. In England and Wales in 1998, for example, there were over 23 000 confirmed cases of human salmonellosis and over 58 000 cases of *Campylobacter* infection reported (Anonymous, 1999a). Evidence from cohort studies of patients visiting general practitioners suggests that the true figure for these infections may be up to 10 times higher. The World Health Organization (WHO) estimates that 1% of all the population of developed countries is infected with *Campylobacter* each year (Notermans, 1994).

A wide variety of foods have been identified as vehicles of infection in both outbreaks and sporadic cases of food-borne disease. Processed foods play an important part in the dissemination of certain pathogenic bacteria, although infections are almost always the result of some mistake or breakdown in the manufacturing process. This problem is exacerbated by the fact that certain pathogens can have low infectious doses. This is especially true when foods are high in fat, to protect the bacteria from gastric acidity, and the consuming populations belong to vulnerable groups such as the very young, the old, those already ill or the immunocompromised. This is now well understood for *Campylobacter jejuni* (Robinson, 1981) and *Escherichia coli* O157 : H7 (Buchanan & Doyle, 1997; Smith, 1997), particularly in children, who will have a lower gastric acidity, an immature immune response and who may have a reduced protective effect from resident gut microflora (Lehmacher, Bockemühl & Aleksic, 1995). There is also increasing recognition that certain *Salmonella* serotypes can cause infection in low numbers, even in apparently healthy people (D'Aoust, 1994; Anonymous, 1997). Table 1 provides details of outbreaks of salmonellosis in which the vehicles of infection were processed foods apparently containing low numbers of *Salmonella* spp.

The continuing high incidence of food-borne disease internationally clearly

Table 1. Examples of food-poisoning outbreaks attributed to low levels of *Salmonella* contamination (adapted from D'Aoust, 1994).

Food vehicle	Serotype	Estimated no. of cells ingested	Reference
Chocolate	*Salmonella eastbourne*	<100	Craven *et al.* (1975)
Chocolate	*Salmonella napoli*	10–100	Greenwood & Hooper (1983)
Chocolate	*Salmonella typhimurium*	≤10	Kapperud *et al.* (1990)
Cheddar cheese	*Salmonella heidelberg*	100	Fontaine *et al.* (1980)
Cheddar cheese	*Salmonella typhimurium*	1–10	D'Aoust (1985)
Savoury maize snack	*Salmonella agona*	2–45	Killalea *et al.* (1996)
Hamburger	*Salmonella newport*	10–100	Fontaine *et al.* (1978)
Paprika-flavoured potato chips	Various types	4–45	Lehmacher, Bockemühl & Aleksic (1995)

indicates a need to better protect public health by preventing/removing bacterial contamination from our food. There are several ways by which this goal can be achieved, such as by preventing raw ingredients from becoming contaminated. In the UK, approximately 30% of chickens on retail sale are contaminated with *Salmonella* spp. and over 75% will be *Campylobacter*-positive (Anonymous, 1998). A better understanding of the epidemiology of these two important human pathogens in poultry meat production would allow on-farm control measures to be identified. Intervention on the farm is the most cost-effective way to reduce the high levels of chicken-associated human infections. As many as 200 birds min^{-1} are killed in modern poultry processing plants and this intensity of production militates against control. Given the ease with which bacteria present on chicken carcases are spread in the domestic kitchen (De Wit, Brockhuizen & Kampelmacher, 1979), it is also unrealistic to expect consumers to be important in the control process, particularly when contamination levels with *Campylobacter* spp. can exceed 10^8 cells per carcase (Hood, Pearson & Shahamat, 1988).

The proper investigation of epidemiology, whether in animal carriers or human cases, requires definitive typing methods so that bacteria can be accurately identified with potential sources. This subject will be discussed later. It is inevitable that factories producing processed foods will continue to receive contaminated raw materials. Food processors have a duty to produce foods which are free from potentially pathogenic bacteria. It is now believed that this is best achieved by carrying out a Hazard and Critical Control Point (HACCP) analysis and, indeed, under UK food law certain food processors are required to do this. An HACCP scheme requires the identification of critical points and these can be ones which are vital in ensuring the destruction of pathogenic bacteria. The establishment of an HACCP system requires some initial microbiological examination to identify time/temperature regimens: for example, those which will destroy even the more tolerant of food-borne pathogenic bacteria. In theory, it is possible to operate an HACCP scheme without end product testing. The need to demonstrate due diligence and the fear of causing an infection, which could be financially disastrous for the company concerned, however, mean that much end product testing is still undertaken. One of the difficulties with classic microbiological techniques is the time taken to achieve a definitive 'yes' or 'no' answer (see below). There has thus been a great deal of research activity over the last 10–15 years to identify more rapid methods which could be used before the positive release of perishable products, i.e. where such products are held until tests show them to be pathogen-free. Many of the rapid methods use molecular techniques to detect the pathogens and this subject area will form the basis of the second part of this chapter.

Molecular typing and the control of food-borne disease

Understanding the way in which infection by food-borne pathogens occurs is a key strategy for prevention. Typing, especially molecular subtyping in its many forms, is a powerful investigative tool that makes it possible to identify the source and site of entry of pathogenic bacteria into the food chain and the events that precede human infection. To summarize, these techniques can be used as follows.

1 To enhance our understanding of the ever-changing epidemiology of food-borne pathogens. In the UK, as elsewhere, it is not uncommon to see changes in the dominance of certain *Salmonella* sero- or phage types. New pathogens, such as *E. coli* O157 : H7, will also appear in both the animal and human populations. Typing plays a vital role in monitoring the presence of these bacteria and ultimately in the identification of sources of infection and potential control measures.

2 To monitor any epidemiological changes induced by alterations in food processing. Consumer pressure to reduce preservatives generally, and salt and sugar in particular, is reflected in the significant increase in consumption of chilled foods and foods of low or intermediate (0·6–0·9) water activity (a_w). Concurrent with this, however, has been an emergence of previously unrecognized virulence characteristics which now have a molecular explanation. In particular, chilling is now known to induce filament formation in *Salmonella* spp. where growth continues but without the need for concomitant septation (Phillips, Humphrey & Lappin-Scott, 1998). *Salmonella* filaments are also formed at certain low a_w values (K. Mattick, personal communication). The public health implications of these activities are not yet fully understood but both chilled and dried foods have been vehicles of infection in very large international outbreaks (Table 1).

3 To permit the better targeting of intervention measures. The many possible sources of food contamination or food animal infection require that those of the greatest importance are properly identified. Typing strategies (Table 2), if used

Table 2. An overview of common typing methods.

	Advantage	Disadvantage	Example
Phenotypic	Wide availability	Phenotypic variability	Biochemical
	Good for screening	Lower discrimination	Serological
	Relatively quick	Labour intensive	Phage
	Cheaper		Antibiogram
			Bacteriocin
Genotypic	Restricted availability	Higher discrimination	Plasmid profiling
	Few screening methods,	Same method for different	Various PCRs
	e.g. PCR	species can be expensive	Ribosomal
	Many are laborious	Some enable population analyses	Insertion elements
			PFGE
			RFLP (used in combination
			with many of the above)
			MLEE*

*Multilocus enzyme electrophoresis (MLEE) is classified conveniently with genotypic methods, as the amino acid sequences that determine the electrophoretic mobilities are the result of specific gene sequences.

properly, facilitate this process and can allow the proper targeting of intervention measures. The definitive identification of pathogenic bacteria also allows assessments of the efficacy of preventative measures by providing data on changes in infection or contamination rates.

An overview of typing

In human or animal medicine, it is customary to identify sources of infection by typing the bacterial isolates from the infected individual and from potential reservoirs of infection. If typing shows that the bacteria are either identical or very closely related, the source of infection is believed to be confirmed. When typing of *Salmonella* spp. was first undertaken, it was a relatively simple procedure and was largely based on serology. As the importance of *Salmonella* spp. as human pathogens increased and the epidemiology became more complicated, more and more detailed typing was required. The usefulness of these approaches in the protection of processed foods is discussed below.

There are several general reviews of molecular typing methods (e.g. Böttger, 1996; Farber, 1996; Olive & Bean, 1999; Struelens *et al.*, 1996; van Belkum, 1994; Milch, 1998) and many more species-specific ones (see below), but for this chapter it will be useful to provide an overview. There are two groups of methods, phenotypic and genotypic (Table 2), and the recent explosion in the variety of tests available has been almost exclusively in the latter group. For example, many probes are now available that target virulence gene sequences. Also, automation has been applied to, for example, ribotyping (reviewed by Odumeru *et al.*, 1999), which, although expensive, is useful for fast, high-volume throughput. Notably, ribotyping is not applicable to *E. coli* O157 strains at present. Of the newer techniques, amplified fragment length polymorphism-PCR (AFLP-PCR) is attracting a lot of attention and methods now exist for *Salmonella* (B. Mackey, personal communication), *Campylobacter* (Duim *et al.*, 1999) and *E. coli* O157 (Arnold *et al.*, 1999). AFLP-PCR is a very promising method but, as yet, not rapid. Automated high-speed sequencers now exist, however, and analytical software continues to improve. AFLP-PCR is a powerful, discriminatory tool that may become essential to the formulation of intervention strategies for newly emerging pathogens and HACCP analysis.

Each method group has its advantages and disadvantages (Table 2) and the defining quality of a particular method is that of epidemiological usefulness. There is no one predominant method, particularly in a European context, where a hierarchical approach using a combination of phenotypic typing and genotypic subtyping has proved particularly useful for various bacteria (Domingue *et al.*, 1997; Milch, 1998; Willshaw *et al.*, 1997), as long as caution is applied in interpreting results obtained by the different methods (Preston, 1999). The recent establishment of international and national electronic networks for the recruitment of typing data applicable to food-borne communicable disease has proved invaluable in enhancing large-scale surveillance of both raw food materials and finished products. Examples of these are the European-based Enter-Net (http://www.b3e.jussieu.fr/ceses/eurosurv) for enteropathogens, and the American Pulse-Net (CDC, Atlanta) for *E. coli* O157 strains. Ultimately though, each food-borne pathogen may also require a unique approach to typing and this is illustrated by the examples below for three food-borne pathogens of great current importance.

Salmonella

Serotyping, where antibodies against somatic 'O' and flagella 'H' antigens are used, still forms the basis for the identification of *Salmonella* spp. This is usually supplemented by determining the phage type (PT), and the serotype *Salmonella typhimurium* has over 200 PTs. Recent advances in molecular techniques have allowed an even greater level of discrimination and one PT can be further subdivided into different groups based on the analysis of plasmids or plasmid-associated genes such as the *Salmonella* plasmid virulence (*spv*) gene, insertion sequences in the chromosome, and DNA RFLP methods, e.g. PFGE. There have been a number of excellent reviews on this topic (Threlfall, Hampton & Ridley, 1998; see also Landeras, González-Hevia & Mendoza, 1998). In one vigorous comparison, PFGE appeared more discriminatory than plasmid analysis and ribotyping (Ridley, Threlfall & Rowe, 1998).

Molecular subtyping has revealed the existence of dominant *Salmonella enteritidis* clones associated with both poultry (chicken and egg) transmission and gastroenteritis (Landeras, González-Hevia & Mendoza, 1998). It has also been of great value in characterizing the noticeable increase in outbreaks, sometimes on an international scale, associated with foods of intermediate or low a_w, such as chocolate (*Salmonella napoli*), snack bars (*Salmonella agona*) and dried milk (*Salmonella anatum*) (Anonymous, 1995, 1997; Gill *et al.*, 1983; Greenwood & Hooper, 1983; Rowe *et al.*, 1987). Without the existence of sophisticated typing methods, it would not have been possible to identify accurately the sources of infection, which ultimately led to control measures and a concomitant reduction in human salmonellosis.

Campylobacter jejuni

Campylobacter jejuni is the most commonly reported food-borne pathogen in developed countries (Phillips, 1995). The majority of human cases are sporadic, which for a long time posed a problem in the establishment of the events leading to contamination of the food chain. Trends in *Campylobacter* epidemiology, however, are now emerging. This is due to the development and application of typing methods similar to those used for *Salmonella* spp. (Gibson & Owen, 1998; Steele *et al.*, 1998). These approaches have made it possible to demonstrate that *Campylobacter* spp. found in animals such as chickens and cattle are identical to those from human cases. The next stage is to use typing to identify sources and routes of infection in the key food animals (Nielsen, Engberg & Madsen, 1997).

Campylobacter spp. in poultry provide a good example of the usefulness of typing as an aid to preventing food contamination. Poultry flocks can be colonized with *Campylobacter* spp. from a number of potential reservoirs. These include contaminated drinking water, colonized parent birds possibly allowing vertical transmission and a contaminated environment outside the poultry house; the latter is believed to be the most important. A range of phenotypic and genotypic tests have confirmed that identical isolates of *C. jejuni* were obtained from chickens and the area around the broiler house (D. Newell, personal communication). This confirmation will make it easier to convince the poultry industry to adopt intervention measures. Boot dipping in strong phenolic disinfectant before people entered broiler houses was found to either delay or prevent flock colonization with *C. jejuni* (Humphrey, Henley & Lanning, 1993). A reduction in the colonization of the live animal and the associated fall in carcase contamination rates will ease some of the pressure on food processors who receive raw chicken meat.

Future needs aimed at reducing the contamination of the food supply are

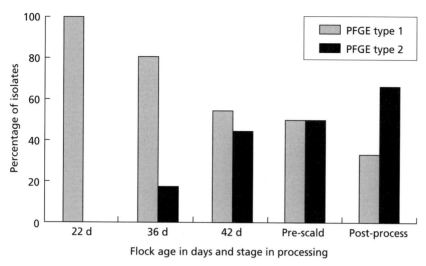

Fig. 1. Use of PFGE to show the changes in predominant *C. jejuni* subtypes during a chicken flock growing period and during processing.

numerous. High-resolution typing, such as that afforded by PFGE, needs to become more widespread to help clarify epidemiological blind spots. The finding that not all human serotypes are found in poultry suggests that there are significant sources of infection for humans yet to be identified. For example, the epidemiology of non-*C. jejuni/coli* that are zoonotically important must be thoroughly examined. Also, *Campylobacter* spp. are found in natural water sources throughout the year (Altekruse *et al.*, 1999), but their presence initially was found not to correlate strongly with indicator organisms for faecal contamination (Carter *et al.*, 1987). When typing was applied, however, water and human isolates were found to share subtypes (Bolton *et al.*, 1987).

Molecular typing also makes it possible to monitor changes in *Campylobacter* subpopulations either during the life of the flock or when birds are processed. Recent work in Exeter used differences in PFGE profiles to follow changes in the *C. jejuni* population of broiler flocks during the growing period and processing (Fig. 1).

E. coli O157:H7

This is often referred to as a newly emergent pathogen and characteristics such as its production of verocytotoxins, low infecting dose and relatively high case fatality rate in young and elderly victims have led it to be classified in the UK as a containment Category 3 pathogen (reviewed by Bolton & Aird, 1998; Buchanan & Doyle, 1997; Karmali, 1989; Lansbury & Ludlam, 1997; Smith, 1997). In England and

Wales, a peak of over 1000 cases was reported in 1997 (Anonymous, 1999b, c). The majority of cases are sporadic and this has influenced the approach in typing investigations to unravel the epidemiology of *E. coli* O157 at both the phenotypic and, especially, at the molecular level. After hierarchical clustering by biochemical tests, serotyping, phage typing and antibiograms, molecular subtyping is applied. This consists of using molecular probes to examine isolates for virulence factor sequences in their DNA. These will include *eae*, which encodes intimin, and sequences encoding verotoxins (VT1 and VT2). PCR-based tests include the subtyping of VT2 genes and PFGE can provide final, high-level discrimination (Domingue *et al.*, 1997; Grif *et al.*, 1998; Willshaw *et al.*, 1997). This combined approach with a molecular emphasis has established several important facts relevant to improving food safety.

For currently unknown reasons, the dominant *E. coli* O157:H7 clone in England and Wales is PT2 expressing verocytotoxin type 2, i.e. PT2 VT2. In contrast, PT49 strains are now isolated infrequently from foods whereas a few years ago they were common. Once the reasons for these population changes have been established, and this will undoubtedly be achieved with molecular-based techniques, realistic intervention strategies can be targeted.

The main reservoir of *E. coli* O157 is cattle and sheep, where for animals older than 6 months carriage is usually asymptomatic (see reviews cited above and Chapman *et al.*, 1997). Entry of organisms into the food chain is typically by faecal contamination. This route has been shown to exist not only in post-slaughter processing but also prior to this, so that incidents have been associated with contamination of a wide variety of foods including fruit juice, dairy products and potatoes stored in contaminated peat. Contamination of a public water supply and direct transmission from animals to humans and from humans to humans have also been reported (reviewed by Adak *et al.*, 1996; Lansbury & Ludlam, 1997; Parry *et al.*, 1998; see also Jones & Roworth, 1996; Trevena *et al.*, 1999). The latter route obviously has implications for food handlers in all types of settings. Universality may not apply in the natural history of the typical human infection, making the emphasis upon precautions different from country to country. For example, in the USA, the major risk factor is the consumption of undercooked beefburgers, and an abattoir-based zero tolerance HACCP analysis is favoured, as exemplified by the large-scale recall of raw minced beef from a source implicated in an outbreak (LaBudde, 1997). In the UK, undercooked beefburgers are also important but cross-contamination from raw to cooked meat would seem to be the greater risk factor (G. K. Adak, personal communication). This was shown by the large butcher-associated outbreak in central Scotland in 1991 (Reid, 1997). The heightened awareness following an investigation of this important outbreak (Pennington, 1997) has led to the inclusion of molecular subtyping in the refinement of existing HACCP systems and the monitoring of meat retail outlets (Anonymous,

1999c). Also, other nationwide surveillance has permitted identification of, for example, the point of breakdown in producing contaminated unpasteurized cheese and also enabled public warnings (Anonymous, 1999d).

Another area for concern is the natural history of non-O157 verocytotoxin-producing *E. coli* strains that have caused food-borne human infection and which are known to be zoonotically important. This area needs to be quickly and thoroughly examined as indicated by the serotype O11 salami-associated outbreak in Australia (Goldwater & Bettelheim, 1998). For food microbiologists, the situation is hampered currently by poor awareness of the potential importance of these serogroups and also the paucity of commercial kits for their detection.

Prevention at the molecular level (molecular intervention) of food-borne disease

The use of probiotic flora in the gastrointestinal tract of food animals to limit colonization of food-borne pathogens shows promise as an on-farm intervention strategy. There is an ever-increasing choice of molecular probes that target DNA and RNA sequences: for example, 16S rRNA, which is a molecule that has severe constraints placed upon its mutation rate because of its functional importance, i.e. it can be regarded as a molecular clock (Woese, 1987). Because of this, 16S rRNA is genus-specific and therefore can be utilized in detection and identification methods. This technique has many potential applications and is being used to explore the genetic diversity and activity of various gut flora micro-communities, so allowing an understanding of, for example, their identities, ecological niches, inter-relationships, functionality and specific gene(s) expression, all as influenced by dietary factors (Ricke & Pillai, 1999). This hopefully will lead to greater use of probiotics and reduced usage of growth promoters such as antibiotics.

Rapid methods and the detection/isolation of food-borne pathogens

This subject area will be discussed largely with the detection/isolation of *Salmonella* spp. as the example.

Methods to detect Salmonella in food

Methods for the isolation and detection of *Salmonella* were first developed for clinical specimens because of widespread human and animal disease. Later, when food became apparent as an important factor in the chain of salmonellosis, microbiologists naturally applied the same methods to food samples. For many years, the detection of *Salmonella* in foods was based on the application of culture methods originally designed to detect this pathogen in clinical specimens. As more became known about the low infectious dose of *Salmonella*, it was apparent that clinical methods were no longer suitable for foods. New methods had to be developed that could detect as few as a single *Salmonella* cell. The detection of such low levels was made even more difficult with the realization that debilitated cells, which could be present as a consequence of food processing or storage, could also cause infection. Without some provision to rejuvenate these injured cells, the selective conditions required in the culture isolation procedure could render the bacteria undetectable. Such an occurrence could have serious consequences. Legislation in most countries, therefore, states that relevant products must be demonstrated to be free of *Salmonella*. This is done by testing a representative sample, typically 25 g, for the presence or absence of the bacterium.

Conventional culture methods

Five basic steps are common to most culture methods for the detection, isolation and identification of *Salmonella* in foods: (i) pre-enrichment of the sample in a nutritious non-selective broth; (ii) subsequent enrichment in a selective broth, allowing *Salmonella* to proliferate while suppressing the growth of competing species of bacteria; (iii) isolation of pure cultures of *Salmonella*, by streaking onto a selective agar; (iv) presumptive identification of positive isolates and biochemical characterization; and (v) definitive serological confirmation (D'Aoust, 1981; Andrews, 1985; Fricker, 1987). These five basic steps, first submitted to the Association of Official Analytical Chemists (AOAC) for approval in 1967, still make up the conventional culture method. All new methods that seek AOAC approval are judged against these criteria.

Rapid methods

The primary disadvantage of the procedures mentioned above is the time required to determine whether a sample contains *Salmonella*. A sample cannot be reported as negative until the fourth day after initiation of analysis. Samples presumptively positive for *Salmonella* on the fourth day require several more days for confirmation of identification. The added expense of this lengthy period on food processing and warehousing has stimulated an enormous interest in faster methods. Since 1973, 20 new methods have been granted AOAC approval (Andrews, 1994), and there is an even larger number of publications comparing different pre-enrichment broths, different selective enrichment broths and modified protocols for the detection of *Salmonella* (for reviews see Kitchfield, 1973; D'Aoust, 1981; Fricker, 1987; Feng, 1992; Blackburn, 1993). Over the past two decades, there have been many significant advances in rapid methods (Table 3), for example ELISAs, electrophoresis methods, nucleic acid probes, PCRs, with test times now as short as 30 min (for reviews see Feng, 1992; Blackburn, 1993). Despite promising results in laboratory-based experiments, when used in real food enrichments even the most

Table 3. Rapid methods for the detection or identification of pathogenic bacteria in foods. Data taken from Swaminathan & Feng (1994).

General test area	Specific development
1 Miniaturized biochemical assays	API
	Micro-ID
	Enterotube
	GPI and GNI cards
2 Physicochemical assays	Impedance
	Bioluminescence
	Fluorescence
	GLC
3 Antibody-based tests	Immunofluorescence
	ELISA
	Latex coagglutination
4 Nucleic acid-based tests	DNA probes
	In vitro amplification (PCR)

sensitive commercially available detection systems require at least 10^4 target cells (ml enrichment broth)$^{-1}$ to generate reliable results (Jay, 1992).

Rapid detection and identification to strain level is an obvious, highly desirable goal in food microbiology, particularly for HACCP style enforcement. The ideal method would allow the direct examination of food samples without the need for time-consuming culture. This requires the detection of a live cell or a cellular component which is either present or active when the cell is alive or absent or inactive when the cell is dead. This is clearly a difficult target to achieve and this part of the chapter discusses developments in this field of microbiology. There are a wide range of rapid tests available; Table 3 outlines the various methods which are either available or are being developed.

At the molecular level, developing the correct sensitivity and specificity for an assay is problematic, particularly for PCR-based technology. Here the ultimate aim is to target and amplify a gene sequence in an organism whose viability needs to be determined and which may be present in low numbers within a food matrix (Scheu, Berghof & Stahl, 1998). Nucleic acid extraction strategies are very important also (Grant, Dickinson & Kroll, 1993). The enhancement of cultural methods by utilizing specific antibodies via immunomagnetic separation (IMS) is well established (Chapman *et al.*, 1997; Willshaw *et al.*, 1997). IMS is now becoming more common for enhancing PCR-based pathogen detection. This may help overcome the various problems above (Scheu, Berghof & Stahl, 1998). Also, throughput can be increased significantly by the trend towards microtitre plate formats for PCR reactions, and convenience is available now with the advent of bead format PCR kits.

Sensitivity of rapid methods

The new methodologies can be divided into two broad categories. There are those designed to detect/isolate viable cells and those which detect cellular components such as DNA. There are concerns that the latter methods may detect either dead cells or DNA released from them and that cultural techniques may lack the required sensitivity.

Detection of cellular components

Conventional culture techniques, as discussed earlier, can be long-winded, cumbersome and expensive. How much better it would be if it were possible to detect pathogenic bacteria by the direct application of a test to a food sample. *Salmonella* spp., and presumably other food-borne pathogens, appear to have unique DNA sequences which could allow detection. For example, Olsen *et al.* (1991) demonstrated that *S. typhimurium* has a DNA sequence which is not present in either *E. coli* or *Citrobacter freundii*. They developed both a DNA probe (Olsen *et al.*, 1991) and a PCR method (Aabo *et al.*, 1993). With these techniques they demonstrated that the molecular methods compared well to conventional culture when used on enrichment broths. The above studies were designed to obviate the need for selective culture and serological/biochemical identification. There are other examples, and various automated PCR-based systems are now available. Most of them require an enrichment period, which delays results, but sensitivity can be of the order of 3 c.f.u. per 25 g food (Chen *et al.*, 1997). Other new approaches include flow cytometry (Goodridge, Chen & Griffiths, 1999; Pyle, Broadaway & McFeters, 1999). These methods have faced the criticism that they are still too slow for the rapid pace of modern processed food production methods. How much more sensible it would be if one could intervene earlier in the culture process or test foods directly. For example, the incorporation of molecular probes into plasticized membranes used in food packaging to detect virulence factors such as verocytotoxins has been cited as an example of an on-the-spot 'litmus test' that the food industry needs (Stevens & Cheng, 1996).

It should be borne in mind that, although the above tests may be relatively inexpensive at the individual sample level, collectively they may represent a considerable financial burden. It has yet to be properly established whether the consumer is prepared to pay the higher prices the 'tested' foods may command.

Sublethal injury and rapid culture techniques

There is an understandable desire to reduce the time taken to confirm the presence or absence of food-borne pathogens. Unfortunately, bacteria in foods, particularly foods which have been processed, will often be sublethally damaged by the treatments the foods have received. For example, in *Salmonella*, as in other food-borne pathogens, sublethal injury has a number of manifestations. The most important of these is an increased sensitivity to selective agents, which non-injured cells more easily resist. Thus, while *Salmonella* cells in faecal specimens will grow well in the presence of selenite, because they are essentially uninjured, cells in foods will not. Resistance to selective agents will be regained if cells are given the opportunity to repair the cellular damage responsible for the increased sensitivity, usually damage to the outer membrane in Gram-negative bacteria (Hurst *et al.*, 1973; Calcott & Macleod, 1975). This is usually achieved by incubating samples in a non-selective medium such as buffered peptone water. The recovery step is vital to the success of food testing for pathogenic bacteria whether by rapid or conventional methods. It is perhaps unfortunate that relatively little attention has been paid to this step in the development of rapid methods. Injured bacteria can have very long recovery times, even under conditions believed by microbiologists to be ideal. Heat-stressed *Pseudomonas fluorescens* had a lag time of 9 h before regaining resistance to selective agents (McCoy & Ordal, 1979) and Dabbah, Moats & Mattick (1969) demonstrated that lag times could be as long as 48–72 h. Mackey & Derrick (1982) calculated lag times of *Salmonella* injured by different types of stress and showed that, in a population of cells with a lag time of 9 h, some cells required up to 14 h to regain resistance to high salt concentrations, thus demonstrating the heterogeneity of the extent of injury within a population. This type of result confirmed the fact that the measured lag time for a population was likely to reflect only the least severely injured cells that repaired and multiplied first. Such long recovery periods, and here highly damaged cells of *Campylobacter coli* can take up to 4 d to recover (K. Martin, personal communication), run counter to the desire for more rapid detection/isolations.

There have been a number of attempts to shorten recovery but few were successful. D'Aoust & Maishment (1979) reported that pre-enrichment for 6 h in various non-selective media failed to identify about half of low- and high-moisture foods contaminated with *Salmonella*. Later, D'Aoust, Sewell & Jean (1990) reported that shorter incubation times, 3–8 h, in non-selective enrichment media did not result in effective resuscitation of stressed organisms, and gave unacceptably high numbers of false negative results. To overcome the problems found by these workers, a detailed understanding of the functioning of pre-enrichment is required. It also has to be recognized that, within a population of *Salmonella* spp., individual cells can have markedly different recovery times, even when the population as a whole has been subjected to the same stress. This is illustrated in Fig. 2 with a heat-damaged population of *S. typhimurium*, where recovery was measured using a Bioscreen (Stephens *et al.*, 1997). In order for methods to become faster, either detection systems have to become more sensitive or enrichment culture has to be made more effective. In particular, it is important to gain a better understanding of sublethal injury and recovery so that culture conditions which minimize the effects of the former and which speed the latter can be identified.

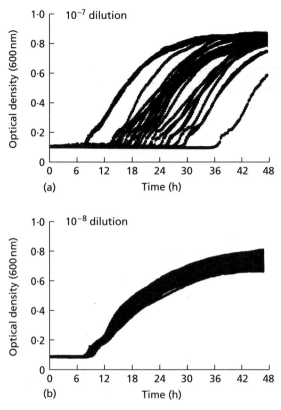

Fig. 2. Recovery and growth of (a) heat-injured and (b) undamaged *S. typhimurium* (CMCC 3073) at different dilution levels in a typical pre-enrichment medium measured with the Bioscreen analyser. Data taken from Stephens *et al.* (1997) and reproduced with permission from the Society for Applied Microbiology.

Are they living, or are they dead?

Certain raw ingredients of processed foods, like meat, poultry, milk and vegetables, will be frequently contaminated with food-borne pathogens. Food treatment processes are designed not to remove these bacteria but to kill them. This means that the finished product will contain dead cells and material released from them. The most rapid of the rapid methods use systems for the detection of DNA or RNA. Within the bacterial chromosome there will be sequences of DNA which are likely to be unique to that organism. Identification and amplification of these sequences by PCR potentially provides detection/identification methods. There are two problems with this approach. The first is interference from food matrices in which the bacteria are enmeshed (Werners *et al.*, 1991). It is possible, however, to remove bacteria from food samples by IMS and this, coupled to PCR testing, may offer an attractive alternative to conventional culture (Grant, Dickinson & Kroll, 1993).

A greater problem with DNA detection is the fact that DNA is heat-stable. Thus intact DNA will be present in processed foods. Duprey *et al.* (1997) also demonstrated that DNA both in a free form and within dead cells of *Salmonella* could survive for considerable periods in sea water. One possible way to overcome this problem is to include a culture step (Werners *et al.*, 1991) as a means of detecting viable cells. Not only will this increase the length of the test, but it may not be sufficient to allow the growth of more severely damaged cells, as discussed earlier.

An alternative nucleic acid target to determine viability would be an RNA molecule that exists as many copies in a cell and which exhibits rapid turnover. mRNA may prove to be a suitable target as it has a reported half-life of 2 min or less, and PCRs have been published for the detection of viability in *E. coli* (Sheridan *et al.*, 1998) and the thermophilic *Campylobacter* spp. (Sails *et al.*, 1998). The rate of degradation of the mRNA is influenced by the method of inactivation, however, and this has to be considered when assessing food-processing conditions (Sails *et al.*, 1998).

Conclusions

Molecular techniques can play an important part in the prevention of food contamination by assisting in the identification of sources and routes of infection in food animals and by their use as rapid methods of detection for key food-borne pathogens.

Typing, in its many forms, has proved invaluable in identifying sources of infection in outbreaks and in following the changing patterns of dominance of food-borne and other pathogenic bacteria. Typing is particularly useful in the context of an outbreak where the evidence of similarity between bacteria can be supported by food consumption patterns and relationships between cases in terms of time and geography. Typing becomes more tenuous in individual sporadic cases. Investigations of these events are hampered by the fact that bacteria do not have markers of individuality and are only very rarely confined to one discrete geographical area or to one food contamination event. Thus, although a bacterium from an implicated source can be shown by typing methods to be identical to that isolated from a case, in the absence of powerful epidemiological evidence it can be difficult to prove a link. Typing is being increasingly challenged in the courts where the ubiquity of certain bacterial strains casts doubt on the identification of vehicles or causes of infection.

Rapid methods are also not without difficulties and many are currently too insensitive to ensure food safety, although much progress is being made in this area. Cultural methods are attractive because they enable the organisms to be visualized. They have also been fundamental to microbiology for over 100 years. The traditional methods will always have a usefulness but it is likely that they will be supplemented by more and more molecular methodologies.

Rapid methods may well have an important future role in food safety microbiology. At present, however, there is insufficient faith in these tests in the scientific community for the classical culture techniques to be abandoned. It is likely, given the conservatism of science, that the molecular techniques will supplement traditional methods rather than replace them.

References

Aabo, S., Rasmussen, O. F., Rosser, L., Sorensen, P. D. & Olsen, J. E. (1993). *Salmonella* identification by the polymerase chain reaction. *Molecular and Cellular Probes* 7, 171–178.

Adak, G. K., Wall, P. G., Smith, H. R., Cheasty, T., Bolton, F. J., Griffin, M. A. S. & Rowe, B. (1996). PHLS begins a national case control study of *Escherichia coli* O157 infection in England. *Communicable Disease Report* 6, R144–R146.

Altekruse, S. F., Stern, N. J., Fields, P. I. & Swerdlow, D. L. (1999). *Campylobacter jejuni* – an emerging foodborne pathogen. *Emerging Infectious Diseases* 5, 28–35.

Andrews, W. H. (1985). A review of culture methods and their relation to rapid methods for the detection of *Salmonella* in foods. *Food Technology* 39, 77–82.

Andrews, W. H. (1994). Update on validation of microbiological methods by AOAC International. *Journal of AOAC International* 77, 925–931.

Anonymous (1995). An outbreak of *Salmonella agona* due to contaminated snacks. *Communicable Disease Report* 5, 29–32.

Anonymous (1997). *Salmonella anatum* infection in infants linked to dried milk. *Communicable Disease Report* 7, 33–36.

Anonymous (1998). The House of Commons Agriculture Committee. 4th Report: *Food Safety*, vol. 1. Report and Proceedings of the Committee. London: The Stationery Office.

Anonymous (1999a). *Salmonella* infections, England & Wales; common gastrointestinal infections, England & Wales. *Communicable Disease Report* 9, 10–11.

Anonymous (1999b). VTEC O157 phage type 21/28 infection in North Cumbria: update. *Communicable Disease Report* 9, 105.

Anonymous (1999c). Outbreak of VTEC O157 infection in East Sussex. *Communicable Disease Report* 9, 219, 222.

Anonymous (1999d). *Escherichia coli* O157 associated with eating unpasteurised cheese. *Communicable Disease Report* 9, 113, 116.

Arnold, C., Metherell, L., Willshaw, G., Maggs, A. & Stanley, J. (1999). Predictive fluorescent amplified fragment length polymorphism analysis of *Escherichia coli*. *Journal of Clinical Microbiology* 37, 1274–1279.

van Belkum, A. (1994). DNA fingerprinting of medically important micro-organisms by use of PCR. *Clinical Microbiology Reviews* 7, 174–184.

Blackburn, C. W. (1993). Rapid and alternative methods for the detection of salmonellas in foods. *Journal of Applied Bacteriology* 75, 199–214.

Bolton, F. J. & Aird, H. (1998). Verocytotoxin-producing *Escherichia coli* O157: public health and microbiological significance. *British Journal of Biomedical Science* 55, 127–135.

Bolton, F. J., Coates, D., Hutchinson, D. N. & Godfree, A. F. (1987). A study of thermophilic campylobacters in a river system. *Journal of Applied Bacteriology* 62, 167–176.

Böttger, E. C. (1996). Approaches for identification of micro-organisms. *ASM News* 62, 247–250.

Buchanan, R. L. & Doyle, M. P. (1997). Foodborne disease significance of *Escherichia coli* O157 : H7 and other enterohemorrhagic *E. coli*. *Food Technology* 51, 69–76.

Calcott, P. H. & Macleod, R. A. (1975). The survival of *E. coli* from freeze–thaw damage:

permeability barrier damage and viability. *Canadian Journal of Microbiology* **21**, 1724–1732.

Carter, A. M., Pacha, R. E., Clark, G. W. & Williams, E. A. (1987). Seasonal occurrence of *Campylobacter* spp. and their correlation with standard indicator bacteria. *Applied and Environmental Microbiology* **53**, 523–526.

Chapman, P. A., Sididons, C. A., Cerdan-Malo, A. T. & Harkin, M. A. (1997). A 1-year study of *Escherichia coli* O157 in cattle, sheep, pigs and poultry. *Epidemiology and Infection* **119**, 245–250.

Chen, S., Yee, A., Griffiths, M., Wu, K. Y., Wang, C.-N., Rahn, K. & De Grandis, S. A. (1997). A rapid, sensitive and automated method for detection of *Salmonella* species in foods using AG-9600 AmpliSensor Analyser. *Journal of Applied Microbiology* **83**, 314–321.

Craven, P. C., Mackel, D. C., Baine, W. B. & 8 other authors (1975). International outbreak of *Salmonella eastbourne* infection traced to contaminated chocolate. *Lancet* **1**, 788–793.

Dabbah, R., Moats, W. A. & Mattick, J. F. (1969). Factors affecting resistance to heat and recovery of heat-injured bacteria. *Journal of Dairy Science* **52**, 608–614.

D'Aoust, J.-Y. (1981). Update on pre-enrichment and selective enrichment conditions for detection of *Salmonella* in foods. *Journal of Food Protection* **44**, 369–374.

D'Aoust, J.-Y. (1985). Infective dose of *Salmonella typhimurium* in cheddar cheese. *American Journal of Epidemiology* **122**, 717–720.

D'Aoust, J.-Y. (1994). *Salmonella* and the international food trade. *International Journal of Food Microbiology* **24**, 11–31.

D'Aoust, J.-Y. & Maishment, C. (1979). Pre-enrichment conditions for effective recovery of *Salmonella* in foods and feed ingredients. *Journal of Food Protection* **42**, 153–157.

D'Aoust, J.-Y., Sewell, A. & Jean, A. (1990). Limited sensitivity of short (6 h) selective enrichment for detection of foodborne *Salmonella*. *Journal of Food Protection* **53**, 562–565.

De Wit, J. C., Brockhuizen, G. & Kampelmacher, E. H. (1979). Cross-contamination during the preparation of frozen chickens in the kitchen. *Journal of Hygiene* **83**, 27–32.

Domingue, G., Smith, H. R., Willshaw, G., Cheasty, T. & Rowe, B. (1997). Genotyping of food, animal and human isolates of vero-cytotoxin producing strains of *Escherichia coli* O157 (O157 VTEC). In *Proceedings of the Fourth International Meeting on Bacterial Epidemiological Markers*, p. 20. Elsinore, Denmark. Presentation number S203.

Duim, B., Wassenaar, T. M., Rigter, A. & Wagenaar, J. (1999). High-resolution genotyping of *Campylobacter* strains isolated from poultry and humans with amplified fragment length polymorphism fingerprinting. *Applied and Environmental Microbiology* **65**, 2369–2375.

Duprey, E., Caprais, M. P., Derrien, A. & Fach, P. (1997). *Salmonella* DNA persistence in natural sea waters using PCR analysis. *Journal of Applied Microbiology* **82**, 507–510.

Farber, J. M. (1996). An introduction to the Hows and Whys of molecular typing. *Journal of Food Protection* **59**, 1091–1101.

Feng, P. (1992). Commercial assay systems for detecting foodborne *Salmonella*: a review. *Journal of Food Protection* **55**, 927–934.

Fontaine, R. E., Arnon, S., Martin, W. T., Vernon, T. M., Gangarosa, E. J., Farmer, J. J., Moram, A. B., Silliker, J. H. & Decker, D. L. (1978). Raw hamburger: an interstate common source of human salmonellosis. *American Journal of Epidemiology* **107**, 36–45.

Fontaine, R. E., Cohen, M. L., Martin, W. T. & Vernon, T. M. (1980). Epidemic salmonellosis from cheddar cheese: surveillance and prevention. *American Journal of Epidemiology* **111**, 247–253.

Fricker, C. R. (1987). A review: the isolation of *Salmonellas* and *Campylobacters*. *Journal of Applied Bacteriology* **63**, 99–116.

Gibson, J. R. & Owen, R. J. (1998). *Campylobacter* infections: species identification and typing. In *Methods in Molecular Medicine*, vol. 15, *Molecular Bacteriology: Protocols and Clinical Applications*, pp. 407–418. Edited by N. Woodford & A. P. Johnson. New Jersey: Humana Press.

Gill, O. N., Sockett, P. N., Bartlett, C. L. R. & Vaile, M. S. B. (1983). Outbreak of *Salmonella napoli* infection caused by contaminated chocolate bars. *Lancet* **i**, 574–577.

Goldwater, P. N. & Bettelheim, K. A. (1998). New

perspectives on the role of *Escherichia coli* O157 : H7 and other enterohaemorrhagic *E. coli* serotypes in human disease. *Journal of Medical Microbiology* **47**, 1039–1045.

Goodridge, L., Chen, J. & Griffiths, M. (1999). Development and characterisation of a fluorescent-bacteriophage assay for detection of *Escherichia coli* O157 : H7. *Applied and Environmental Microbiology* **65**, 1397–1404.

Grant, K. A., Dickinson, J. H. & Kroll, R. G. (1993). Specific and rapid detection of foodborne bacteria with rRNA sequences and the polymerase chain reaction. In *New Techniques in Food and Beverage Microbiology* SAB Technical Series No. 31, pp. 147–162. Edited by R. G. Kroll, A. Gilmour & M. Sussman. London: Blackwell Scientific Publications.

Greenwood, M. H. & Hooper, W. L. (1983). Chocolate bars contaminated with *Salmonella napoli*: an infectivity study. *British Medical Journal* **286**, 1394.

Grif, K., Karch, H., Schneider, C. & 7 other authors (1998). Comparative study of five different techniques for epidemiological typing of *Escherichia coli* O157. *Diagnostic Microbiology and Infectious Disease* **32**, 165–176.

Hood, A. M., Pearson, A. D. & Shahamat, M. (1988). The extent of surface contamination of retailed chickens with *Campylobacter jejuni* serogroups. *Epidemiology and Infection* **100**, 17–25.

Humphrey, T. J., Henley, A. & Lanning, D. G. (1993). The colonisation of broiler chickens with *Campylobacter jejuni*: some epidemiological investigations. *Epidemiology and Infection* **110**, 601–607.

Hurst, A., Hughes, A., Beare-Rogers, J. L. & Collins-Thompson, D. L. (1973). Physiological studies on the recovery of salt tolerance by *Staphylococcus aureus* after sublethal heating. *Journal of Bacteriology* **116**, 901–907.

Jay, J. M. (1992). Determining micro-organisms and their products in foods. In *Modern Food Microbiology*, 3rd edn, pp. 95–190. Edited by J. M. Jay. New York: Chapman & Hall.

Jones, I. G. & Roworth, M. (1996). An outbreak of *Escherichia coli* O157 and campylobacteriosis associated with contamination of a drinking water supply. *Public Health* **110**, 277–282.

Kapperud, G., Gustavsen, S., Hellesnes, I., Hansen, A. H., Lassen, J., Hirn, J., Jahkola, M., Montenegro, M. A. & Helmuth, R. (1990). Outbreak of *Salmonella typhimurium* infection traced to contaminated chocolate and caused by a strain lacking the 60-megadalton virulence plasmid. *Journal of Clinical Microbiology* **28**, 2597–2601.

Karmali, M. A. (1989). Infection by verocytotoxin producing *E. coli*. *Clinical Microbiology Reviews* **2**, 15–28.

Killalea, D., Ward, L. R., Roberts, D. & 12 other authors (1996). International epidemiological and microbiological study of outbreak of *Salmonella agona* infection from a ready to eat savoury snack – I: England and Wales and the United States. *British Medical Journal* **313**, 1105–1107.

Kitchfield, J. H. (1973). *Salmonella* and the food industry – methods for isolation, identification and enumeration. *CRC Critical Reviews in Food Technology* **3**, 415–456.

LaBudde, R. A. (1997). The facts of the matter, in re: The Hudson Foods O157 : H7 outbreak. *International Food Safety News* **6**, 3–11.

Landeras, E., González-Hevia, M. A. & Mendoza, M. C. (1998). Molecular epidemiology of *Salmonella* serotype Enteritidis. Relationships between food, water and pathogenic strains. *International Journal of Food Microbiology* **43**, 81–90.

Lansbury, L. E. & Ludlam, H. (1997). *Escherichia coli* O157: lessons from the past 15 years. *Journal of Infection* **34**, 189–193.

Lehmacher, A., Bockemühl, J. & Aleksic, S. (1995). Nationwide outbreak of human salmonellosis in Germany due to contaminated paprika and paprika-powdered potato chips. *Epidemiology and Infection* **115**, 501–511.

McCoy, D. R. & Ordal, Z. J. (1979). Thermal stress of *Pseudomonas fluorescens* in complex media. *Applied and Environmental Microbiology* **37**, 443–448.

Mackey, B. M. & Derrick, C. M. (1982). The effect of sublethal injury by heating, freezing, drying and gamma-radiation on the duration of the lag phase of *Salmonella typhimurium. Journal of Applied Bacteriology* **53**, 243–251.

Milch, H. (1998). Advances in bacterial typing methods. *Acta Microbiologica et Immunologica Hungarica* **45**, 401–408.

Nielsen, E. M., Engberg, J. & Madsen, M. (1997). Distribution of serotypes of *Campylobacter jejuni* and *C. coli* from Danish patients, poultry, cattle, and swine. *FEMS Immunology and Medical Microbiology* **19**, 47–56.

Notermans, S. (1994). Epidemiology and surveillance of *Campylobacter* infections. In *Report on a WHO Consultation on Epidemiology and Control of Campylobacteriosis*, pp. 35–44. Bilthoven, The Netherlands, 25–27 April 1994: The World Health Organization.

Odumeru, J. A., Steele, M., Fruhner, L., Larkin, C., Jiang, J., Mann, E. & McNab, W. B. (1999). Evaluation of accuracy and repeatability of identification of foodborne pathogens by automated bacterial identification systems. *Journal of Clinical Microbiology* **37**, 944–949.

Olive, D. M. & Bean, P. (1999). Principles and applications of methods for DNA-based typing of microbial organisms. *Journal of Clinical Microbiology* **37**, 1661–1669.

Olsen, J. E., Aabo, S., Nielsen, E. O. & Nielsen, B. B. (1991). Isolation of a *Salmonella*-specific DNA hybridisation probe. *APMIS* **99**, 114–120.

Parry, S. M., Salmon, R. L., Willshaw, G. A. & Cheasty, T. (1998). Risk factors for and prevention of sporadic infections with verocytotoxin (shiga toxin) producing *Escherichia coli* O157. *Lancet* **351**, 1019–1022.

Pennington, H. (1997). *Report on the Circumstances Leading to the 1996 Outbreak of Infection with E. coli O157 in Central Scotland, the Implications for Food Safety and the Lessons to be Learned.* London: The Stationery Office.

Phillips, C. A. (1995). Incidence, epidemiology and prevention of foodborne *Campylobacter* species. *Trends in Food Science Technology* **6**, 83–87.

Phillips, L. E., Humphrey, T. J. & Lappin-Scott, H. M. (1998). Chilling invokes different morphologies in two *Salmonella enteritidis* PT4 strains. *Journal of Applied Microbiology* **84**, 820–826.

Preston, M. A. (1999). Interpretation of molecular typing results for bacterial species or subtypes with limited genetic diversity. *Canadian Journal of Infectious Disease* **10**, 83.

Pyle, B. H., Broadaway, S. C. & McFeters, G. A. (1999). Sensitive detection of *Escherichia coli* O157 : H7 in food and water by immunomagnetic separation and solid-phase laser cytometry. *Applied and Environmental Microbiology* **65**, 1966–1972.

Reid, D. (1997). Second SCIEH Verocytotoxigenic *E. coli* Workshop, 31st January 1997. *SCIEH Weekly Report* (supplement) **1**, no.97/13, p. 1.

Ricke, S. C. & Pillai, S. D. (1999). Conventional and molecular methods for understanding probiotic bacteria functionality in gastrointestinal tracts. *Critical Reviews in Microbiology* **25**, 19–38.

Ridley, A. M., Threlfall, E. J. & Rowe, B. (1998). Genotypic characterisation of *Salmonella enteritidis* phage types by plasmid analysis, ribotyping, and pulsed-field gel electrophoresis. *Journal of Clinical Microbiology* **36**, 2314–2321.

Robinson, D. A. (1981). Infective dose of *Campylobacter jejuni* in milk. *British Medical Journal (Clinical Research Edition)* (May 16) **282**, 1584.

Rowe, B., Begg, N. T., Hutchinson, D. N., Dawkins, H. C., Gilbert, R. J., Jacob, M., Hales, B. H., Rae, F. A. & Jepson, M. (1987). *Salmonella ealing* infections associated with consumption of infant dried milk. *Lancet* **ii**, 900–903.

Sails, A. D., Bolton, F. J., Fox, A. J., Wareing, D. R. A. & Greenway, D. L. A. (1998). A reverse transcriptase polymerase chain reaction assay for the detection of thermophilic *Campylobacter* spp. *Molecular and Cellular Probes* **12**, 317–322.

Scheu, P. M., Berghof, K. & Stahl, U. (1998). Detection of pathogenic and spoilage micro-organisms in food with the polymerase chain reaction. *Food Microbiology* **15**, 13–31.

Sheridan, G. E. C., Masters, C. I., Shallcross, J. A. & Mackey, B. M. (1998). Detection of mRNA by reverse transcription-PCR as an indicator of viability in *Escherichia coli* cells. *Applied and Environmental Microbiology* **64**, 1313–1318.

Smith, H. R. (1997). Verocytotoxin-producing *Escherichia coli* O157: cause for concern. *Society for General Microbiology Quarterly* 24, 54–57.

Steele, M., McNab, B., Fruhner, L., DeGrandis, S., Woodward, D. & Odumeru, J. A. (1998). Epidemiological typing of *Campylobacter* isolates from meat processing plants by pulsed-field gel electrophoresis, fatty acid profile typing, serotyping, and biotyping. *Applied and Environmental Microbiology* 64, 2346–2349.

Stephens, P. J., Joynson, J. A., Davies, K. W., Holbrook, R., Lappin-Scott, H. M. & Humphrey, T. J. (1997). The use of an automated growth analyser to measure recovery times of single heat-injured *Salmonella* cells. *Journal of Applied Microbiology* 83, 445–455.

Stevens, R. & Cheng, Q. (1996). A 'litmus test' for molecular recognition using artificial membranes. *Chemistry & Biology* 3, 113–120.

Struelens, M. J., Bauernfeind, A., van Belkum, A. & 14 other authors (1996). Consensus guidelines for appropriate use and evaluation of microbial epidemiologic typing systems. *Clinical Microbiology & Infection* 2, 2–11.

Swaminathan, B. & Feng, P. (1994). Rapid detection of foodborne pathogenic bacteria. *Annual Review of Microbiology* 48, 401–426.

Threlfall, E. J., Hampton, M. D. & Ridley, A. M. (1998). Application of molecular methods to the study of infections caused by *Salmonella* spp. In *Methods in Molecular Medicine*, vol. 15, *Molecular Bacteriology: Protocols and Clinical Applications*, pp. 355–368. Edited by N. Woodford & A. P. Johnson. New Jersey: Humana Press.

Trevena, W. B., Willshaw, G. A., Cheasty, T., Domingue, G. & Wray, C. (1999). Zoonotic transmission of verocytotoxin-producing *Escherichia coli* O157 from farm animals to humans within Cornwall and West Devon. *Communicable Disease and Public Health* 2, 263–268.

Werners, K., Delfgou, E., Soetoro, P. S. & Notermans, S. (1991). A successful approach for detection of low numbers of enterotoxigenic *Escherichia coli* O157 in minced meat by using the polymerase chain reaction. *Applied and Environmental Microbiology* 57, 1914–1919.

Willshaw, G. A., Smith, H. R., Cheasty, T., Wall, P. G. & Rowe, B. (1997). Verocytotoxin-producing *Escherichia coli* O157 outbreaks in England and Wales, 1995: phenotypic methods and genotypic subtyping. *Emerging Infectious Diseases* 3, 561–565.

Woese, C. R. (1987). Bacterial evolution. *Microbiological Reviews* 51, 221–271.

Is global clean water attainable?

Jamie Bartram and José Hueb

Water, Sanitation and Health Programme, World Health Organization, Geneva, Switzerland

Introduction

The benefits of improved water and sanitation include both health and non-health effects. The direct health benefits are related to two contrasting roles of water: that of disease vector when it carries pathogens; and that of health mediator through its use in personal and domestic hygiene. Indirect effects related to health include, for example, improved quality of life and decreased expenditure on medical expenses. Non-health effects include time savings for productive activity or education.

The fact that the health impact of inadequate water supply services, especially in the developing world, has never been established is recognized (Troaré, 1992) and recent work highlights unrecognized health burdens elsewhere—both from apparently 'good quality' supplies (Payment *et al.*, 1991, 1997) and from outbreaks of disease (Ford & Colwell, 1996).

Disease burden from unsafe, inadequate water supply

The disease burden associated with inadequate access to safe drinking water and the health impact of improved water supply are distinct, although the difference is sometimes overlooked. There are numerous identified problems in undertaking assessments (World Bank, 1976; Blum & Feachem, 1983; Esrey, Feachem & Hughes, 1985; Esrey *et al.*, 1991) and as a result a limited body of evidence is available. The most recent estimates of the global burden of disease (GBD) suggest that around 6% of the global disease burden is linked to 'basic hygiene' (water, sanitation, food, hygiene behaviour). However, these estimates do not take account of more severe infectious outcomes that can be transmitted or prevented through safe drinking-water supply (such as infectious hepatitis or typhoid); neither do they take account of water-washed disease, water contact, vector-borne disease (schistosomiasis, malaria) or non-infectious diseases (arsenicosis, fluorosis). Murray & Lopez (1996) suggest a declining relative importance for some of the basic hygiene diseases. This trend may not seem to be in line with other authoritative sources that suggest an

Table 1. Percentage reductions in diarrhoeal morbidity rates from improvements in one or more components of water and sanitation (after Esrey *et al.*, 1991).

	All studies		Rigorous studies	
	n*	Reduction (%)	n*	Reduction (%)
Water and sanitation	7/11	20	2/3	30
Sanitation	11/30	22	5/18	36
Water quality and quantity	22/43	16	2/22	17
Water quality	7/16	17	4/7	15
Water quantity	7/15	27	5/10	20
Hygiene	6/6	33	6/6	33

*The first figure relates to the number of studies suitable for use in calculating the median; the second the total number of studies considered.

'unequivocal increase in infectious disease attributable to microbiologically unsafe drinking water' (Ford & Colwell, 1996). It appears that the trend prediction was based upon a (questionable) assumption of increasing sanitation coverage. Such GBD estimates relate to present-day disease burden and not to the gains already achieved and 'held in check' by good water and sanitation management. The reviews of health impact studies by Esrey, Feachem & Hughes (1985) and Esrey *et al.* (1991), like that of Cairncross (1990a), suggest that water supply and sanitation improvements can reduce the overall incidence of infant and child diarrhoea substantially, in the range of 15–36% for single or combined interventions, and total infant and child mortality substantially. Thus through an exhaustive critical appraisal of published health impact studies, Esrey, Feachem & Hughes (1985) and Esrey *et al.* (1991) reported the results summarized in Table 1.

Generalized analyses of this type may undervalue the importance of water quality, because of the ability of piped water supplies to propagate *outbreaks* of disease. Large-scale outbreaks such as those caused by *Cryptosporidium* in Milwaukee, USA (Mackenzie *et al.*, 1994), hepatitis in India (Ramalingaswami & Purcell, 1988) and the outbreaks of cholera in Latin America (Pan-American Health Organization, 1995) attest to its importance and are unaccounted for in such studies. The effect of quality during extreme disruptions, whether seasonal or related to natural disasters, may also be underestimated (Ramalingaswami & Purcell, 1988).

Drinking-water quality

Concern for the microbiological quality of drinking water centres primarily upon human health. Whilst some microbiological factors may be of relevance for other reasons, such as taste and odour nuisance, biofilm development and some aspects of corrosion, it is public health which dominates thinking concerning the microbiological quality of drinking water.

A shift of interest and emphasis in water quality from microbiological to chemical aspects was seen in relatively recent decades (Farland & Gibb, 1993). This reflected a belief in many wealthier countries that microbiological quality problems were those associated with underdevelopment and were generally conquered. This perspective was fuelled by reductions in the number of detected outbreaks of water-borne disease in countries and supply types where high standards of service quality were largely maintained. More recently, there has been increasing concern regarding microbiological quality amongst these same nations. This has resulted especially from the following (adapted from Bartram, 1998):

• the recognition of 'new' pathogens which defied (then) accepted treatment practice in some countries, *Giardia* and more recently *Cryptosporidium* providing the most obvious examples (Mackenzie *et al.*, 1994; Solo-Gabriele & Neumeister, 1996);

• the failure to make further advances in the control of outbreaks of water-borne disease, and reported increases in the numbers of outbreaks detected (Ford & Colwell, 1996);

• recognition that ability to detect water-borne disease outbreaks, even with relatively advanced surveillance systems, is relatively poor, and that if greater resources are deployed in surveillance then the sensitivity of detection will increase;

• a nascent literature regarding the potentially significant contribution of 'properly' treated and distributed drinking water to the background rates of some diseases (Payment *et al.*, 1991, 1997), supported by evidence related to hepatitis A (Lagarde *et al.*, 1995);

• the general observation of very low infectious doses for several important pathogens and particularly *Giardia lamblia* (Regli *et al.*, 1991), *Cryptosporidium parvum* (DuPont *et al.*, 1995) and some viruses (Rose & Gerba, 1991; Haas *et al.*, 1993).

• a line of questioning regarding the significance of analytical results indicating compliance that may represent 1 c.f.u. of a 'faecal indicator bacteria' in 100 ml or 1 in 10 000 ml (see, for example, Gale, 1996), especially since systems of 'good' quality (e.g. in which most samples show <1 c.f.u. ml^{-1} of the selected indicator) show large numbers in a few samples (Pipes, Ward & Ahn, 1977; Christian & Pipes, 1983; Pipes & Christian, 1984; Gale, 1996).

Reasonable questions are therefore being raised regarding the 'safety' of even that fraction of water consumed by a small privileged minority of the population of the globe that meets prevailing standards.

In addition to the overall global importance of the microbiological quality of drinking water, it should be recalled that microbiological contamination is a special problem in small supplies. Whilst definitions of what constitutes a 'small community supply' vary greatly, data from across the world indicate that such supplies—however locally defined—have a high rate of failure to comply with prevailing microbiological standards. This is a challenge shared by wealthy and poorer countries (Bartram, 2000a).

It might reasonably be concluded that the microbiological quality of drinking water, alongside the need to expand coverage with improved supply, remain the principal drinking-water-related issues for human health. Microbiological quality is rising on the agenda in wealthier nations whilst retaining its prime place elsewhere.

Drinking-water supply

Information concerning the global population with access to 'safe' drinking-water is collected periodically by the World Health Organization (WHO) and the United Nations Fund for Children (UNICEF). The WHO/UNICEF 'Joint Monitoring Programme' (JMP) for the water supply and sanitation sector is the only available authoritative source of information on global water supply and has reported on status, progress and trends in the sector (WHO, 1984, 1986, 1990, 1992; WHO/UNICEF/WSSCC, 1996). A year 2000 assessment is in preparation. Tables 2 and 3 describe the progress and status of coverage with safe drinking-water supply in developing countries as last reported by JMP in 1996.

The data used in the WHO/UNICEF JMP are collected through questionnaires completed by appropriate national authorities, supplemented in the most recent assessment by data from other sources (notably cluster surveys). No consistent attempt was made to determine coverage in the industrially developed countries. Whilst coverage in these areas was recognized to be incomplete, they were judged to represent a lower relative public health priority. JMP has begun to collect data from these countries also and the year 2000 assessment will therefore include worldwide coverage information.

A pragmatic definition of 'safe' drinking water was adopted for JMP that required access to an improved source. Recognizing different approaches to water supply, countries selected definitions of 'access to an improved source' that varied

Table 2. Development of coverage with 'safe' drinking-water supply 1990–2000 (population in billions).

	1990	2000
World total population	5·27	6·06
Population served	4·13	4·94
Population unserved	1·13	1·11

Source: WHO/UNICEF/WSSCC (2000) (in preparation).

Table 3. Twenty-five countries where half or more of the total population had no 'safe' drinking-water supply in 1994.

Country	Percentage without safe drinking water		
	Urban	Rural	Total
Afghanistan	61	95	88
Central African Republic	82	82	82
Chad	52	83	76
Zaire	63	77	73
Papua New Guinea	16	83	72
Haiti	63	77	72
Madagascar	17	90	71
Liberia	42	92	70
Angola	31	85	68
Mozambique	83	60	68
Sierra Leone	42	79	66
Uganda	53	68	66
Vietnam	47	68	64
Mali	64	62	63
Myanmar	64	61	62
Lao PDR	60	61	61
Nigeria	37	74	61
Swaziland	59	56	57
Iraq	–	–	56
Nepal	34	59	56
Zambia	36	73	57
Malawi	48	56	55
Sri Lanka	57	53	54
Benin	59	47	50
Sudan	34	55	50

Source: based on data from WHO/UNICEF/WSSCC (1996).

between piped supply within dwellings to an improved well within 30 min walk. The year 2000 assessment defines access to an improved source when such sources are within 1000 m and ensure a minimum of 20 l per capita per day.

Improvements in assessment approaches

Coverage assessment, that is a division of populations into 'haves' and 'have nots', represents the most commonly used measure of water supply adequacy. 'Coverage' generally refers to access to an organized water supply of some type, whether an improvement or not on traditional sources. However, the important differences relate to the nature and quality of provision and the adequacy of these with respect to basic needs — and principally needs for health. Being 'covered' therefore does not necessarily imply the adequacy, reliability, convenience, acceptability or use of the supply.

Some workers have noted that contaminated—but more accessible—water sources are frequently used because convenience is more important to consumers than quality (Smith, 1991). Others have often reported that 'collected' volumes do not reflect domestic water use as some uses such as washing and laundry may be undertaken on site.

There is no clear health gain-related basis for simple coverage monitoring. Nevertheless, there is good reason to believe that means of provision (level of service) have a link to health. Thus, for example: (a) different means of provision lead to increased or decreased use of water for personal and domestic hygiene and evidence has been presented to relate, for instance, transmission of some pathogens with service level directly: McJunkin (1982) for *Shigella*, Esrey *et al.* (1991) for *Ascaris* and diarrhoeal disease, and Edungbola *et al.* (1988) for *Dracunculus*; (b) certain means of provision are associated with greater or lesser difficulty in maintaining water quality (regardless of the quality of water at source) (typically because of the need for or practice of successive manipulations); and (c) different levels of service are associated with different quantity use characteristics and, by implication, increased use for personal and domestic hygiene and thereby health (Cairncross, 1990b).

The progressively changing concept from 'who has water?' to 'how do people obtain water?' reflects developments in sector thinking during recent decades. Until the 1970s, those concerned with public health engineering tended to emphasize the need for a comprehensive solution to water supply: either nothing should be done or a filtered and chlorinated supply to multiple taps should be provided (Bradley, 1993). During the period leading up to and throughout the International Drinking-Water Supply and Sanitation Decade (IDWSSD), 1981–1990, increasing attention was paid to intermediate interventions (Anonymous, 1990; IRC, 1995). The effect of this on reducing the unit cost of establishing supply—hopefully— enabled improved supply to be provided to a greater number of persons. Proposals have been made for replacing coverage with more attainable and health-related indicators (Bartram, 2000a).

What is 'safe' water?

The concept of microbiologically safe drinking water is simple to comprehend— implying security against outbreaks of disease transmitted by water and that water quality would not contribute to background rates of disease. However, its application in analytical terms is more complex. The complexity has several dimensions:

• For whom is water intended to be safe? The healthy 'normal' population? The immunocompromised (e.g. HIV positive, or on immunosuppressive therapy), the young and the elderly? The resident or visiting population?

• What is the real public health significance of infection with a pathogen typically causing mild, self-limiting disease (such as mild diarrhoea), the consequences of which may be greater for a malnourished than a well-nourished child?

• Is 'safe' or 'clean' an absolute or relative concept—for instance, depending upon the relative importance of drinking water when compared to other routes of transmission of disease?

• Where is water judged as being safe or otherwise? At the well where it is collected? In the household where it is stored before use? At the limit of responsibility of a formal supply agency? At a user's tap?

• How applicable are occasional point measurements of a parameter (such as bacterial indicators of faecal contamination) known to vary widely and rapidly and for which short-term changes may have significant public health consequences?

• What is the role of positive 'measures of safety' rather than 'measures of contamination'? Measures of safety may include, for example, assessment of the safeguards in place to prevent access of contaminants, or the existence of 'multiple barriers' in a treatment chain.

WHO, in its *Guidelines for Drinking-Water Quality* (WHO, 1993), defines its Guideline Values for potentially hazardous water constituents—which provide a basis for assessing drinking-water quality—as 'the concentration of a constituent that does not result in any significant risk to the health of a consumer over a lifetime of consumption'.

This definition was developed when particular concern related to chemical hazards—especially those associated with more severe health outcomes—and reflects this. However, many microbial hazards encountered in drinking water are associated with relatively mild and often self-limiting health effects and these effects may arise following short-term exposure. Both the terms 'significant risk' and 'lifetime of exposure' merit reflection in their application to microbial hazards, especially since the 'constituent' measured, such as faecal indicator bacteria, in the routine monitoring of drinking water is rarely the direct cause of ill health.

Safe water for who?

Very substantial variations in response occur on human exposure to water-borne pathogens, depending on, for example, age and immune status. Thus, for example, the impact of hepatitis A virus following initial infection in childhood may be relatively mild but where initial infection occurs in adults the course of disease may be more severe. For hepatitis E virus (HEV), mortality rates amongst pregnant women are reported to be as high as 20–40%. Recent interest in *Cryptosporidium* has been fuelled in part because it leads to potentially life-threatening infection in HIV/AIDS patients, but generally to mild, self-limiting disease amongst healthy individuals. The developing interest in *Mycobacterium avium* complex (MAC) is fuelled by similar factors.

The increasing prevalence of sensitive sub-groups, particularly those who are immunocompromised, in addition to the young, elderly and pregnant, unites wealthier and poorer countries. The issue has been especially highlighted because of HIV/AIDS, for which the majority of affected persons are resident in the least developed or developing nations. In some countries, other causes (notably immunosuppressive therapy) are also significant. Whether targets (standards) for water quality supplied through public means should aim to protect sensitive sub-populations is an emerging issue. Important questions remain regarding, for example, the variability within human populations and whether we understand the nature of the water quality requirements of sensitive populations. In turn, these raise questions, not only for conventional, piped water supply, but also regarding the suitability of alternative risk management approaches such as consumption of bottled water.

The issue of sensitive sub-populations applies at a global (international) level as well as local levels. One example is provided by infectious hepatitis. Travellers from areas of low endemicity (and not previously exposed), such as many of the industrialized countries, to areas where infectious hepatitis is more common may therefore be at increased risk.

Comparing health outcomes: a 'tolerable' disease burden?

In its definition of drinking-water 'safety', WHO indicates that 'The judgement of safety—or what is an acceptable level of risk in particular circumstances—is a matter in which society as a whole has a role to play. The final judgement as to whether the benefit resulting from the adoption of any of the Guideline Values . . . justifies the cost is for each country to decide.' (WHO, 1993).

There is increasing recognition—especially amongst the policy-making and scientific communities—of the concept of 'acceptable' risk. The term 'tolerable' risk is preferred by some workers to recognize that the risk is not truly acceptable but may be tolerated, either absolutely, or in deference to greater or more highly perceived priorities.

Different agencies are exploring what might constitute a tolerable disease burden. WHO calculates its Guideline Values for genotoxic carcinogens as equivalent to 1 in 100 000 excess lifetime risk of cancer. The United States Surface Water Treatment Rule is concerned with minimizing health risks from pathogenic micro-organisms occurring in surface water sources and originally established a goal that treatment should ensure that fewer than one person in 10 000 per year became ill from exposure to the protozoan *Giardia* in drinking water, and this was assumed to be protective of other diseases at the time.

However, microbiological hazards — even those typically thought of as mild and self-limiting — may carry unwanted sequelae of greater public health significance. Bennett *et al.* (1987), for example, indicate case fatality rates from 0·0001 to 6% for a range of water-borne diseases and, for example, 1% for miscellaneous enteric infections. Accepting a tolerable disease burden of one excess death per 100 000 over a lifetime of exposure (i.e. comparable to that employed for genotoxic carcinogens) would suggest a tolerable incidence (of miscellaneous enteric infections) of around one per 100 population per year. Apart from death, it has been suggested that infections which are known or are likely to be spread by the water-borne route may be associated with kidney failure and brain damage (*Escherichia coli* O157: Boyce, Swerdlow & Griffin, 1995), reactive arthritis (*Salmonella, Campylobacter* and *Yersinia*: Maki-Ikola & Granfors, 1992; Granfors *et al.*, 1989); Guillain–Barré and Miller–Fisher syndromes (*Campylobacter*: Mishu, Ilyas & Koski, 1993; Jacobe, Endtz & Mech, 1995), miscarriages, diabetes and heart disease (various viruses: Axelsson *et al.*, 1993; Frisk *et al.*, 1992; Hyypia, 1993) and stomach cancer (following chronic infection with *Helicobacter pylori*: Parsonnet, Hansen & Rodriguez, 1994).

Comparing, contrasting and valuing diverse and sometimes additive adverse health outcomes in a way meaningful to policy makers have proven difficult and application of the concept of tolerable disease burden to regulations therefore impeded by the lack of an 'exchange unit'. Several potential units such as DALYs (Disability-adjusted Life Years) are now increasingly employed and this impediment increasingly overcome, enabling distinct health end points to be accommodated in assessing water-associated microbial hazards.

The increasing need to understand what constitutes an 'acceptable' or 'tolerable' risk concerns microbiologists especially because many of the disease outcomes they are concerned with are frequently self-limiting and may be perceived as 'mild'. However, the question is not one which microbiologists may effectively engage alone. Whilst quantified risk assessment has much to offer in describing outcomes, interdisciplinary work engaging, for example, social scientists is required to understand what is a tolerable risk and how it varies within and between societies.

Safe water for consumers?

The quality of water supplied may be different to that consumed (Feachem *et al.*, 1978; Shiffman *et al.*, 1978) and yet there is good reason to believe that quality deterioration in the household is sometimes the principal source of faecal contamination (El Hennawy, 1987; Beg & Mahmood, 1991). Water quality interventions at household level have been demonstrated to be effective in both water quality improvements and in terms of epidemiologically measured health impact. Addressing quality from the viewpoint of the recipient is therefore important.

WHO has suggested that 'safe water implies protection of water sources as well as proper transport and storage within the home' (Anonymous, 1995), thereby supporting the idea that monitoring extends from source to point of use. However, most monitoring schemes are not orientated so as to include investigation of household-level contamination. For example, sampling procedures may specify that taps are cleaned and water left to flush for a significant period before sampling. Where service quality (especially continuity) is poor, or where the means of provision do not provide water within the dwelling, this does not provide a reliable indication of the quality of water as consumed—which is that which has stagnated and been influenced by tap materials, attached debris and subject to regrowth.

Across the body of drinking-water-related research, relatively limited effort is devoted to the *health* significance of quality changes in piped water distribution systems. Such changes may relate to regrowth, recurrent recontamination (for example, associated with incomplete system integrity, cross-connections) or sporadic events (such as back-siphonage or mains repair). The use of a residual disinfectant such as chlorine would be expected to neutralize minor intrusion of chlorine-sensitive micro-organisms into distribution mains. In such circumstances, the use of chlorine might mask the detection of problems by reducing or eliminating the indicators most commonly used in monitoring. Massive intrusion of contaminated water through cross-connections, or leaks, may not be neutralized effectively, even if high levels of free residual chlorine are the norm, although sporadic contamination in particular may be undetected by monitoring programmes.

Measuring contaminants or measuring safety?

Prevailing standards for microbiological quality of drinking water originate from the early 20th century when they were developed to monitor the treatment processes (slow sand filtration, chlorination) that had proven successful in controlling bacterial pathogens causing diseases such as cholera and typhoid. These problems are still real with important consequences in large parts of the globe and such standards and their application for this purpose remain valid.

These standards are based upon the concept of 'faecal indicator organisms', generally bacteria, ideally of no direct public health relevance, present in large numbers of faeces, otherwise environmentally rare and which are unable to multiply in the environment and are at least as resistant to environmental conditions as the pathogens that may also be present in faeces. Faecal indicator theory then presupposes that the absence of these indicators implies the absence of pathogens. These organisms provided a simple indicator of sanitary status and measure of process efficiency in relation to the (bacterial) agents of disease then preoccupying reformers.

However, the concept of faecal indicators has progressively developed such that faecal indicator organisms have adopted the role of end product quality indicators in their own right, despite the evident absence of an adequate scientific justification for this. This development has paralleled developments related to approaches to risk assessment for chemicals in drinking water.

For the great majority of chemicals occurring in drinking water, health effects arise from long-term exposures, and acute exposure is of limited health significance. Assessment of exposure to chemicals in drinking water may typically be reasonably undertaken through achievable sampling strategies towards finished waters. Furthermore, meaningful management intervention is often feasible within a time frame of action following detection of a change in chemical quality of water. Microbiological quality in contrast varies both rapidly and widely and both temporally and spatially. Short-term exposure can lead to severe, acute adverse health effects. Health effects may arise before the results of analyses become available. In the light of water quality variability, it is doubtful whether short-term deterioration is readily detected by present approaches to water quality testing, and there is greater doubt that any detected deterioration could be associated with detectable health outcomes. Fundamental questions therefore exist regarding the primacy of end product quality standards in ensuring the microbiological safety of drinking-water supplies.

This present status of much drinking-water quality legislation, emphasizing end product quality, should be contrasted with much earlier legislation, which was led by good practice and safety.

Attention has begun to focus on the potential for application of an approach analogous to Hazard Assessment Critical Control Point Analysis (HACCP) to drinking-water supply (Brian, 1992; WHO, 1997). HACCP has been developed and applied extensively in the food industry over the past two decades and is analogous to the idea of 'sanitary inspection' as was once common practice in the water supply industry. In their classic textbook, Prescott & Winslow (1931) state that 'the first attempt of the expert called in to pronounce upon the character of a potable water should be to make a thorough sanitary inspection'. Nevertheless, very limited systematic recent literature is available (examples include Jensen, 1966, 1967; WHO, 1976, 1985, 1997; Lloyd & Bartram, 1991). The following are common to these approaches (i.e. to sanitary inspection and HACCP):
• recognition that microbiological contamination varies widely and rapidly and is therefore difficult to estimate by analytical regimes unless sampling density and frequency are high (with attendant cost implications and logistical problems);
• acceptable levels are zero or very low and therefore monitoring fails to identify the degree of acceptability;
• in the light of the above, some reliance must be placed upon the susceptibility of systems to contamination; and

• implicit recognition of the multiple barrier principle.

WHO has consistently advocated 'sanitary inspection' as a vital and complementary measure to microbiological and residual disinfectant water quality analysis (WHO, 1976, 1985, 1993, 1997) and their complementarity has also been emphasized by Lloyd & Bartram (1991) and Lloyd & Helmer (1991). These two tools for investigating water quality are distinct. One relates to the susceptibility of a water supply system to contamination and the other is an instantaneous measure of actual contamination (Lloyd & Bartram, 1991).

How microbiological risks are managed

Risk management is a relatively recent discipline in which developments are still occurring rapidly. Various model schemes for risk management have been developed and published, most of which have some common elements. These include most frequently the need for an information base to support decision-making, the need to recognize that decisions are made on the basis of available (i.e. generally less than ideal) information, the need to compare and value multiple and differing outcomes of various alternatives, and the need for broad participation in all stages. One example is included in Fig. 1, which describes the process as circular (thereby showing the feedback from policy evaluation which may otherwise be seen as a 'dead end') and with communication as a central, two-way process.

Whilst risk management may be readily perceived as a highly rational process, it should be recognized that the scientific basis for many of its elements is often weak.

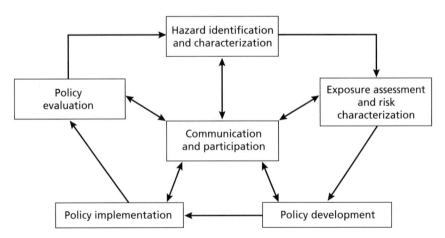

Fig. 1. Risk management cycle (adapted from Soby, Simpson & Ives, 1993).

More importantly, no scientific assessment will support effective risk management if it fails to address the perceptions and priorities of the society concerned. A discussion of the factors involved in the various stages for one particular hazard (toxic cyanobacteria in water) is provided in Bartram *et al.* (1999).

Whether 'clean water for all' is achievable depends immediately upon the informed individual or community and to a large extent upon policy makers and decision-takers. These individuals act at both a domestic level (concerning, for example, national policy development, standards and their implementation and investments) and also at an international level, both as the organized international community (for example, in relation to international trade in goods and services, regulated through the World Trade Organization (WTO)) and as part of the 'development community', providing support in various forms to assist acceleration of improvements in less economically advantaged countries and regions. Whilst work in environmental health administrations generally seeks to follow a rational approach (reviewed by Steensberg, 1989), it should be recognized that there are many examples where the former has had greater influence than the latter (see, for example, Bennett, 1998).

Despite the evidence for health gain from interventions in water supply and quality, their role as public health interventions has been questioned. Walsh & Warren (1979) in their work on 'selective primary health care' in essence argued that the cost per health improvement (such as infant death averted) was too high for water supply and sanitation to be included in a package of selective 'primary health care'. Their paper generated much debate, most recently in the form of a paper by Varley, Tarvid & Chao (1999), which argued for a level playing field in such assessments—for example, if the costs of constructing hospitals and clinics were not accounted for as part of the health care service then why was the cost of constructing water supplies to be accounted for? Especially since the health services sector was not in the business of water supply *per se*.

Economic feasibility of global clean water

According to Listorti (1999), 'the three disciplines of health, environment and economics have not reinforced each other's efforts' and 'economic analyses have concentrated more on costs than on benefits; while health analyses have concentrated more on solutions of the health care system itself than on preventive measures outside'.

Most infrastructure investments are initially justified on grounds other than health. In the case of water supply, for example, economic and social (non-health) benefits as measured, for example, in terms of 'willingness to pay' justify investment in water supply *per se*, which will accrue substantial direct benefits to health alone. Ensuring water quality is justified largely in terms of health gain (and to a lesser extent aesthetic or acceptability aspects), but the additional cost of building quality aspects into a water supply project is typically a small component of the total.

The economic achievability of 'clean water for all' depends both upon achieving sustainability in existing supply systems and in extending coverage to the substantial residual population without access to 'safe drinking water'.

Further extension of coverage towards the ultimate goal of 'universal access' to safe water and sanitation is technically feasible, although it was estimated in 1990 that to achieve this goal by the year 2000 would require investment in the order of US$50 billion per year, which should be compared to a rate of around US$13 billion per year during the IDWSSD.

As a supportive activity preparatory to the development of the Protocol on Water and Health to the Convention on the Protection and Use of Transboundary Watercourses and International Lakes, a cost–benefit analysis was undertaken. This study looked at the costs of bringing water supply and quality in the countries of central and eastern Europe and the newly independent states to the standards prevailing in western Europe in 1998. The costs amounted to some US$125 billion. This is equivalent to around US$30 per capita per annum (range 10–76 across the total population) and represented 1–2·5% of per capita GDP.

Information regarding overall levels of investment in water supply and sanitation is contradictory. Apparently sound sources claim that either after the end of the Decade overall investment in water supply and sanitation declined, due to donor fatigue, economic recession and competition with other development sectors and other factors (Anonymous, 1995); or the level of investment has been growing (Gibbons *et al.*, 1996). In Peru, for example, overall social expenditure (including investments in water supply and sanitation) decreased from 49 to 12.5 US$ per capita between 1980 and 1993 (Zucchetti, 1994). Lowered overall investment may to some extent be countered by lower unit costs, for instance, as achieved during the IDWSSD (IRC, 1995).

Maintaining the gains already achieved

In developing countries, construction of new facilities (in order to quickly make up the large sector deficit by providing services to those without adequate facilities) is typically given higher priority than ensuring the continual functioning of existing systems. As a result, many water supply facilities collapse or work at a fraction of their capacity. External financial assistance, whilst small in comparison to national contributions (Le Moigne, 1996), has a large influence on how national finances are directed (Okun & Lauria, 1991). Most external finance is provided for the development of *new* facilities (Robinson, 1991). This places the system intended to ensure continued functioning of existing infrastructure under greater stress (Taylor, 1993).

The deterioration of the accumulated water supply infrastructure—of increasing importance as the total 'capital' increased (Bays, 1992)—became a widely recognized issue towards the end of the IDWSSD. It created challenges to management and potential opportunities for the transmission of water-borne disease. The need to increase attention to the sustainability of water supply service provision generally and to operation and maintenance in particular is considered to be one of the principal lessons that may be learned from the Decade (IRC, 1995; Gibbons *et al.*, 1996).

In many large urban areas, where the cost of producing water is normally high, the unaccounted-for water has been reported as more than 50% of the water produced (WHO, 1991, 1994). The reduction of unaccounted-for water alone would facilitate access to water supply services for the urban poor, particularly those living in peri-urban areas. Technology selection may lead to repeated breakdown or short lifespan of systems, or the systems may be too sophisticated for the water agency to maintain them. These factors may be aggravated by the long-standing tradition of some governments and external support agencies to consider water and sanitation as being a free social service for all, the costs of which are not borne by the user.

It has been proposed that there are certain overriding principles that should be adopted if water supply and sanitation services are to be sustainable (WHO, 1991, 1994).

• The provision of water is a service which requires a service-orientated attitude by the agencies involved. To ensure sustainability, water exploitation should be on a financially sound and cost-effective basis, subject to legal and regulatory controls to ensure conservation, protection and wise utilization of water.

• The supply of water to consumers should normally be based on *effective demand*, which can be defined as the standard of service that the users are willing to maintain, operate and finance. The effective demand has to satisfy the priorities of the community at large.

• Water supply systems should be managed and operated following the principles of good business practice. The form of management will vary depending on the local situation. The responsible agency should be autonomous from government but manage the system under technical, financial and administrative guidelines set by national governments. The agency should be transparent and accountable to its consumers.

• Sanitation is undervalued and emphasis is required on sanitation development and for forging closer links between water supply and environmental sanitation (solid and liquid waste management) in the planning of new programmes.

• The concerns of government to satisfy the basic needs of the disadvantaged segments of the population should be recognized. Governments may require agencies to provide service at lifeline tariffs for such groups or institute temporary subsidies to promote public health and economic development.

Rural and small community supplies have special problems

A large majority (80%) of the estimated 1·2 billion people lacking access to a safe water supply at the end of the IDWSSD lived in rural areas (Appleton, 1995).

Figures in the range 40–60% of point source infrastructure being non-functional are often reported (Troaré, 1992) and reliable sources have quoted figures as high as 80%. One UNICEF study estimated 80% of rural water supply infrastructure in India and Asia to be 'non-operational' (Watters, 1990), and an official of the Swedish Foreign Ministry is quoted as saying that in the Swedish rural water programme in Tanzania and Kenya 'after 15 years of quite considerable investment in piped water and mechanical pumping equipment, hardly more than 10% of our installations are still in use' (cited in Pickford, 1991). Work in the region of Tolima, Colombia, showed that although water quality was extremely poor and considerable infrastructure for water treatment existed, no water treatment plant in the study area (which included both urban and rural communities) was operational (Bartram *et al.*, 1991).

Infrastructure failure relates not only to breakdown (i.e. interruptions to supply), but to the fact that small and private supplies, especially in rural areas, are more frequently contaminated. Failure of supply and contamination are often connected (Bartram, 2000a).

Technical challenges to achieving global clean water

The goal of universal provision of safe drinking water directly to households involves abstraction from surface water or groundwater resources. The importance of desalination is increasing, especially in water-scarce regions. Abstracted water is then typically treated through physical, chemical and/or biological means to eliminate or reduce to safe levels contaminants in the source, or a combination of these. Finally, the treated water is distributed, generally through pipes, to individual households and must be protected against quality deterioration throughout this.

There is a traditional preference for groundwater sources in many parts of the globe. Technically, groundwaters are generally viewed as being of more stable quality and of better microbiological quality than their surface water counterparts because of the attenuation capacity of surface (e.g. soil) layers and aquifer materials themselves. However, extensive experience has highlighted that this 'good quality' may be readily prejudiced if adequate protection measures (e.g. at a well head) are not in place. Some evidence also exists for viral contamination of well-protected deep aquifers with the suggestion that those organisms have migrated large distances in aquifers. On a global scale, ongoing overexploitation of aquifers and irreversible quality deterioration will place finite limits on the availability of water from this type of source.

Surface waters such as rivers and lakes are extensively exploited as drinking-water sources worldwide. Their contribution to drinking-water supply varies almost from 0 to 100%, depending primarily upon the availability of safe, accessible groundwaters. Surface waters are generally considered to be open to contamination. They vary widely in quality and are subject to rapid changes in quality. Surface waters are also subject to overexploitation but this may be more readily reversible than for groundwaters. The dependability of surface waters as drinking-water sources may decrease with predicted global climate changes.

Drinking-water treatment generally follows the 'multiple barrier' principle and the degree of treatment applied relates to source water quality. More polluted water sources are therefore more costly to treat than less polluted sources and increasing reliance on degraded water resources is likely to increase the costs of supply. Increasing recognition of environmentally robust protozoa, such as *Giardia* and *Cryptosporidium*, as water-borne pathogens has led to *Cryptosporidium* in particular serving as the point of reference in evaluating treatment efficiency for surface waters. The importance of a short-term decrease in efficiency (such as that occurring following filter backwash) for public health is increasingly recognized but inadequately understood.

Microbiologically safe drinking water entering distribution systems may deteriorate in quality as a result of recontamination or regrowth. Knowledge of the extent and health impact of both is limited (see also 'Safe water for consumers?', p. 45).

Around 20% of the world's population lacks access to an improved water source and a far greater population do not benefit from a piped water supply. Whilst the goal of extending safe, piped supply is generally accepted, there is less consensus concerning intermediate solutions, especially for water quality improvement within households. There is accumulating evidence that improving and safeguarding water quality in the home can contribute significantly to reducing rates of disease. Experience with large-scale promotion of household treatment in developing countries is limited.

Environmental constraints to achieving global clean water

The concept of 'safe water' has for many years been seen to relate principally to drinking water and the importance of drinking-water supply and quality—for both health and development—remains unquestioned.

Nevertheless, lessons from recent decades suggest that water cannot be effectively managed in discrete compartments, especially when these compartments are associated with specific uses, each with distinct and conflicting demands and impacts upon water quality and quantity. Attempts to provide safe drinking-water supply in isolation of the management of wider water resource issues are unsustainable.

Managing water for health depends upon ensuring that the water resources themselves remain healthy in terms of both quality and quantity of water and increasing emphasis is placed upon integrated management of freshwater resources. It is widely agreed, for example, that freshwater should be managed as a finite and vulnerable resource and management action should be integrated into the framework of national economic, social and environmental policy.

Current water use patterns across much of the globe are unsustainable because of two principal types of pressures: (a) quantitative, relating to increased demands, poor allocation and water wastage; and (b) qualitative, relating to the use of freshwater resources as a means for carriage, treatment and final disposal of society's wastes.

Water use continues to increase due to a combination of population growth and increased individual usage. Developments in industrial economies in water-use and water-savings devices, and in reuse practices may eventually contribute to controlling demand. Great potential remains to reduce wastage in drinking-water supply, where around 50% of water may be unaccounted for, and in agriculture, which accounts for around 70% of all water abstractions. Nevertheless, by the year 2025, it is projected that as much as two-thirds of the world's population will live in areas of water stress, including countries as diverse as Mauritius, India, Iran, Zimbabwe, Togo and Poland.

In freshwater systems, water pollution inhibits the use of, or increases the costs of, beneficial uses of water such as for drinking-water supply, irrigation and aquaculture. Human health may be affected in the transmission of water-borne disease or through the accumulation of pathogens in foodstuff from aquaculture, irrigated agriculture or shellfish or marine fisheries. Both freshwater and coastal areas are used for recreation, and the health significance of sewage pollution of recreational waters is increasingly recognized (WHO, 1998).

A number of major influences will limit the effectiveness of measures to improve the management of water resources and to increase access to water for health and development. These include population growth (and consequential urbanization, industrialization and agricultural intensification), a general deterioration in water reuse availability and quality and, potentially, global climate change (Bartram, 2000b).

Population growth

The most dramatic single pressure on a global scale results from population growth. Global population growth was nearly 90 million in 1996 — higher than at any previous time in human history. It is anticipated that the rate of increase will peak around the year 2000 and slow thereafter, although the overall population of the globe is still likely to double before the middle of this century.

Population growth demands significant resources simply to enable maintenance of population coverage with access to safe drinking water. The effect of population growth on the environment more generally may have equally if not more profound implications for water management. It will create additional demands in most water-using sectors and exacerbate tensions arising between sectors with competing legitimate demands for water, for example, for consumption and for food production, and between states, especially those sharing water resources.

Much population growth will occur in low-income countries. The link between population growth and poverty leads to exploitation of resources that may be precarious or already degraded and may thereby impact upon resource security.

Urbanization

Cities worldwide are growing rapidly, especially in developing countries. The proportion of the global population living in cities will increase from around 45% to around 62% by the year 2025.

Urban demands on limited resources may increase water supply costs as more remote sources are exploited, and may contribute to long-term and possibly irreversible water resource depletion. Urban centres are sources of pollution affecting both resources used by cities and downstream users.

Global climate change

Global climate change is an area of intensive debate and one in which very significant uncertainty remains, and is likely to remain for some time. Nevertheless, an increased frequency of extreme weather events — floods and droughts — is recognized with severe implications for water resources and infrastructure, and significant impact upon human health and well-being.

Extreme weather events may damage or destroy infrastructure (dams, water supply) and cause contamination of domestic supplies or previously uncontaminated resources. Drought may cause shortages in water availability, which may lead to increased water supply contamination through 'rationing' or low pressure in supply systems, through decreased 'self-purification' (e.g. decreased dilution) in water courses or by forcing exploitation of 'new' and potentially more contaminated sources.

References

Anonymous (1990). *Global Consultation on Safe Water and Sanitation for the 1990s, Background Papers.* Secretariat for the Global Consultation on Safe Water and Sanitation for the 1990s. New Delhi. India.

Anonymous (1995). *Community Water Supply and Sanitation: Needs, Challenges and Health Objectives.* Report by the Director General to the 48th World Health Assembly, WHO Document A48/INF.DOC/2, p. 4. Geneva: World Health Organization.

Appleton, B. (1995). The financial and institutional implications of the operation and maintenance of rural water supply and sanitation schemes. In *Integrated Rural Management,* vol. II, *Technical Documents Prepared for the Second Technical Consultation,* pp. 53–60. Geneva: World Health Organization.

Axelsson, C., Bondestam, K., Frisk, G., Bergstrom, S. & Diderholm, H. (1993). Coxsackie B virus infections in women with miscarriage. *Journal of Medical Virology* 39, 282–285.

Bartram, J. (1998). Policy and administrative issues. In *Molecular Technologies for Safe Drinking-water,* pp. 1–16 (Anonymous). Paris: OECD.

Bartram, J. (2000a). Effective monitoring of small drinking-water supplies. In *Providing Safe Drinking-Water in Small Systems,* pp. 353–366. Edited by J. Cotruvo, G. Craun & N. Hearne. Boca Raton: Lewis Publishers.

Bartram, J. (2000b). Future perspectives and international cooperation. In *Security of Public Water Supplies. Proceedings of a NATO International Conference on Water and Security, Hungary.* Edited by R. Deininger, P. Literathy & J. Bartram. New York: Kluwer.

Bartram, J., Suarez, M., Quiroga, E. & Galvis, G. (1991). Environmental monitoring and institutional roles in post-disaster development. In *Appropriate Development for Basic Needs. Proceedings of the Conference on Appropriate Development for Survival — the Contribution of Technology,* pp. 265–268. Edited by D. P. Maguire. London: Thomas Telford.

Bartram, J., Burch, M., Falconer, I. R., Jones, G. & Kuiper-Goodman, T. (1999). Situation assessment, planning and management. In *Toxic Cyanobacteria in Water,* pp. 179–208. Edited by I. Chorus & J. Bartram. London: E&FN Spon.

Bays, L. (1992). Urbanisation and birth-rate thwart global water progress. In *Water Technology International 1992,* pp. 11–13. Edited by M. Munro. London: Century Press.

Beg, A. & Mahmood, A. (1991). *Water Decontamination Technologies Developed and Disseminated by PCSIR.* Karachi, Pakistan: PCSIR.

Bennett, J. V. (1998). Housewives, urban protest and water policy in Monterrey, Mexico. In *Water — The Key to Socio-economic Development and Quality of Life (Proceedings of the 8th Stockholm Water Symposium 10–13 August 1998, Stockholm, Sweden).* Stockholm: Stockholm International Water Institute.

Bennett, J. V., Holmberg, S. D., Rogers, M. F. & Solomon, M. L. (1987). Infectious and parasitic diseases. In *Closing the Gap: the Burden of Unnecessary Illness.* Edited by R. W. Amler & H. B. Dull. New York: Oxford University Press.

Blum, D. & Feachem, R. G. (1983). Measuring the impact of water supply and sanitation investments on diarrhoeal diseases: problems of methodology. *International Journal of Epidemiology* 12, 357–365.

Boyce, T. G., Swerdlow, D. L. & Griffin, P. M. (1995). *Escherichia coli* O157 : H7 and the hemolytic-uremic syndrome. *New England Journal of Medicine* **333**, 364–368.

Bradley, D. (1993). Environmental and health problems of developing countries. In *Environmental Change and Human Health*, Ciba Foundation Symposium 175. Edited by Lake, Rock & Ackill. New York: Wiley.

Brian, S. C. (1992). *Hazard Analysis Critical Control Point Evaluation: a Guide to Identification of Hazards and Assessing Risks Associated with Food Preparation and Storage.* Geneva: World Health Organization.

Cairncross, S. (1990a). Health impacts in developing countries: new evidence and new prospects. In *Institution of Water and Environmental Management Engineering for Health, Technical Papers of Annual Symposium, 27–28 March 1990, UMIST, Manchester (paper 1)*. London: Institution of Water and Environmental Management.

Cairncross, S. (1990b). Health aspects of water and sanitation. In *Community Health and Sanitation*, pp. 30–34. Edited by C. Kerr. London: Intermediate Technology Publications.

Christian, R. R. & Pipes, W. O. (1983). Frequency distribution of coliforms in water distribution systems. *Applied and Environmental Microbiology* **45**, 603–609.

DuPont, H. L., Chappel, C. L., Sterling, C. R., Okhuysen, P. C., Rose, J. B. & Jakubowski, W. (1995). The infectivity of *Cryptosporidium parvum* in healthy volunteers. *New England Journal of Medicine* **332**, 855–859.

Edungbola, L. D., Watts, S. J., Alabi, T. O. & Bello, A. B. (1988). The impact of the UNICEF-assisted rural water project on the prevalence of guinea-worm disease in Asa, Kwara State, Nigeria. *American Journal of Tropical Medicine and Hygiene* **39**, 79–85.

El Hennawy (1987). Water quality in relation to water resources and supply systems in Alexandria. In ORDEV, WHO, UNDP and UNICEF, *Proceedings of the National Water Quality Seminar*, vol. II, *Alexandria, 23–26 November 1987*. Alexandria: World Health Organization/EMRO.

Esrey, S., Feachem, R. & Hughes, J. (1985). Interventions for the control of diarrhoeal diseases among young children: improving water supplies and excreta disposal facilities. *Bulletin of the World Health Organization* **63**, 757–772.

Esrey, S. A., Potash, J. B., Roberts, L. & Schiff, C. (1991). Effects of improved water supply and sanitation on ascariasis, diarrhoea, dracunculiasis, hookworm infection, schistosomiasis and trachoma. *Bulletin of the World Health Organization* **69**, 609–621.

Farland, W. H. & Gibb, H. J. (1993). Perspective on balancing chemical and microbial risks of disinfection. In *Safety of Water Disinfection: Balancing Chemical and Microbial Risks*, pp. 3–10. Edited by G. F. Craun. Washington, DC: ILSI Press.

Feachem, R. G., Burns, E., Cairncross, S., Cronin, A., Cross, P., Curtis, D., Khan, M. K., Lamb, D. & Southall, H. (1978). *Water, Health and Development: an Interdisciplinary Evaluation.* London: Tri-Medical Books.

Ford, T. E. & Colwell, R. C. (1996). *A Global Decline in Microbiological Safety of Water: a Call for Action.* Washington, DC: American Academy of Microbiology.

Frisk, G., Nilsson, E., Tuvemo, T., Friman, G. & Diderholm, H. (1992). The possible role of Coxsackie A and echo viruses in the pathogenesis of type I diabetes mellitus studied by IgM analysis. *Journal of Infection* **24**, 13–22.

Gale, P. (1996). Coliforms in the drinking-water supply: what information do the 0/100ml samples provide? *Journal of Water Science and Technology—Aqua* **45**, 155–161.

Gibbons, G., Edwards, L., Gross, A., Lee, S. M. & Noble, B. (1996). *The UNDP-World Bank Water and Sanitation Programme, Annual Report July 1994–June 1995.* Washington: International Bank for Reconstruction and Development.

Granfors, K., Jalkanen, R., von Essen, R. & 7 other authors (1989). *Yersinia* antigens in synovial fluid cells from patients with reactive arthritis. *New England Journal of Medicine* **320**, 216–221.

Haas, C. N., Rose, J. B., Gerba, C. & Regli, S. (1993). Risk assessment of virus in drinking water. *Risk Analysis* **13**, 545–552.

Hyypia, T. (1993). Etiological diagnosis of viral heart disease. *Scandinavian Journal of Infectious Diseases* **88**, 25–31.

IRC (1995). *Water and Sanitation for All: a World Priority*, vol. 2: *Achievements and Challenges*. The Hague: Ministry of Housing, Spatial Planning and the Environment.

Jacobe, B. C., Endtz, H. P. & Mech, G. G. A. (1995). Serum anti-GQ1b antibodies recognise surface epitopes on *Campylobacter jejuni* from patients with Miller-Fisher syndrome. *Annals of Neurology* **37**, 260–264.

Jensen, P. (1966). *Drinking water examinations: drinking water supply in Danish rural districts* (English summary of Danish text). PhD thesis, Kobenhaven Costers Bogtrykkeri, Copenhagen.

Jensen, P. (1967). Examination of water supply and drinking water. *Danish Medical Bulletin* **14**, 273–280.

Lagarde, E., Joussemet, M., Latialade, J. & Fabre, G. (1995). Risk factors for hepatitis A infection in France: drinking tap water may be of importance. *European Journal of Epidemiology* **11**, 145–148.

Le Moigne, G. (1996). Change of emphasis in World Bank lending. In *New World Water 1996*, pp. 17–21. Edited by S. Aubry. London: Sterling Publications.

Listorti, J. A. (1999). Is environmental health really a part of economic development—or only an afterthought. *Environment and Urbanisation* **11**, 89–100.

Lloyd, B. & Bartram, J. (1991). Surveillance solutions to microbiological problems in water quality control in developing countries (Proceedings of the IAWPRC Symposium on Health-Related Water Microbiology, Tubingen, April 1990). *Water Science and Technology* **24**, 61–75.

Lloyd, B. & Helmer, R. (1991). *Surveillance of Drinking Water Quality in Rural Areas.* London: Longman.

McJunkin, E. (1982). *Water and Human Health*. Washington, DC: US Agency for International Development.

Mackenzie, W. R., Hoxie, N. J., Proctor, M. E. *et al.* (1994). A massive outbreak in Milwaukee of *Cryptosporidium* infection transmitted through the public water supply. *New England Journal of Medicine* **331**, 161–167.

Maki-Ikola, O. & Granfors, K. (1992).

Salmonella-triggered reactive arthritis. *Lancet* **339**, 1096–1098.

Mishu, B., Ilyas, A. A. & Koski, C. L. (1993). Serologic evidence of previous *Campylobacter* infection in patients with the Guillain–Barré syndrome. *Annals of Internal Medicine* **118**, 947–953.

Murray, J. L. M. & Lopez, A. D. (1996). *The Global Burden of Disease*. Harvard School of Public Health, World Health Organization and World Bank.

Okun, D. A. & Lauria, D. T. (1991). Capacity building for water sector management: an international initiative for the 1990s. In *A Strategy for Water Sector Capacity Building. Proceedings of the UNDP Symposium, Delft 3–5 June, 1991*. Edited by G. J. Alaerts, T. L. Blair & F. J. A. Hartvelt. Delft: International Institute for Hydraulic and Environmental Engineering.

Pan-American Health Organization (1995). Cholera in the Americas. *Epidemiology Bulletin* **16**, 11–13.

Parsonnet, J., Hansen, S. & Rodriguez, L. (1994). *Helicobacter pylori* infection and gastric lymphoma. *New England Journal of Medicine* **330**, 1267–1271.

Payment, P., Richardson, L., Siemiatycki, J., Dewar, M., Edwardes, M. & Franco, E. (1991). A randomised trial to evaluate the risk of gastrointestinal disease due to consumption of drinking water meeting current microbiological standards. *American Journal of Public Health* **81**, 703–708.

Payment, P., Siemiatycki, J., Richardson, J., Gilles, R., Framnco, E. & Prevost, M. (1997). A prospective epidemiological study of gastrointestinal health effects due to the consumption of drinking-water. *International Journal of Environmental Health Research* **7**, 5–31.

Pickford, J. (1991). Water and sanitation. In *Appropriate Development for Basic Needs. Proceedings of the Conference on Appropriate Development for Survival—the Contribution of Technology*, pp. 145–159. Edited by D. P. Maguire. London: Thomas Telford.

Pipes, W. O. & Christian, R. R. (1984). Estimating mean coliform densities of water distribution systems. *Journal of the American Water Works Association* **76**, 60–64.

Pipes, W. O., Ward, P. & Ahn, S. H. (1977). Frequency distributions for coliform bacteria in water. *Journal of the American Water Works Association* **69**, 644–647.

Prescott, S. C. & Winslow, C. E. A. (1931). *Elements of Water Bacteriology with Special Reference to Sanitary Water Analysis*, 5th edn. New York: Wiley; London: Chapman & Hall.

Ramalingaswami, V. & Purcell, R. H. (1988). Waterborne non-A, non-B hepatitis. *Lancet* **12**, 571–573.

Regli, S., Rose, J. B., Haas, C. N. & Gerba, C. P. (1991). Modelling the risk from *Giardia* and viruses in drinking water. *Journal of the American Water Works Association* 76–84.

Robinson, R. (1991). Maintenance. In *Appropriate Development for Basic Needs. Proceedings of the Conference on Appropriate Development for Survival — the Contribution of Technology*, pp. 203–218. Edited by D. P. Maguire. London: Thomas Telford.

Rose, J. B. & Gerba, C. P. (1991). Use of risk assessment for development of microbial standards. *Water Science and Technology* **24**, 29–34.

Shiffman, M. A., Schneider, R., Faigenblum, J. M., Helms, R. & Turner, A. (1978). Field studies on water, sanitation and health education in relation to health status in Central America. *Progress in Water Technology* **11**, 143–150.

Smith, P. (1991). Rural water supply: best placed? *World Water and Environmental Engineering*, October, 40.

Soby, B. A., Simpson, A. C. D. & Ives, D. P. (1993). *Integrating Public and Scientific Judgements into a Toolkit for Managing Food-Related Risks. Stage 1: Literature Review and Feasibility Study.* University of East Anglia, Norwich, UK. Cited in T. O'Riordan. *Environmental Science for Environmental Management.* Harlow: Longman Scientific and Technical.

Solo-Gabriele, H. & Neumeister, S. (1996). US outbreaks of cryptosporidiosis. *Journal of the American Water Works Association* September 76–86.

Steensberg, J. (1989). *Environmental Health Decision Making.* Copenhagen: Almquist and Wiksell International.

Taylor, P. (1993). Global perspectives on drinking water: water supplies in Africa. In *Safety of Water Disinfection: Balancing Chemical and Microbial Risks*, pp. 21–28. Edited by G. F. Craum. Washington: ILSI Press.

Troaré, A. (1992). Water for the people — community water supply and sanitation. In *International Conference on Water and the Environment: Development Issues for the 21st Century: Keynote Speakers.* Geneva: World Meteorological Organization.

Varley, R. C. G., Tarvid, J. & Chao, D. N. W. (1999). A reassessment of the cost-effectiveness of water and sanitation interventions in programmes for controlling childhood diarrhoea. *Bulletin of the World Health Organization* **76**, 617–631.

Walsh, J. A. & Warren, K. S. (1979). Selective primary health care: an interim strategy for disease control in developing countries. *New England Journal of Medicine* **301**, 967–974.

Watters, G. (1990). The interface between engineering and medicine. In *Proceedings of the Annual Symposium of the Institute of Water and Environmental Management, Manchester, UK, 27–28 March 1990*, pp. 6.1–6.18. London: Institute of Water and Environmental Management.

WHO (1976). *Surveillance of Drinking-Water Quality.* WHO Monograph Series No. 63. Geneva: World Health Organization.

WHO (1984). *The International Drinking Water Supply and Sanitation Decade — Review of National Baseline Data* (as at 31 December 1980). WHO offset publication No. 85. Geneva: World Health Organization.

WHO (1985). *Guidelines for Drinking-Water Quality*, vol. 3: *Drinking-water Quality Control in Small-Community Supplies.* Geneva: World Health Organization.

WHO (1986). *The International Drinking Water Supply and Sanitation Decade — Review of National Progress* (as at 31 December 1983). WHO/EHE/CWS/90.16. Geneva: World Health Organization.

WHO (1990). *The International Drinking Water Supply and Sanitation Decade — Review of Decade Progress* (as at December 1988). WHO offset publication no. 85. Geneva: World Health Organization.

WHO (1991). *Proceedings of the Advisory Committee Meeting of the Operation and Maintenance Working Group, Geneva, March 1991.* Report WHO/CWS/91.06. Geneva: World Health Organization.

WHO (1992). *The International Drinking Water Supply and Sanitation Decade — End of Decade Review* (as at December 1990). Geneva: World Health Organization.

WHO (1993). *Guidelines for Drinking-Water Quality,* vol. 1: *Recommendations,* 2nd edn. Geneva: World Health Organization.

WHO (1994). *Operation and Maintenance of Urban Water Supply and Sanitation Systems: a Guide for Managers.* Geneva: World Health Organization.

WHO (1997). *Guidelines for Drinking-Water Quality,* vol. 3: *Drinking Water Quality Control in Small-Community Supplies,* 2nd edn. Geneva: World Health Organization.

WHO (1998). *Guidelines for Safe Recreational Water Environments,* vol. 1: *Coastal and Freshwaters, Draft for Consultation.* Geneva: World Health Organization.

WHO/UNICEF/WSSCC (1996). *Water Supply and Sanitation Sector Monitoring Report 1996* (sector status as of 31 December 1994). WHO/EOS/96.15. Geneva: World Health Organization.

World Bank (1976). *Measurement of the Health Benefits of Investments in Water Supply.* Report no. PUN 20. Washington, DC: The World Bank.

Zucchetti, A. (1994). Drinking water supplies in Peru: microbiological surveillance and good disinfection practices. *Society for General Microbiology Quarterly* 21, 13–15.

Vaccine development: past, present and future

Coenraad F.M. Hendriksen

*National Institute of Public Health and the Environment (RIVM), PO Box 1, 3720 BA
Bilthoven, The Netherlands*

Introduction

One of the most effective ways to fight infectious diseases is the use of immunopro-
phylaxis. Its impact was already recognized as early as in the 6th century. A variety
of vaccines are now at our disposal, both in human and veterinary health care, and
infectious diseases such as tetanus, diphtheria and poliomyelitis no longer constitute a threat to health in the industrial countries. Many more are under development, not only against infectious diseases, but also against parasitic diseases and
cancers. This chapter will describe the various periods in vaccine development,
from the empiric approach in the very beginning to the rational scientifically based
development of conventional vaccines at the end of the 19th century. Finally, the
introduction of modern methodologies in vaccine development, such as recombinant DNA technology and DNA immunization, will be discussed. The history of
the use of adjuvant products to augment the immune response induced by the
vaccine, as well as the novel adjuvant products being developed, will also be
addressed.

Vaccine development: the early beginning (<1800)

The first attempts to vaccinate humans were made in China as long ago as the 6th
century. An early Chinese medical text, '*The Golden Mirror of Medicine*', lists several
forms of inoculation against smallpox (also called variolation), such as blowing
powdered scabs of human smallpox lesions into the nose, and stuffing a piece of
cotton smeared with the contents of smallpox vesicles into the nose (Plotkin &
Plotkin, 1988). In Turkey, the powdered scabs were administered by skin scarifica-
tion. Variolation was introduced into Britain in 1721 by Lady Mary Wortly Mon-
tague, the wife of the British ambassador in Turkey. The method became popular at
the intercession of King George I, after a public inoculation-challenge experiment
in a group of prisoners and orphans. Although the treatment was often effective,
results were erratic, and 2–3% of those treated died of smallpox obtained from the
variolation itself (Parish, 1965) or from a contamination of the product such as

61

with syphilis. In the 18th century, 'variolation' was also used to combat cattle plague, a disease which caused major losses in Europe at the time (De Vries, 1994), and measles. The British researcher Francis Home wrote about measles in 1758: 'I thought I should do no small service to mankind if I could render this disease more mild and safe in the same way as the Turks have taught us to mitigate the smallpox'.

The first attempt to control an infectious disease by means of a deliberate systematic inoculation with cowpox is attributed to Edward Jenner. After 25 years of study, Jenner published his findings on cowpox and smallpox in 1798 (Jenner, 1798). This publication also contained the results of his successful experiment in 1796 in which an 8-year-old boy, James Phipps, was inoculated with cowpox material and subsequently challenged with smallpox virus. In the 19th century, Pasteur introduced the term 'vaccination' (*vacca* = cow) as a homage to Jenner. It might be interesting to note that based on studies of the virus genome it is now believed that the 'cowpox' virus was in fact the poxvirus of the horse.

The understanding of pathogenicity (1800–1880)

During the 82 years that elapsed between Jenner's treatise on Variolae Vaccinae and the development of the first vaccine based on scientific research — the fowl cholera (*Pasteurella multocida*) vaccine discovered by Pasteur in 1880 — the field of microbiology was not dormant. Generally speaking, the emphasis was on the development of comparative pathology and the gaining of insight into the aetiology of infectious diseases. The prevailing belief that miasmas were the causal agent was replaced by the understanding that diseases are caused by a living organism (*contagium vivum*). In 1840, the German pathologist and anatomist Jacob Henle (1809–1885) published his work on contagious diseases, in which he demonstrated that living organisms and not miasmas were the cause of disease. Henle concluded that for establishing the association between causal agent and infection, it is essential to culture the micro-organism outside a living host. Koch's postulates, published in 1884, have been of historic importance regarding the study of infectious diseases. These postulates stated the criteria to be met in relating a micro-organism to a given infection. The postulates included the isolation of the micro-organism in pure culture, followed by the introduction of the pure culture into a suitable experimental animal. This must result in the induction of typical clinical signs in the experimental animal. Koch's postulates gained general acceptance in microbiology and helped to lay the foundation for the intensive search for prophylactic vaccination and therapeutic serum treatment (Hendriksen, 1996).

Vaccine development: the beginning of the scientific approach
(1880–1900)

It was Pasteur who laid the foundation for a rational and experimental approach to vaccine development, in which animal experiments played an important role. He discovered in 1880 that, when chickens were injected with old cultures of *P. multocida* (the causative agent of fowl cholera), very few clinical symptoms developed. A subsequent injection of fresh virulent cultures into the same animals had no adverse effects. He drew two fundamental conclusions from this: (a) the virulence of germs can be attenuated by storing the culture for some time; and (b) these cultures can be used to induce artificial immunity.

The vaccine against fowl cholera thus became the first experimentally developed vaccine, soon followed by vaccines against anthrax (1881) and rabies (1885).

Since Pasteur's pioneering work, immunization as a means of combating infectious diseases was taken up and extended by several researchers. In particular, the discovery of diphtheria antitoxin by von Behring and Kitasato in 1890 (von Behring & Kitasato, 1890) facilitated further research. It was one of the first immunotherapies where the direct result of a dramatic decline of mortality and morbidity could be demonstrated. With a mortality rate of up to around 40%, diphtheria (also known as 'the strangling angel of children') was one of the most feared epidemic diseases of childhood (Grundbacher, 1992). In addition, it was obvious to the public that experimental animal work had laid the foundation for effectively combating this disease (see Table 1), a fact that influenced the emotional discussion at that time on the use of experimental animals. Furthermore, many common routine procedures in the quality control of immunobiologicals arose from diphtheria research.

With the discovery of the early vaccines in the last decades of the 19th century, it seemed a matter of time before such dreaded diseases as tuberculosis, typhus,

Table 1. Animal experiments in the development of diphtheria treatment and prevention (reproduced from Hendriksen, 1996, with permission).

Development	Year	Name	Animal species*
Isolation of the causal micro-organism, *Corynebacterium diphtheriae*	1884	Loeffler	Pigeon, chicken, rabbit, guinea pig
Production of the exotoxin	1884	Roux & Yersin	Various animal species, GUINEA PIG
Demonstration of the therapeutic value of antitoxin	1890	Behring	GUINEA PIG, dog, mouse, rat, various other species
Large-scale production of antitoxin	1894	Roux & Martin	Dog, sheep, goat, HORSE, cow
Toxin–antitoxin mixtures for active immunization	1913	Behring	GUINEA PIG
Diphtheria toxoid	1923	Ramon	Various animal species

*The animal species finally chosen is given in upper case.

typhoid fever and syphilis could be controlled. However, by 1910, it became evident that these and many other diseases would not soon be amenable to such human control (Trager, 1990), due to little knowledge of fundamental immunology and technical limitations. For instance, although the viral vaccines against smallpox and rabies were among the first to be developed, the real breakthrough of viral vaccine development came only after the 1950s with the introduction of the large-scale use of tissue culture technology (Ruitenberg & Van Mourik, 1991).

Vaccine development: the conventional (traditional) approach (1900–1975)

From about 1900 to 1950, vaccine-related research became the domain of pathology, immunology and cell culture technology. Emphasis was placed on the understanding of disease processes, for instance regarding routes of transmission, and on the nature of the immune response (antibody specificity and antibody–antigen interactions, complement fixation, humoral and cellular immunity, etc.). Attention was also given to improving the immune response by the use of adjuvant products (see later).

A number of new vaccines became available in this period, all based on the traditional technique of attenuation or inactivation of the virulent micro-organism or product thereof. Ramon was successful in detoxifying diphtheria toxin and tetanus toxin by treatment with formaldehyde in 1923 and 1927, respectively. The pertussis vaccine was first used in 1923 and influenza and mumps vaccines were produced in the 1940s by culturing of the virus in chicken embryos.

Although Roux had already taken the first steps towards *in vitro* culture of cells in 1885, it was not until the discovery of penicillin by Fleming in 1928 that cell culture conditions could be improved and cell cultures could be adopted for virus vaccine production. Therefore, it was only in 1949 that the paper by Enders, Weller & Robbins was published describing the successful cultivation of the Lansing strain of poliovirus in cultures of non-nervous human tissues, which provided the breakthrough for the development of a polio vaccine (Robbins, 1988). In 1955, Sabin was successful in producing the attenuated (oral) polio vaccine and, in 1960, Salk produced the inactivated polio vaccine. Cell culture vaccines against the major virus diseases in humans became available in the 1960s and early 1970s.

The new generations of vaccines (1975–2000)

The value of conventional vaccine production as described above has been beyond doubt. It resulted in a generation of vaccines that significantly reduced the incidence of diphtheria and whooping cough and led to the eradication of

smallpox and the imminent eradication of polio. Nevertheless, it was felt that the conventional production approach suffered from several drawbacks (Lerner,1983):
• attenuated vaccine strains can mutate, resulting in an increase in virulence or loss of immunogenicity;
• with inactivated vaccines, there is a risk of incomplete inactivation, and several vaccine-related accidents have occurred, such as the Cutter incident due to incomplete inactivation of the inactivated poliomyelitis vaccine in 1955 (Hendriksen, 1996);
• certain micro-organisms, such as the hepatitis B virus, cannot be cultured by conventional techniques;
• the cells and medium used for production of viral vaccines in particular are a potential reservoir of contaminating micro-organisms;
• working with pathogenic vaccine strains carries a health risk;
• the cost of assuring the quality of the current vaccines is high; and
• it is sometimes difficult to separate toxic components from the vaccine micro-organisms, so that vaccination may be accompanied by adverse reactions. This was one of the main reasons for the decline in pertussis immunization in the UK and in many other countries in the mid-1970s.

It is in part because of these shortcomings that other production techniques have been researched, using the new advances from immunology, molecular biology and biochemistry. For example, antigens are now being identified and purified with the help of monoclonal antibodies. Biochemistry has advanced our knowledge of the structure of antigens and has enabled the coupling of antigens to carrier molecules in order to elicit T-cell responses. Immunology has provided information on the critical factors in B- and T-cell stimulation and in the generation and activation of memory cells (Ada, 1988). Recombinant DNA techniques are employed to produce specific antigens. With this fundamental knowledge, a rational approach is being taken to devise novel vaccine production strategies such as subunit vaccines, synthetic vaccines and (polynucleotide) DNA vaccines. In using the protein carrier technique, it is now possible to produce effective vaccines against polysaccharide antigens (glycoconjugate vaccines). Some characteristics of the new generation of vaccines are discussed below.

Subunit vaccines

These products are based on the isolation by conventional biochemical procedures or by recombinant DNA techniques of the relevant immunogenic components of the vaccine micro-organism. Several subunit pertussis vaccines are currently on the market based on detoxified pertussis toxin, eventually combined with some other antigens, such as filamentous haemagglutinin (FHA) or outer-membrane proteins. This has resulted in a reduction of the toxicity of the vaccine. By recombinant DNA techniques it has been possible to produce a hepatitis B vaccine based on the immunogenic surface antigens.

Synthetic vaccines

Advances in biochemistry have increased the interest in polypeptide vaccines produced by chemical synthesis. The appeal for this approach is that it permits the manufacture of chemically well-defined products on an industrial scale. However, experience has shown that, though the antigen peptides can be synthesized, it is very difficult to obtain the specific configuration of the peptide chain necessary for immunogenicity. The pioneering system has been performed using the foot-and-mouth disease virus (FMDV). The viral peptides related to protective immunity have been identified (the capsid protein VP1) and have been synthesized in the laboratory. However, so far these antigens have conferred only limited protection in the natural host (Sobrino *et al.*, 1999). Advantages and limitations of the synthetic FMDV in comparison to the conventional vaccines are illustrated in Table 2.

Glycoconjugate vaccines

By covalently linking weak (T-cell independent) antigens like polysaccharides to protein carriers it is possible to induce T-cell dependency. Conjugate technology has enabled the development of glycoconjugate vaccines, such as for *Haemophilus influenzae* type b (Hib), *Streptococcus pneumoniae* and *Neisseria meningitidis*. These vaccines are now routinely used to eliminate infant diseases like meningitis, pneumonia and otitis media (Madore, 1996; Madore, Strong & Eby, 1999).

Table 2. Conventional and synthetic vaccines: advantages and limitations. Adapted from Sobrino *et al.* (1999), and reproduced with permission.

Conventional, chemically inactivated whole virus vaccines
+ Good immunogenicity
− Possibility of virus release during vaccine production
− Stability (requirement of cold chain)
− Limited duration of immunity (revaccination every 6–12 months)
− Antigenic variability (need for selection of vaccine strains)
− Poor distinction between infected/vaccinated animals

Synthetic and subunit vaccines
+ Innocuous
+ Chemically defined
+ Stable
+ Possibility of inducing lasting immunity
+ Possibility of inducing wider protection spectra
+ Distinction between infected/vaccinated animals
− Low immunogenicity

DNA (polynucleotide) vaccines

The strategy in DNA vaccination differs from the more traditional antigen-based vaccines in that the 'vaccine' antigen is synthesized *in vivo* after direct immunization of its encoding sequences, generally by intramuscular injection. This method of vaccination provides a stable and long-lived source of protein antigen and has been shown to elicit both humoral and cell-mediated immune responses. In laboratory animals, DNA immunization has provided effective immunity to a variety of pathogens and also has proven to be useful against non-infectious diseases. There are still several hurdles that need to be overcome, such as improving delivery and potency (Tuteja, 1999). Safety concerns are also raised about the long-term effects of antigen expression, the development of anti-DNA antibodies and the potential expression of toxic antigens.

A general disadvantage of the novel generations of vaccines is that these products are often based on highly purified, small molecular antigens that are generally weakly immunogenic and, therefore, require an adjuvant to evoke the desired immune response (Gupta *et al.*, 1993).

The history of adjuvant development

Adjuvants (Latin: *adjuvare* = to help) are compounds that, when used in combination with specific vaccine antigens, augment the resultant immune response (Newman & Powell, 1995). Although adjuvants have already been used for a long time, the introduction of the new vaccine production technologies, such as subunit and conjugate vaccines, has resulted in an increased interest in adjuvants.

The history of adjuvant products started with the work of the French veterinarian Ramon in 1925 (Ramon, 1925). Ramon studied the immune response against diphtheria and tetanus toxoid and showed that it was possible to increase protective immunity by administering the toxoid together with tapioca, pyogenic bacteria, agar, starch oil and even breadcrumbs. A major drawback, however, was the development of large abscesses at the injection site. One year later, in 1926, Glenny *et al.* (1926) investigated the use of a number of precipitating and adsorbing aluminium salts, such as $AlK(SO_4)_2$, $AlPO_4$ and $Al(OH)_3$. A significant increase in the antigenicity of the toxoid, with only minimal adverse effects, could be demonstrated.

A milestone in adjuvant research was the work performed by Freund and others in 1937 on the activity of mineral (paraffin) oil mixed with killed mycobacteria (Freund, Casals & Hosmer, 1937). This product (named Freund's complete adjuvant or FCA) appears to be one of the most potent adjuvants described so far. Unfortunately, it was too toxic for use in humans. Instead, the mineral oil without mycobacteria, also named Freund's incomplete adjuvant or FIA, has been used in human vaccines, in particular influenza and inactivated poliomyelitis vaccine, for many years. However, the use of FIA in human vaccines was discontinued in the early 1970s because of reported side effects, such as sterile abscesses, ulcers and delayed-type local reactions (Gupta *et al.*, 1993; Stewart-Tull, 1995). However, it is still used as an adjuvant in veterinary vaccines.

Currently, the only adjuvants that are accepted for use in human vaccines are the aluminium salts.

The mode of action of adjuvant products

Since the first reports on adjuvants, extensive studies have been conducted to elucidate the mechanism of immune potentiation. Glenny, for instance, showed in 1931 by studies in guinea pigs that the adjuvants delayed the adsorption of the antigens. The animals were injected intracutaneously with an alum-precipitated toxoid or with toxoid alone. Several days after immunization, the portion of the skin containing the site of injection was excised and an emulsion thereof was injected into naïve guinea pigs. Successful immunizations of the recipient animals could be demonstrated only among those that had received emulsions from the alum-toxoid-injected guinea pigs, whereas both groups of donors became immune (Lindblad, 1995).

The benchmark adjuvants for the study of the mode of action have been the aluminium salts and oil products (FCA and FIA). From these studies, we now know that a number of mechanisms are involved, such as: (1) the formation of a depot of antigen at the site of administration which is slowly released; (2) the presentation of antigen to immunocompetent cells; (3) the activation of complement; and (4) the induction of different cytokines (Chedid, 1985). We also know that the mechanism is in some way related to the type of adjuvant. So, aluminium salts in particular induce Th2 helper cells and stimulate the production of IgE and IgG1, whereas FCA enhances a mixed Th1/Th2 response.

The finding that adjuvants can not only be used to augment the immune response but also to modify the response and/or to elicit a specific activity has led to the development of many new types of products.

The new generation of vaccine adjuvants

The main goal of the conventional adjuvants was to support the conventional vaccines in inducing a high and long-lasting humoral response. However, the new vaccines or combinations of vaccines frequently have a highly specific activity, thereby requiring specific adjuvants. Vaccines are now being developed to induce mucosal immunity as a first line defence. Cholera toxin, for instance, was shown to augment the local mucosal immunity to orally administered antigens (Gupta *et al.*, 1993).

Conventional vaccines are generally administered several times in order to induce long-lasting protective activity. As the number of vaccines increases in paediatric immunization programmes, there is a need to reduce the frequency of injection. Biodegradable microspheres have been studied to solve the problem of multi-injection immunization schedules. Vaccine antigens encapsulated into the microsphere formulations are delivered and released to the immune system more slowly than are soluble antigens. This might lead to vaccine formulations that only require one single immunization (Newman & Powell, 1995). Antigen-specific cellular immune responses can be induced using adjuvants such as the saponin Quil A (a natural glycoside derived from the bark of *Quillaja saponaria*) and immune-stimulating complexes (ISCOMs). These products have been used in experimental HIV vaccines (Bomford, 1995). The working mechanism of ISCOMs is related to encapsulation of the antigen in cage-like structures of saponin/cholesterol/phospholipids. Liposomes have a similar mode of action.

A new generation of adjuvant products might be found in the group of cytokines that regulate immune interactions. In fact, the traditional adjuvants exert their activity by the non-specific induction of several cytokines (Dong, Brunn & Ho, 1995). Based on this, it is hypothesized that the selective use of cytokines should effectively and directly improve the immunogenicity of the weak antigens while minimizing the side effects of the conventional adjuvants. To date, cytokine research is still faced with conflicting data and further studies will be needed.

Other adjuvant types being studied include the non-ionic block copolymers, the bacterial cell wall products such as muramyl dipeptide (MDP) or lipid A. Detailed information on these products can be found in several reference volumes such as those published by Stewart-Tull (1995) and Newman & Powell (1995).

Conclusions

Immunization as a way to prevent infectious diseases has a history of almost 14 centuries. The scientific approach to vaccine development started with the work of Pasteur and Koch at the end of the 19th century. Since then, numerous vaccines, both human and veterinary, have been registered.

The impact of vaccination programmes has been dramatic. Smallpox is now eradicated, polio is on the verge of eradication and the incidence of most of the important paediatric diseases has declined significantly.

Until very recently, vaccine production was based on conventional approaches: the attenuation or inactivation of virulent micro-organisms or products thereof. However, these vaccines have a number of disadvantages related to safety and efficacy. In addition, problems are encountered in developing vaccines to specific diseases, such as AIDS and malaria. With the advent of novel technology coming from molecular biology and biochemistry, new vaccine production approaches are now within reach. The approaches include subunit vaccines, glycoconjugate vaccines, synthetic vaccines and recently also DNA vaccines. These novel products might be characterized as safe, without a risk for reversion to virulence. The novel techniques are now being exploited to develop improved versions of existing vaccines and new vaccines against emerging pathogens, tumours or autoimmune diseases (Mahon *et al.*, 1999). Also, vectors for recombinant DNA vaccine production now range from micro-organisms to plants.

The novel vaccines are often based on highly purified, small molecular antigens that are generally weakly immunogenic. Therefore, the use of an adjuvant is essential to evoke the desired immune response. There is also a need for adjuvants that not only augment, but also modulate the humoral and cell-mediated immune response. This has resulted in an increased interest in the development of novel adjuvants.

It is believed that our future health care policy will be significantly influenced by the novel vaccine production technologies, shifting the balance from therapeutics to prophylactics.

References

Ada, G. L. (1988). What to expect of a good vaccine and how to achieve it. *Vaccine* 6, 77–79.

von Behring, E. & Kitasato (1890). Uber das Zustandekommen der Diphtherie-Immunität und der Tetanus-Immunität bei Tieren (Production of diphtheria immunity and tetanus immunity in animals). *Deutsche Medizinische Wochenschrift* 49, 113.

Bomford, R. (1995). Adjuvants in AIDS vaccines. In *The Theory and Practical Application of Adjuvants*, pp. 353–363. Edited by D. E. S. Stewart-Tull. Chichester: Wiley.

Chedid, L. (1985). Adjuvants of immunity. *Annales de l'Institut Pasteur Immunology* 136D, 283–291.

De Vries, J. (1994). De bestrijding van de runderpest in Friesland gedurende de 18e eeuw (Combatting cattle plague in Friesland in the 18th century). *Argos* 10, 315–323.

Dong, P., Brunn, C. & Ho, R. J. Y. (1995). Cytokine as vaccine adjuvants: current status and potential applications. In *Vaccine Design. The Subunit and Adjuvant Approach*, pp. 625–639. Edited by M. F. Powell & M. J. Newman. New York: Plenum.

Enders, J. F., Weller, T. H. & Robbins, F. C. (1949). Cultivation of the Lansing strain of poliomyelitis virus in cultures of various human embryonic tissue. *Science* 109, 85–87.

Freund, J., Casals, J. & Hosmer, E. P. (1937). Sensitization and antibody formation after injection of tubercle bacilli and paraffin oil. *Proceedings of the Society for Experimental Biology* 37, 509–513.

Glenny, A. T., Pope, C. G., Waddington, H. & Wallace, U. (1926). Immunological notes XVII to XXIV. *Journal of Pathology* 29, 31–40.

Grundbacher, F. J. (1992). Behring's discovery of diphtheria and tetanus antitoxins. *Immunology Today* 13, 188–189.

Gupta, R. K., Relyveld, E. H., Lindblad, E. B., Bizzini, B., Ben-Efraim, S. & Gupta, C. K. (1993). Adjuvants—a balance between toxicity and adjuvanticity. *Vaccine* 11, 293–306.

Hendriksen, C. F. M. (1996). A short history of the use of animals in vaccine development and quality control. *Developments in Biological Standardization* 86, 3–10.

Jenner, E. (1798). *An Inquiry into the Causes and Effects of the Variolae Vaccinae*. London: Samson Low.

Lerner, R. A. (1983). Synthetic vaccines. *Scientific American* 248, 48–56.

Lindblad, E. B. (1995). Aluminium adjuvants. In

The Theory and Practical Application of
Adjuvants, pp. 21–37. Edited by D. E. S.
Stewart-Tull. Chichester: Wiley.

Madore, D. V. (1996). Impact of immunization
on *Haemophilus influenzae* type b disease.
Infectious Agents and Disease 5, 8–20.

Madore, D. V., Strong, N. & Eby, R. (1999). Use of
animal testing for evaluating glycoconjugate
vaccine immunogenicity. *Developments in
Biological Standardization* 101, 49–56.

Mahon, B. P., Moore, A., Johnson, P. A. & Mills,
K. H. (1999). Approaches to new vaccines.
Critical Reviews in Biotechnology 18, 257–282.

Newman, M. J. & Powell, M. F. (1995). *Vaccine
Design. The Subunit and Adjuvant Approach.*
New York: Plenum.

Parish, H. J. (1965). *A History of Immunization.*
London: E & S Livingstone.

Plotkin, S. L. & Plotkin, S. A. (1988). A short
history of vaccination. In *Vaccines*, pp. 1–7.
Edited by S. A. Plotkin & E. A. Mortimer.
Philadelphia: W. B. Saunders.

Ramon, G. (1925). Sur la production de
l'antitoxine diphterique (Production of
diphtheria antitoxin). *C R Academie des
Sciences (Paris)* 93, 506.

Robbins, F. C. (1988). Polio-historical. In
Vaccines, pp. 98–114. Edited by S. A. Plotkin &
E. A. Mortimer. Philadelphia: W. B. Saunders.

Ruitenberg, E. J. & Van Mourik, P. C. (1991).
Production of biological products and the use
of laboratory animals. In *Animals in
Biomedical Research, Replacement, Reduction
and Refinement: Present Possibilities and Future
Trends*, pp. 69–79. Edited by C. F. M.
Hendriksen & H. B. W. M. Koëter. Amsterdam:
Elsevier.

Sobrino, F., Blanco, E., Garcia-Briones, M. & Ley,
V. (1999). Synthetic peptide vaccines: foot-
and-mouth disease virus as a model.
Developments in Biological Standardization
101, 39–43.

Stewart-Tull, D. E. S. (1995). Freund-type
mineral oil adjuvant emulsions. In *The Theory
and Practical Application of Adjuvants*, pp.
1–20. Edited by D. E. S. Stewart-Tull.
Chichester: Wiley.

Trager, J. A. (1990). History of immunology.
Immunology Today 11, 182–183.

Tuteja, R. (1999). DNA vaccines: a ray of hope.
*Critical Reviews in Biochemistry and Molecular
Biology* 34, 1–24.

Live attenuated vectors: have they delivered?

Myron M. Levine, James E. Galen, Eileen Barry, Carol Tacket, Karen Kotloff, Oscar Gomez-Duarte, Marcela Pasetti, Thames Pickett and Marcelo Sztein

Center for Vaccine Development, University of Maryland School of Medicine, Baltimore, MD 21201, USA

Introduction

An area of vaccinology that has generated considerable excitement during the past decade is the use of attenuated strains of bacteria and viruses to carry protective antigens of foreign, unrelated pathogens and deliver those foreign antigens to the immune system, thereby resulting in a relevant, protective response (Dougan, Hormaeche & Maskell, 1987; Levine *et al.*, 1990). Initially, the strategy was confined to engineering the attenuated strains to express foreign proteins and polysaccharide antigens directly. More recently, live vectors, in particular bacteria, have been modified so that they carry DNA vaccine plasmids (i.e. foreign genes on plasmids under the control of eukaryotic expression promoters) (Darji *et al.*, 1997; Fennelly *et al.*, 1999; Pasetti *et al.*, 1999). In this way, the bacteria deliver the DNA vaccine plasmids to phagocytic antigen-presenting cells such as macrophages, and the foreign antigen is then expressed by the eukaryotic cells themselves.

An array of viruses such as vaccinia (Moss, 1996), other poxviruses (e.g. canarypoxvirus) (Belshe *et al.*, 1998) and adenovirus (Gonin *et al.*, 1996) have been investigated as live vectors. Similarly, a wide variety of bacteria have been utilized as live vectors. These have included attenuated strains derived from wild-type pathogens such as *Salmonella typhi* (Barry *et al.*, 1996; Chatfield *et al.*, 1992a; Galen *et al.*, 1997; Tacket *et al.*, 1997a) and *Salmonella typhimurium* (Chatfield *et al.*, 1992b; Schodel *et al.*, 1996), *Shigella* (Noriega *et al.*, 1996a, b), *Vibrio cholerae* (Butterton *et al.*, 1997; Ryan *et al.*, 1997a, b), *Listeria monocytogenes* (Frankel *et al.*, 1995), *Mycobacterium bovis* (bacille Calmette–Guérin, BCG) (Edelman *et al.*, 1999) and *Yersinia enterocolitica* (Igwe *et al.*, 1999; Morris, Tacket & Levine, 1992). However, the bacterial live vectors have also included non-pathogenic strains derived from normal flora such as *Streptococcus gordonii* (Medaglini *et al.*, 1995; Pozzi *et al.*, 1992) or other sources such as *Lactobacillus acidophilus* (Maassen *et al.*, 1999; Zegers *et al.*, 1999) and *Lactococcus* (Robinson *et al.*, 1997; Steidler *et al.*, 1998). This chapter will confine itself to a discussion of *Salmonella* and *Shigella* strains as examples of bacterial live vectors derived by attenuation of otherwise pathogenic parent bacteria, and will consider their use both for prokaryotic expression of foreign antigens and as live delivery systems for eukaryotic expression plasmids (i.e. DNA vaccine plasmids).

The array of immune responses elicited by *Salmonella* and *Shigella*

Infection with wild-type *Salmonella* and *Shigella* or with attenuated strains used as live oral vaccines is capable of stimulating every arm of the immune system. Serum IgG antibodies (Hohmann *et al.*, 1996; Tacket *et al.*, 1992a, b, 1997a, b), secretory IgA (SIgA) intestinal antibodies (Forrest *et al.*, 1991; Forrest, Shearman & LaBrooy, 1990; Viret *et al.*, 1999), proliferation of γ-interferon-secreting CD4$^+$ lymphocytes (Sztein *et al.*, 1994), specific MHC I-restricted CD8$^+$ cytotoxic lymphocytes (CTL) (Sztein *et al.*, 1995) and specific antibody-dependent mononuclear cell killing responses (Tagliabue *et al.*, 1985, 1986) have all been described in humans and in animal models. Depending on the type of immune response necessary to prevent infection by the pathogen of interest, *Salmonella* and *Shigella* live vectors encoding heterologous protective antigens can elicit the relevant immune response, be it humoral or cell-mediated (Barry *et al.*, 1996; Chatfield *et al.*, 1992b; Galen *et al.*, 1997; Gonzalez *et al.*, 1994, 1998; Hess *et al.*, 1996; Noriega *et al.*, 1996a).

Studies in animal models with *Salmonella* live vectors expressing foreign antigens

Wild-type *S. typhimurium* is highly pathogenic for mice. Following oral inoculation, *Salmonella* strains target the microfold (M) cells that overlie the gut-associated lymphoid tissue (GALT) and are then passed to macrophages in the underlying GALT. Lymphatic drainage leads, via the thoracic duct, to a primary bacteraemia that seeds the organs of the reticuloendothelial system (spleen, liver, bone marrow, etc.) and, after a moderate incubation period, results in a systemic illness. In many of these features, *S. typhimurium* in mice shares a resemblance to typhoid fever in humans caused by the human host-restricted pathogen *S. typhi*. The mouse model of *S. typhimurium* has been used in vaccine development to study the degree of attenuation that various precise mutations confer upon wild-type *S. typhimurium* (Chatfield *et al.*, 1992c; Curtiss & Kelly, 1987; Hoiseth & Stocker, 1981; Miller, Kukral & Mekalanos, 1989). Typically, mortality is used as the readout and mutants that no longer cause mortality (or for which the lethal dose is raised by 4–5 logs) are then assessed for immunogenicity and for their ability to protect against challenge with a lethal dose of wild-type *S. typhimurium*. Several of these homologous mutations were subsequently introduced into wild-type *S. typhi* and the resultant strains were evaluated in clinical studies to assess whether these mutations adequately attenuated *S. typhi* for humans (Hohmann *et al.*, 1996; Levine *et al.*, 1987; Tacket *et al.*, 1992a, b, 1997a, b).

The attenuated *S. typhimurium* vaccine strains have been used extensively to study their utility as live vector vaccines in mice. The most popular *S. typhimurium* strains are those that harbour mutations in the following genes. (1) *cya* (encoding adenylate cyclase) and *crp* (encoding the cAMP receptor protein): these genes comprise a global regulatory system that affects multiple virulence and housekeeping genes (Curtiss & Kelly, 1987). (2) *aroA, aroC* or *aroD*: these genes encode enzymes involved in the biosynthesis of aromatic metabolites, including folate, enterochelin and the aromatic amino acids (Dougan *et al.*, 1988). Such mutants are dependent on substrates (*para*-aminobenzoic acid and 2,3-dihydroxybenzoate) that are not available to the bacteria in sufficient concentration once they have invaded systemically. (3) *phoP, phoQ*: these genes constitute another regulatory system that controls genes allowing *Salmonella* to adapt to survival within phagolysosomal vacuoles in macrophages (Miller *et al.*, 1989). (4) *htrA*: this gene encodes a stress response protein that functions as a serine protease (Chatfield *et al.*, 1992c).

Scores of studies have been carried out in the mouse model with *S. typhimurium* carrying foreign antigens encoded either on plasmids or by genes integrated into the chromosome. The perspective gained from the extensive experience with *Salmonella* live vectors in the mouse model is that this vaccinology strategy has been extraordinarily successful. Proteins and polysaccharide antigens from other genera of bacteria, viruses and parasites have been expressed in *Salmonella* and oral immunization with the live vectors has elicited appropriate specific serum antibody, SIgA mucosal antibody or cell-mediated immune responses that have conferred protection against challenge with wild-type organisms or toxins (Aggarwal *et al.*, 1990; Ascon *et al.*, 1998; Chatfield *et al.*, 1992b; Clements *et al.*, 1986; Gomez-Duarte *et al.*, 1998; Karem *et al.*, 1997; Nayak *et al.*, 1998; Sadoff *et al.*, 1988; Toebe *et al.*, 1997; Xu *et al.*, 1997).

Experience in human clinical trials with attenuated *S. typhi* and *S. typhimurium* expressing foreign antigens

Despite the extensive, overwhelmingly positive experience in mice with attenuated *Salmonella* live vectors expressing foreign antigens, only a few Phase I and II clinical trials have been carried out in humans using attenuated *S. typhi* live vectors expressing foreign antigens. These trials have given variable results, none being overly positive (Black *et al.*, 1987; DiPetrillo *et al.*, 1999; Gonzalez *et al.*, 1994; Herrington *et al.*, 1990; Tacket *et al.*, 1997a; C. O. Tacket, J. Galen, M. B. Sztein, G. Losonsky, T. L. Wynant, J. Nataro, S. S. Wasserman, R. Edelman, S. Chatfield, G. Dougan & M. M. Levine, unpublished; Tramont *et al.*, 1984). In some clinical trials, the *S. typhi* live vectors have simply failed to elicit a detectable immune response to the foreign antigen, despite immunogenicity of the same (or homologous *S. typhimurium*) constructs when tested in preclinical animal models. On the other hand, a few clinical trials have resulted in the elicitation of both serum and cell-mediated immune responses to the foreign antigen, including antigens such as the circumsporozoite protein (CSP) of *Plasmodium falciparum* (Gonzalez *et al.*, 1994) that are known to be poorly immunogenic in humans.

One Phase I clinical trial has been carried out with an attenuated *S. typhimurium* live vector expressing *Helicobacter pylori* urease A and B (Angelakopoulos & Hohmann, 1999). This construct was not well tolerated as two of six subjects developed headaches. Nevertheless, three of six subjects manifested immune responses to the urease foreign antigen.

Parameters that affect the immunogenicity of foreign antigens expressed by attenuated *Salmonella*

A careful review of the few clinical trials in humans and of the numerous mouse model studies using *S. typhimurium* and other *Salmonella* vectors allows a systematic analysis to be undertaken of the various factors that influence the ability of *S. typhi* constructs to function as oral live vector vaccines in humans. These are discussed in the ensuing paragraphs.

Choice of S. typhi strain

The choice of which attenuated *Salmonella* strain to use is critical. Although several candidate strains may all be acceptably well tolerated, they may differ notably in their immunogenicity both with respect to eliciting responses to the live vector itself (i.e. anti-*Salmonella*) and to the foreign antigen. In general, the strength of the immune responses to the vector and the heterologous antigen coincide together. For example, Ty21a, the well-tolerated licensed live oral typhoid vaccine strain, is only modestly immunogenic. Thus it is not a good choice to serve as a live vector. Depending on the attenuating mutations, strains may differ in whether they preferably elicit antibody or cell-mediated responses. For example, *Salmonella* strains that harbour mutations in the *phoPQ* operon, which limits survival within the phagolysosomes of macrophages, appear to preferentially elicit antibody responses (Benyacoub *et al.*, 1999). Strains harbouring mutations in *aro* genes and *htrA* appear to stimulate both Th1 and Th2 type responses.

Choice of protective foreign antigens

While the flexibility of the *Salmonella* system is considerable, it nevertheless has limitations. For example, if a protective antigen is a virus protein that is usually produced by a eukaryotic cell and post-translational modifications such as glycosylation contribute to the correct folding of the antigen in order for it to stimulate specific antibodies, *Salmonella* would not be the preferred system for direct expression of such an antigen. *Salmonella* would not be able to glycosylate such a protein. In contrast, even if such a virus protein is ordinarily found glycosylated *in vivo*, if the critical immune response is cell-mediated (and therefore not conformation-dependent), *Salmonella* may function quite well.

Because of the broad immune response that *Salmonella* strains are able to elicit, there is much interest in utilizing *Salmonella* to deliver antigens of parasites to the immune system (Levine *et al.*, 1997). Expressing these eukaryotic proteins has been generally challenging and has often required protein engineering in which highly hydrophobic regions, usually found on the N-terminus or C-terminus, are deleted. When such regions are removed, these eukaryotic proteins can usually be expressed.

Optimization of codon usage

Attempts to express genes from highly unrelated organisms have illustrated the importance of codon usage. This has been true in expressing genes encoding eukaryotic proteins in bacteria such as *Salmonella* and *Escherichia coli*, in expressing bacterial genes in higher plants, and even in expressing proteins from Gram-positive bacteria such as *Clostridium tetani* in *Salmonella*. Therefore, in order to improve direct expression by *Salmonella*, the codon usage of the foreign gene should be optimized for *Salmonella* preferences.

Selection of a promoter

The choice of promoter to control the expression of a foreign gene in *Salmonella* has proven to be critical in affecting the immunogenicity of the construct. For some antigens that are well tolerated by *Salmonella*, the use of a powerful constitutive promoter that achieves high-level continuous expression of the antigen is advantageous (Galen *et al.*, 1997). However, the expression of certain antigens has a deleterious impact on the live vector, taking a metabolic toll that alters its growth curve *in vitro*, and on its colonizing ability *in vivo* (Coulson, Fulop & Titball, 1994). For such antigens, the strategy that has become popular is to delay the onset of full expression until *Salmonella* has successfully reached a protected and immunologically relevant site, such as within M cells or within the phagolysosomes of macrophages. Accordingly, promoters that are activated by signals present in such environments have been used with variable success. Among these 'in *vivo*-activated' promoters are those which regulate expression of *nir15* (activated by low redox potential) (Chatfield *et al.*, 1992b; Gomez-Duarte *et al.*, 1995), *ompC* (activated by isosmolar conditions), *pagC* (activated by conditions in the phagolysosome) (Dunstan, Simmons & Strugnell, 1999) and *htrA* (activated by various environmental stresses) (Everest *et al.*, 1995). The behaviour of the promoter under *in vitro* conditions does not always predict results in animal models or human clinical trials.

Stabilization of foreign genes

There are two strategies to stabilize foreign genes encoding antigens of interest: by chromosomal integration or by a plasmid stabilization system. Integration into the chromosome assures stability but provides only a single copy of the gene and in some instances is technically difficult to accomplish (Gonzalez *et al.*, 1994; Hone *et al.*, 1988). In contrast, plasmid stabilization allows one to have the benefits of multi-copy plasmids. The use of a 'balanced lethal system' assures that bacteria that carry the plasmid survive, whereas those that lose it die (post-segregational killing) (Nakayama, Kelly & Curtiss, 1988). The best known balanced lethal system is based on deleting *asd*, a gene that encodes an enzyme necessary for the synthesis of diaminopimelic acid (DAP), a substrate that Gram-negative bacteria must have in order to synthesize their cell walls (Nakayama, Kelly & Curtiss, 1988). Inactivation of *asd* is lethal unless DAP, which is not present in mammalian tissues, is provided. A plasmid that carries *asd* plus the gene encoding the foreign antigen is introduced into a *Salmonella* strain harbouring a chromosomal deletion in *asd*. This plasmid complements the *asd* mutation *in trans* and the plasmid becomes stabilized because the bacterium would die if the plasmid was lost. One deficiency of such a balanced lethal system alone is that, whereas it assures maintenance of the plasmid in an individual bacterium, it does not account for equitable distribution of the plasmid to daughter cells when replication occurs. This problem was recently solved by Galen *et al.* (1999), who added to the plasmid of a balanced lethal system partitioning loci that enhance the probability that upon replication both daughter cells will inherit the plasmid.

Site of expression

The site of accumulation of the expressed foreign antigen within *Salmonella* can greatly affect immunogenicity. Some antigens, such as fragment C of tetanus toxin, have proven to be highly immunogenic when accumulated as cytoplasmic antigens (Chatfield *et al.*, 1992b; Galen *et al.*, 1997). Other antigens are immunogenic only when amassed in the periplasmic space (where conditions may foster correct folding to achieve conformational epitopes), or are more immunogenic when accumulated on the surface of the bacterium. Yet other antigens must be secreted out of the bacteria in order to be immunogenic. This is not only true for stimulating antibody but also for eliciting cell-mediated immune responses. Hess *et al.* (1996) have shown that the outright secretion of foreign antigen by *Salmonella* using the Haemolysin A secretion system results in significantly enhanced CTL responses.

It is important to emphasize that, as yet, none of the clinical trials in humans with attenuated *S. typhi* live vectors has incorporated all these optimizations. It is anticipated that, if future clinical trials utilize optimized live vector constructs that address the parameters discussed above, enhanced immune responses will be observed.

Co-expression of adjuvants and cytokines

It has been clearly demonstrated in animal models that one can adjuvant and modulate the immune response towards a preferred Th1 or Th2 bias by inclusion of co-expressed cytokines or biological adjuvants in the live vector (Carrier *et al.*, 1993; Chen *et al.*, 1998; Denich *et al.*, 1993; Dunstan, Ramsay & Strugnell, 1996; Whittle *et al.*, 1997). For example, interleukin 4 (IL-4) and IL-5 expressed by attenuated *Salmonella* increase the serum and mucosal antibody responses (Denich *et al.*, 1993; Whittle *et al.*, 1997).

Cholera toxin and the related heat-labile enterotoxin (LT) of enterotoxigenic *E. coli* are powerful adjuvants that augment the serum and mucosal antibody responses to co-administered antigens. Mutant LT molecules have been devised that retain the adjuvanticity property but are virtually devoid of secretogenic activity with respect to intestinal mucosa. There is some evidence that secretion of mutant LT by attenuated *Salmonella* can enhance immune responses to the bacteria (Covone *et al.*, 1998).

Attenuated *Shigella* as live vector expressing foreign antigens

Attenuated strains of *Shigella* expressing fimbrial colonization factors and mutant LT of enterotoxigenic *E. coli* have been highly successful in eliciting mucosal SIgA (in tears) and serum IgG antibodies in guinea pigs immunized intranasally with two spaced doses of the live vector (Noriega *et al.*, 1996a, b).

Attenuated *Shigella* and *Salmonella* as live vectors carrying DNA vaccine plasmids

One of the most startling observations in the field of vaccinology was made in the 1990s when it was reported that immune responses and protection can be elicited in animal models by inoculating with 'naked DNA'. In current usage, a DNA vaccine consists of a plasmid with the gene of interest placed under the control of a eukaryotic promoter (such as the cytomegalovirus immediate early promoter) to drive transcription in mammalian cells, a polyadenylation signal at the 3′ end of the insert to stabilize mRNA and ensure translation, and an origin of replication (e.g. from simian virus 40 or Epstein–Barr virus) that allows some rounds of plasmid replication within eukaryotic cells. When such plasmids enter antigen-processing cells such as dendritic cells and macrophages, the eukaryotic cell machinery actually produces the foreign protein. An advantage is that, for virus proteins in particular, the protein receives all the appropriate post-translational modifications.

In early studies, naked DNA was inoculated intramuscularly or adsorbed to gold particles and inoculated intradermally via a gene gun. However, it was subsequently demonstrated that mucosal immunization with attenuated *Shigella* and *Salmonella* could also deliver DNA vaccine plasmids directly to cells in which eukaryotic expression of the foreign protein can occur, resulting in immune responses (Anderson *et al.*, 2000; Darji *et al.*, 1997; Fennelly *et al.*, 1999; Pasetti *et al.*, 1999; Sizemore, Branstrom & Sadoff, 1995). Since attenuated *Shigella* strains that retain a functional invasion plasmid antigen B (IpaB) have the ability to break out of the phagolysosome and enter the cytosol, they are the preferred live vector for use with DNA vaccine plasmids.

The vaccine development paradigm

Although there may be some disappointment in the modesty of the immune responses seen in the human clinical trials carried out so far with *S. typhi* live vectors (Table 1), such disappointments are commonplace in the transition from animal models into early Phase I clinical trials. Perseverance, commitment and the provision of adequate resources are necessary to move any vaccine along the development path. In this regard, it may be helpful to compare the results of the few clinical trials of attenuated *S. typhi* live vector vaccines with those from the earliest clinical trials with parenterally administered DNA vaccines or with mucosally administered polylactide/polyglycoside microspheres containing protein antigen. Expectations with respect to early clinical trials were similarly high for both the DNA vaccines and mucosal polylactide/polyglycoside microspheres because of the potent immune responses that were earlier observed in animal models. However, with both these types of vaccines, the initial clinical trials were disappointing compared to the expectations. Of the first three non-HIV trials with DNA vaccines (against influenza, hepatitis B and malaria), little or no serum antibody was elicited (Tacket *et al.*, 1999), despite strong serological responses having been observed in animal model studies; the malaria DNA vaccine stimulated CTL responses but these were detected using only one of several assays (Wang *et al.*, 1998). With perseverance, modifications in the vaccines and with the commitment of adequate resources, additional clinical trials were carried out with improved influenza and hepatitis B vaccines; these demonstrated more potent serum antibody responses.

Table 1. Phase I and II clinical trials in humans assessing the safety, immunogenicity and efficacy of attenuated *Salmonella typhi* and *Salmonella typhimurium* constructs expressing foreign protein or polysaccharide antigens.

Salmonella live vector	Mutations	Foreign antigen	Site of foreign gene	Safety	Immune response to foreign antigen	Protection	Reference
S. typhi strains							
Ty21a	*rpoS, galE, Vi-*	*Shigella sonnei* PS	Plasmid	4+	9/9* 9/16†	Variable	Black *et al.* (1987); Herrington *et al.* (1990)
CVD 908	*aroC, aroD*	*P. falciparum* CSP	Chromosome	4+	3/10	NA	Gonzalez *et al.* (1994)
X4632(pYA3167)	*cya, crp, cdt*	Hepatitis B virus core-pre-S	Plasmid‡	4+	0/10	NA	Tacket *et al.* (1997a)
Ty1033§	*phoP, phoQ, purB*	*H. pylori* urease A and B	Plasmid‡	4+	0/11	NA	DiPetrillo *et al.* (1999)
CVD 908-htrA	*aroC, aroD, htrA*	Tetanus fragment C	Plasmid	4+	Low dose 0/3‖ High dose 1/1‖	NA	C. O. Tacket and others, unpublished
S. typhimurium strain							
LH1160	*phoP, phoQ, purB*	*H. pylori* urease A and B	Plasmid‡	2+	3/6	NA	Angelakopoulos & Hohmann (1999)

NA, Not applicable.
*IgA antibody secreting cell response.
†Serum IgA antibody response.
‡Stabilized plasmid.
§Derivative of Ty800.
‖Results in subjects lacking tetanus antitoxin at baseline.

The first clinical trials with mucosal administration of antigens delivered via polylactide/polyglycolide microspheres also showed only modest immunogenicity (Tacket *et al.*, 1994).

Thus it is likely that if sufficient resources are available to pursue clinical trials in a true vaccine development fashion, attenuated *Salmonella* and *Shigella* will prove their worth as carriers of foreign antigens and as deliverers of DNA vaccine plasmids. The advantages of bacterial live vector vaccines, which include extraordinary flexibility to deliver either antigens or DNA plasmids, practicality of oral administration, ability to elicit broad immune responses and economy of manufacture, remain evident and strongly argue for the continued development of such vaccines. Only after clinical trials with optimized constructs are carried out will information be available about the long-term suitability of this vaccinology strategy versus competing ones.

References

Aggarwal, A., Kumar, S., Jaffe, R., Hone, D., Gross, M. & Sadoff, J. (1990). Oral *Salmonella*–malaria circumsporozoite recombinants induce specific CD8+ cytotoxic T-cells. *Journal of Experimental Medicine* 172, 1083–1090.

Anderson, R., Pasetti, M. F., Sztein, M. B., Levine, M. M. & Noriega, F. N. (2000). Δ*guaBA* attenuated *Shigella flexneri* 2a strain CVD 1204 as a *Shigella* vaccine and as a live mucosal delivery system for fragment C of tetanus toxin. *Vaccine* 18, 2193–2202.

Angelakopoulos, H. & Hohmann, E. L. (1999). Human studies of *phoP/phoQ*-deleted *Salmonella* vaccine strains expressing *Helicobacter pylori* urease A/B subunits. In *Proceedings of the 35th Joint Conference on Cholera and other Bacterial Enteric Infections*, US–Japan Cooperative Medical Science Program, December 3–5, 1999, pp. 47–51. Baltimore, MD.

Ascon, M. A., Hone, D. M., Walters, N. & Pascual, D. W. (1998). Oral immunization with a *Salmonella typhimurium* vaccine vector expressing recombinant enterotoxigenic *Escherichia coli* K99 fimbriae elicits elevated antibody titers for protective immunity. *Infection and Immunity* 66, 5470–5476.

Barry, E. M., Gomez-Duarte, O., Chatfield, S., Pizza, M., Rappuoli, R., Losonsky, G. A.,

Galen, J. E. & Levine, M. M. (1996). Expression and immunogenicity of pertussis toxin S1 subunit–tetanus toxin fragment C fusions in *Salmonella typhi* vaccine strain CVD 908. *Infection and Immunity* 64, 4172–4181.

Belshe, R. B., Gorse, G. J., Mulligan, M. J. & 11 other authors (1998). Induction of immune responses to HIV-1 by canarypox virus (ALVAC) HIV-1 and gp120 SF-2 recombinant vaccines in uninfected volunteers. NIAID AIDS Vaccine Evaluation Group. *AIDS* 12, 2407–2415.

Benyacoub, J., Hopkins, S., Potts, A., Kelly, S., Kraehenbuhl, J. P., Curtiss, R., III, De Grandi, P. & Nardelli-Haefliger, D. (1999). The nature of the attenuation of *Salmonella typhimurium* strains expressing human papillomavirus type 16 virus-like particles determines the systemic and mucosal antibody responses in nasally immunized mice. *Infection and Immunity* 67, 3674–3679.

Black, R. E., Levine, M. M., Clements, M. L., Losonsky, G., Herrington, D., Berman, S. & Formal, S. B. (1987). Prevention of shigellosis by a *Salmonella typhi*–*Shigella sonnei* bivalent vaccine. *Journal of Infectious Diseases* 155, 1260–1265.

Butterton, J. R., Ryan, E. T., Acheson, D. W. & Calderwood, S. B. (1997). Coexpression of the B subunit of Shiga toxin 1 and EaeA from enterohemorrhagic *Escherichia coli* in *Vibrio cholerae* vaccine strains. *Infection and Immunity* 65, 2127–2135.

Carrier, M. J., Chatfield, S. N., Dougan, G., Nowicka, U. T. A., O'Callaghan, D., Beesley, J. E., Milano, S., Cillari, E. & Liew, F. Y. (1993). Expression of human IL-1β in *Salmonella typhimurium*. A model system for the delivery of recombinant therapeutic proteins *in vivo*. *Infection and Immunity* 61, 4818–4827.

Chatfield, S. N., Fairweather, N., Charles, I., Pickard, D., Levine, M., Hone, D., Posada, M., Strugnell, R. A. & Dougan, G. (1992a). Construction of a genetically defined *Salmonella typhi* Ty2 *aroA*, *aroC* mutant for the engineering of a candidate oral typhoid-tetanus vaccine. *Vaccine* 10, 53–60.

Chatfield, S., Charles, I., Makoff, A., Oxer, M., Dougan, G., Pickard, D., Slater, D. & Fairweather, N. (1992b). Use of the nirB promoter to direct the stable expression of heterologous antigens in *Salmonella* oral vaccine strains: development of a single-dose oral tetanus vaccine. *Biotechnology* 10, 888–892.

Chatfield, S. N., Strahan, K., Pickard, D., Charles, I. G., Hormaeche, C. E. & Dougan, G. (1992c). Evaluation of *Salmonella typhimurium* strains harbouring defined mutations in *htrA* and *aroA* in the murine salmonellosis model. *Microbial Pathogenesis* 12, 145–151.

Chen, I., Pizza, M., Rappuoli, R. & Newton, S. M. (1998). Effects of the insertion of a nonapeptide from murine IL-1beta on the immunogenicity of carrier proteins delivered by live attenuated *Salmonella*. *Archives of Microbiology* 169, 113–119.

Clements, J. D., Lyon, F. L., Lowe, K. L., Farrand, A. L. & El-Morshidy, S. (1986). Oral immunization of mice with attenuated *Salmonella enteritidis* containing a recombinant plasmid which encodes production of the B subunit of heat-labile *Escherichia coli* enterotoxin. *Infection and Immunity* 53, 685–692.

Coulson, N. M., Fulop, M. & Titball, R. W. (1994). *Bacillus anthracis* protective antigen, expressed in *Salmonella typhimurium* SL 3261, affords protection against anthrax spore challenge. *Vaccine* 12, 1395–1401.

Covone, M. G., Brocchi, M., Palla, E., Dias da Silveira, W., Rappuoli, R. & Galeotti, C. L. (1998). Levels of expression and immunogenicity of attenuated *Salmonella enterica* serovar typhimurium strains expressing *Escherichia coli* mutant heat-labile enterotoxin. *Infection and Immunity* 66, 224–231.

Curtiss, R., III & Kelly, S. M. (1987). *Salmonella typhimurium* deletion mutants lacking adenylate cyclase and cyclic AMP receptor protein are avirulent and immunogenic. *Infection and Immunity* 55, 3035–3043.

Darji, A., Guzman, C. A., Gerstel, B., Wachholz, P., Timmis, K. N., Wehland, J., Chakraborty, T. & Weiss, S. (1997). Oral somatic transgene vaccination using attenuated *S. typhimurium*. *Cell* 91, 765–775.

Denich, K., Börlin, P., O'Hanley, P. D., Howard, M. & Heath, A. W. (1993). Expression of the murine interleukin-4 gene in an attenuated *aroA* strain of *Salmonella typhimurium*: persistence and immune response in BALB/c mice and susceptibility to macrophage killing. *Infection and Immunity* 61, 4818–4827.

DiPetrillo, M. D., Tibbetts, T., Kleanthous, H. & Hohmann, E. L. (1999). Safety and immunogenicity of the *phoP/phoQ Salmonella typhi* expressing *Helicobacter pylori* urease. *Vaccine* 18, 449–459.

Dougan, G., Hormaeche, C. E. & Maskell, D. J. (1987). Live oral *Salmonella* vaccines: potential use of attenuated strains as carriers of heterologous antigens to the immune system. *Parasite Immunology* 9, 151–160.

Dougan, G., Chatfield, S., Pickard, D., Bester, J., O'Callaghan, D. & Maskell, D. (1988). Construction and characterization of vaccine strains of *Salmonella* harbouring mutations in two different *aro* genes. *Journal of Infectious Diseases* 158, 1329–1335.

Dunstan, S. J., Ramsay, A. J. & Strugnell, R. A. (1996). Studies of immunity and bacterial invasiveness in mice given a recombinant *Salmonella* vector encoding murine interleukin-6. *Infection and Immunity* 64, 2730–2736.

Dunstan, S. J., Simmons, C. P. & Strugnell, R. A. (1999). Use of *in vivo*-regulated promoters to deliver antigens from attenuated *Salmonella enterica* var. Typhimurium. *Infection and Immunity* **67**, 5133–5141.

Edelman, R., Palmer, K., Russ, K. G. & 13 other authors (1999). Safety and immunogenicity of recombinant Bacille Calmette–Guérin (rBCG) expressing *Borrelia burgdorferi* outer surface protein A (OspA) lipoprotein in adult volunteers: a candidate Lyme disease vaccine. *Vaccine* **17**, 904–914.

Everest, P., Frankel, G., Li, J., Lund, P., Chatfield, S. & Dougan, G. (1995). Expression of LacZ from the *htrA*, *nirB*, and *groE* promoters in a *Salmonella* vaccine strain: influence of growth in mammalian cells. *FEMS Microbiology Letters* **126**, 97–102.

Fennelly, G. J., Khan, S. A., Abadi, M. A., Wild, T. F. & Bloom, B. R. (1999). Mucosal DNA vaccine immunization against measles with a highly attenuated *Shigella flexneri* vector. *Journal of Immunology* **162**, 1603–1610.

Forrest, B. D., Shearman, D. J. C. & LaBrooy, J. T. (1990). Specific immune response in humans following rectal delivery of live typhoid vaccine. *Vaccine* **8**, 209–211.

Forrest, B. D., LaBrooy, J. T., Dearlove, C. E. & Shearman, D. J. C. (1991). The human humoral immune response to *Salmonella typhi* Ty21a. *Journal of Infectious Diseases* **163**, 336–345.

Frankel, F. R., Hegde, S., Lieberman, J. & Paterson, Y. (1995). Induction of cell-mediated immune responses to human immunodeficiency virus type 1 Gag protein by using *Listeria monocytogenes* as a live vaccine vector. *Journal of Immunology* **155**, 4775–4782.

Galen, J. E., Gomez-Duarte, O. G., Losonsky, G. A., Halpern, J. L., Lauderbaugh, C. S., Kaintuck, S., Reymann, M. K. & Levine, M. M. (1997). A murine model of intranasal immunization to assess the immunogenicity of attenuated *Salmonella typhi* live vector vaccines in stimulating serum antibody responses to expressed foreign antigens. *Vaccine* **15**, 700–708.

Galen, J. E., Nair, J., Wang, J. Y., Tanner, M. K., Sztein, M. B. & Levine, M. M. (1999). Optimization of plasmid maintenance in the attenuated live vector vaccine *Salmonella typhi* strain CVD 908-*htrA*. *Infection and Immunity* **67**, 6424–6433.

Gomez-Duarte, O., Galen, J., Chatfield, S. N., Rappuoli, R., Eidels, L. & Levine, M. M. (1995). Expression of fragment C of tetanus toxin fused to a carboxyl-terminal fragment of diphtheria toxin in *Salmonella typhi* CVD 908 vaccine strain. *Vaccine* **13**, 1596–1602.

Gomez-Duarte, O. G., Lucas, B., Yan, Z. X., Panthel, K., Haas, R. & Meyer, T. F. (1998). Protection of mice against gastric colonization by *Helicobacter pylori* by single oral dose immunization with attenuated *Salmonella typhimurium* producing urease subunits A and B. *Vaccine* **16**, 460–471.

Gonin, P., Oualikene, W., Fournier, A. & Eloit, M. (1996). Comparison of the efficacy of replication-defective adenovirus and Nyvac poxvirus as vaccine vectors in mice. *Vaccine* **14**, 1083–1087.

Gonzalez, C., Hone, D., Noriega, F. & 12 other authors (1994). *Salmonella typhi* vaccine strain CVD 908 expressing the circumsporozoite protein of *Plasmodium falciparum*: strain construction and safety and immunogenicity in humans. *Journal of Infectious Diseases* **169**, 927–931.

Gonzalez, C. R., Noriega, F. R., Huerta, S., Santiago, A., Vega, M., Paniagua, J., Ortiz-Navarrete, V., Isibasi, A. & Levine, M. M. (1998). Immunogenicity of a *Salmonella typhi* CVD 908 candidate vaccine strain expressing the major surface protein gp63 of *Leishmania mexicana* mexicana. *Vaccine* **16**, 1043–1052.

Herrington, D. A., Van de Verg, L., Formal, S. B., Hale, T. L., Tall, B. D., Cryz, S. J., Tramont, E. C. & Levine, M. M. (1990). Studies in volunteers to evaluate candidate *Shigella* vaccines: further experience with a bivalent *Salmonella typhi*–*Shigella sonnei* vaccine and protection conferred by previous *Shigella sonnei* disease. *Vaccine* **8**, 353–357.

Hess, J., Gentschev, I., Miko, D., Welzel, M., Ladel, C., Goebel, W. & Kaufmann, S. H. E. (1996). Superior efficacy of secreted over somatic antigen display in recombinant *Salmonella* vaccine induced protection against listeriosis. *Proceedings of the National Academy of Sciences, USA* **93**, 1458–1463.

Hohmann, E. L., Oletta, C. A., Killeen, K. P. & Miller, S. I. (1996). *phoP*/*phoQ*-deleted *Salmonella typhi* (Ty800) is a safe and immunogenic single-dose typhoid fever vaccine in volunteers. *Journal of Infectious Diseases* 173, 1408–1414.

Hoiseth, S. & Stocker, B. A. D. (1981). Aromatic-dependent *Salmonella typhimurium* are non-virulent and effective as live vaccines. *Nature* 292, 238–239.

Hone, D., Attridge, S., van den Bosch, L. & Hackett, J. (1988). A chromosomal integration system for stabilization of heterologous genes in *Salmonella* based vaccine strains. *Microbial Pathogenesis* 5, 407–418.

Igwe, E. I., Russmann, H., Roggenkamp, A., Noll, A., Autenrieth, I. B. & Heesemann, J. (1999). Rational live oral carrier vaccine design by mutating virulence-associated genes of *Yersinia enterocolitica. Infection and Immunity* 67, 5500–5507.

Karem, K. L., Bowen, J., Kuklin, N. & Rouse, B. T. (1997). Protective immunity against herpes simplex virus (HSV) type 1 following oral administration of recombinant *Salmonella typhimurium* vaccine strains expressing HSV antigens. *Journal of General Virology* 78, 427–434.

Levine, M. M., Herrington, D., Murphy, J. R. & 8 other authors (1987). Safety, infectivity, immunogenicity and *in vivo* stability of two attenuated auxotrophic mutant strains of *Salmonella typhi*, 541Ty and 543Ty, as live oral vaccines in man. *Journal of Clinical Investigation* 79, 888–902.

Levine, M. M., Hone, D., Heppner, D. G., Noriega, F. & Sriwathana, B. (1990). Attenuated *Salmonella* as carriers for the expression of foreign antigens. *Microecology and Therapy* 19, 23–32.

Levine, M. M., Galen, J. E., Sztein, M. B., Beier, M. & Noriega, F. (1997). *Salmonella* expressing protozoal antigens. In *New Generation Vaccines*, 2nd edn, pp. 351–361. Edited by M. M. Levine and others. New York: Marcel Dekker.

Maassen, C. B., Laman, J. D., Bak-Glashouwer, M. J. & 8 other authors (1999). Instruments for oral disease-intervention strategies: recombinant *Lactobacillus casei* expressing tetanus toxin fragment C for vaccination or myelin proteins for oral tolerance induction in multiple sclerosis. *Vaccine* 17, 2117–2128.

Medaglini, D., Pozzi, G., King, T. P. & Fischetti, V. A. (1995). Mucosal and systemic immune responses to a recombinant protein expressed on the surface of the oral commensal bacterium *Streptococcus gordonii* after oral colonization. *Proceedings of the National Academy of Sciences, USA* 92, 6868–6872.

Miller, S. I., Kukral, A. M. & Mekalanos, J. J. (1989). A two-component regulatory system (phoP phoQ) controls *Salmonella typhimurium* virulence. *Proceedings of the National Academy of Sciences, USA* 86, 5054–5058.

Morris, J. G., Tacket, C. O. & Levine, M. M. (1992). Oral carrier vaccines: new tricks in an old trade. *Gastroenterology* 103, 699–702.

Moss, B. (1996). Genetically engineered poxviruses for recombinant gene expression, vaccination, and safety. *Proceedings of the National Academy of Sciences, USA* 93, 11341–11348.

Nakayama, K., Kelly, S. & Curtiss, R. (1988). Construction of an ASD$^+$ expression-cloning vector: stable maintenance and high level expression of cloned genes in a *Salmonella* vaccine strain. *Biotechnology* 6, 693–697.

Nayak, A. R., Tinge, S. A., Tart, R. C., McDaniel, L. S., Briles, D. E. & Curtiss, R., III (1998). A live recombinant avirulent oral *Salmonella* vaccine expressing pneumococcal surface protein A induces protective responses against *Streptococcus pneumoniae. Infection and Immunity* 66, 3744–3751.

Noriega, F. R., Losonsky, G., Wang, J. Y., Formal, S. B. & Levine, M. M. (1996a). Further characterization of $\Delta aroA$, $\Delta virG$ *Shigella flexneri* 2a strain CVD 1203 as a mucosal *Shigella* vaccine and as a live vector vaccine for delivering antigens of enterotoxigenic *Escherichia coli. Infection and Immunity* 64, 23–27.

Noriega, F. R., Losonsky, G., Lauderbaugh, C., Liao, F. M., Wang, M. S. & Levine, M. M. (1996b). Engineered $\Delta guaB$-A, $\Delta virG$ *Shigella flexneri* 2a strain CVD 1205: construction, safety, immunogenicity and potential efficacy as a mucosal vaccine. *Infection and Immunity* 64, 3055–3061.

Pasetti, M. F., Anderson, R. J., Noriega, F. R., Levine, M. M. & Sztein, M. B. (1999).

Attenuated Δ*guaBA Salmonella typhi* vaccine strain CVD 915 as a live vector utilizing prokaryotic or eukaryotic expression systems to deliver foreign antigens and elicit immune responses. *Clinical Immunology* 92, 76–89.

Pozzi, G., Contorni, M., Oggioni, M. R., Manganelli, R., Tommasino, M., Cavalieri, F. & Fischetti, V. A. (1992). Delivery and expression of a heterologous antigen on the surface of streptococci. *Infection and Immunity* 60, 1902–1907.

Robinson, K., Chamberlain, L. M., Schofield, K. M., Wells, J. M. & Le Page, R. W. (1997). Oral vaccination of mice against tetanus with recombinant *Lactococcus lactis. Nature Biotechnology* 15, 653–657.

Ryan, E. T., Butterton, J. R., Zhang, T., Baker, M. A., Stanley, S. L., Jr & Calderwood, S. B. (1997a). Oral immunization with attenuated vaccine strains of *Vibrio cholerae* expressing a dodecapeptide repeat of the serine-rich *Entamoeba histolytica* protein fused to the cholera toxin B subunit induces systemic and mucosal antiamebic and anti-*V. cholerae* antibody responses in mice. *Infection and Immunity* 65, 3118–3125.

Ryan, E. T., Butterton, J. R., Smith, R. N., Carroll, P. A., Crean, T. I. & Calderwood, S. B. (1997b). Protective immunity against *Clostridium difficile* toxin A induced by oral immunization with a live, attenuated *Vibrio cholerae* vector strain. *Infection and Immunity* 65, 2941–2949.

Sadoff, J., Ballou, W. R., Baron, L. & 7 other authors (1988). Oral *Salmonella typhimurium* vaccine expressing circumsporozoite protein protects against malaria. *Science* 240, 336–338.

Schodel, F., Kelly, S., Tinge, S., Hopkins, S., Peterson, D., Milich, D. & Curtiss, R., III (1996). Hybrid hepatitis B virus core antigen as a vaccine carrier moiety. II. Expression in avirulent *Salmonella* spp. for mucosal immunization. *Advances in Experimental Medicine and Biology* 397, 15–21.

Sizemore, D. R., Branstrom, A. A. & Sadoff, J. C. (1995). Attenuated *Shigella* as a DNA delivery vehicle for DNA-mediated immunization. *Science* 270, 299–302.

Steidler, L., Robinson, K., Chamberlain, L., Schofield, K. M., Remaut, E., Le Page, R. W. & Wells, J. M. (1998). Mucosal delivery of murine interleukin-2 (IL-2) and IL-6 by recombinant strains of *Lactococcus lactis* coexpressing antigen and cytokine. *Infection and Immunity* 66, 3183–3189.

Sztein, M. B., Wasserman, S. S., Tacket, C. O., Edelman, R., Hone, D., Lindberg, A. A. & Levine, M. M. (1994). Cytokine production patterns and lymphoproliferative responses in volunteers orally immunized with attenuated vaccine strains of *Salmonella typhi. Journal of Infectious Diseases* 170, 1508–1517.

Sztein, M. B., Tanner, M. K., Polotsky, Y., Orenstein, J. M. & Levine, M. M. (1995). Cytotoxic T lymphocytes after oral immunization with attenuated vaccine strains of *Salmonella typhi* in humans. *Journal of Immunology* 155, 3987–3993.

Tacket, C. O., Hone, D. M., Losonsky, G. A., Guers, L., Edelman, R. & Levine, M. M. (1992a). Clinical acceptability and immunogenicity of CVD 908 *Salmonella typhi* vaccine strain. *Vaccine* 10, 443–446.

Tacket, C. O., Hone, D. M., Curtiss, R., III, Kelly, S. M., Losonsky, G., Guers, L., Harris, A. M., Edelman, R. & Levine, M. M. (1992b). Comparison of the safety and immunogenicity of *aroC,aroD* and *cya,crp Salmonella typhi* strains in adult volunteers. *Infection and Immunity* 60, 536–541.

Tacket, C. O., Reid, R. H., Boedeker, E. C., Losonsky, G., Nataro, J. P., Bhagat, H. & Edelman, R. (1994). Enteral immunization and challenge of volunteers given enterotoxigenic *E. coli* CFA/II encapsulated in biodegradable microspheres. *Vaccine* 12, 1270–1274.

Tacket, C. O., Kelly, S. M., Schodel, F., Losonsky, G., Nataro, J. P., Edelman, R., Levine, M. M. & Curtiss, R., III (1997a). Safety and immunogenicity in humans of an attenuated *Salmonella typhi* vaccine vector strain expressing plasmid-encoded hepatitis B antigens stabilized by the ASD balanced lethal system. *Infection and Immunity* 65, 3381–3385.

Tacket, C. O., Sztein, M. B., Losonsky, G. A. & 7 other authors (1997b). Safety and immune response in humans of live oral *Salmonella typhi* vaccine strains deleted in *htrA* and *aroC,aroD. Infection and Immunity* 65, 452–456.

Tacket, C. O., Roy, M. J., Widera, G., Swain, W. F., Broome, S. & Edelman, R. (1999). Phase 1 safety and immune response studies of a DNA vaccine encoding hepatitis B surface antigen delivered by a gene delivery device. *Vaccine* 17, 2826–2829.

Tagliabue, A., Nencioni, L., Caffarena, A., Villa, L., Boraschi, D., Cazzola, G. & Cavalieri, S. (1985). Cellular immunity against *Salmonella typhi* after live oral vaccines. *Clinical and Experimental Immunology* 52, 242–247.

Tagliabue, A., Villa, L., De Magistiris, M. T., Romano, M., Silvestri, S., Boraschji, D. & Nencioni, L. (1986). IgA-driven T-cell-mediated antibacterial immunity in man after live oral Ty21a vaccine. *Journal of Immunology* 137, 1504–1510.

Toebe, C. S., Clements, J. D., Cardenas, L., Jennings, G. J. & Wiser, M. F. (1997). Evaluation of immunogenicity of an oral *Salmonella* vaccine expressing recombinant *Plasmodium berghei* merozoite surface protein-1. *American Journal of Tropical Medicine and Hygiene* 56, 192–199.

Tramont, E. C., Chung, R., Berman, S., Keren, D., Kapfer, C. & Formal, S. B. (1984). Safety and antigenicity of typhoid-*Shigella sonnei* vaccine (strain 5076-1C). *Journal of Infectious Diseases* 149, 133–136.

Viret, J. F., Favre, D., Wegmuller, B., Herzog, C., Que, J. U., Cryz, S. J., Jr & Lang, A. B. (1999). Mucosal and systemic immune responses in humans after primary and booster immunizations with orally administered invasive and noninvasive live attenuated bacteria. *Infection and Immunity* 67, 3680–3685.

Wang, R., Doolan, D. L., Le, T. P. & 12 other authors (1998). Induction of antigen-specific cytotoxic T lymphocytes in humans by a malaria DNA vaccine. *Science* 282, 476–480.

Whittle, B. L., Smith, R. M., Matthaei, K. I., Young, I. G. & Verma, N. K. (1997). Enhancement of the specific mucosal IgA response *in vivo* by interleukin-5 expressed by an attenuated strain of *Salmonella* serotype Dublin. *Journal of Medical Microbiology* 46, 1029–1038.

Xu, D., McSorley, S. J., Chatfield, S. J., Dougan, G. & Liew, F. Y. (1997). Protection against *Leishmania major* infection in genetically susceptible BALB/c mice by gp63 delivered orally in attenuated *Salmonella typhimurium* (AroA- AroD-). *Immunology* 85, 1–7.

Zegers, N. D., Kluter, E., van Der Stap, H., van Dura, E., van Dalen, P., Shaw, M. & Baillie, L. (1999). Expression of the protective antigen of *Bacillus anthracis* by *Lactobacillus casei*: towards the development of an oral vaccine against anthrax. *Journal of Applied Microbiology* 87, 309–314.

New malaria vaccines:
the DNA–MVA prime–boost strategy

Adrian V.S. Hill

Molecular Immunology Group, Institute of Molecular Medicine, University of Oxford,
John Radcliffe Hospital, Oxford OX3 9DU, UK

Introduction

Advances in the analysis of cellular immune responses and their relevance to protective immunity are leading to a revolution in approaches to vaccine design. Foremost amongst the development in recent years have been the introduction of nucleic acid vaccines and the development of prime–boost strategies. Malaria is amongst the infectious diseases most likely to benefit from these new approaches to inducing cellular immunity and is one of the diseases for which the greatest medical need exists. In this short chapter, I shall describe briefly the rationale for attempts to induce CD8[+] T-cell-mediated immunity to malaria by vaccination and the discovery and advantage of a new prime–boost immunization approach utilizing plasmid DNA priming and boosting of the immune response by recombinant modified vaccinia virus Ankara (MVA). These developments have led to the first trials of DNA vaccines in healthy volunteers in Europe, and the clinical assessment of the prime–boost regime is also under way.

Malaria and CD8[+] T cells

The complex, multi-stage life-cycle of the malaria parasite provides both opportunities and difficulties for vaccine development (Nussenzweig & Long, 1994). It has been known for some time that most but not all malaria antigens exhibit stage specificity, so that different antigens are often needed to induce immunity against the various stages of the parasite's life-cycle. Far more vaccine candidates have progressed to clinical trials for the pre-erythrocytic stage of malaria than for the blood stage or sexual stage. This reflects several factors. Pre-erythrocytic vaccines aim to interrupt the parasite life-cycle either immediately after inoculation of sporozoites by the *Anopheles* mosquito or at the early liver stage of infection that lasts about 5–7 d in humans (Hoffman *et al.*, 1991). During this period, the infected individual is asymptomatic and will not become ill if the liver can be cleared of parasites and blood-stage infection prevented. Thus pre-erythrocytic vaccines, unlike blood-stage vaccines, are potentially useful as vaccines for travellers. More importantly, it has become clear from a variety of epidemiological studies that even a partially effective pre-erythrocytic vaccine could be of great value in reducing malaria morbidity and mortality in endemic areas. This finding contrasts with the simplistic view that such a vaccine would have to be completely effective because if a single sporozoite escaped control, and developed into a blood-stage parasite, disease could ensue. Although this concern is appropriate, there is now much evidence that, at least in individuals with some acquired immunity to malaria in endemic areas, such an attenuated infection is less likely to induce morbidity and mortality.

Several observations support the feasibility of immunization against the pre-erythrocytic stage of malaria. An effective vaccine against this stage exists: the irradiated sporozoite. If large numbers of irradiated sporozoites are administered to animals, complete protection can be induced against subsequent challenge with the same or even different strains of sporozoites (Nussenzweig *et al.*, 1967). Studies in humans with *Plasmodium falciparum*, the parasite responsible for almost all malaria deaths, have confirmed that protective immunity can be induced by these means in humans also (Clyde, 1975). However, to be effective, large numbers of sporozoites need to be administered over a 6 month period by hundreds of mosquito bites, rendering this approach impossible for widespread vaccination. Thus for several years now the challenge has been to reproduce this immunity using a subunit vaccine.

Analysis of the mechanisms of pre-erythrocytic immunity, largely in rodent models, has identified a key role for $CD8^+$ T cells in the irradiated sporozoite vaccination model (Schofield *et al.*, 1987). Although the most important immune effector mechanism varies between host strains, in general $CD8^+$ T cells are the most important effectors (Doolan & Hoffman, 1997). These cells are also induced in humans by natural infection (Hill *et al.*, 1992; Aidoo *et al.*, 1995) and by irradiated sporozoite immunization, and several target antigens have been identified (Wizel *et al.*, 1995). The low levels of malaria-specific $CD8^+$ T cells in the circulation of individuals even in areas holoendemic for malaria (Lalvani *et al.*, 1996; Plebanski *et al.*, 1997) have to date prevented large-scale prospective studies of their relevance to immunity in the field. However, at the same time these low levels have encouraged the view that boosting these with a vaccine might enhance natural immunity to malaria. Consistent with the view that $CD8^+$ T cells are protective is the observation that human leukocyte antigen (HLA) class I type, which determines the specificity and efficacy of $CD8^+$ T cells, influences susceptibility to severe malaria in African children. Gambian children with the HLA-B*5301 allele are less likely to develop severe malaria and $CD8^+$ cytotoxic T cells restricted by this HLA type have been characterized (Hill *et al.*, 1991, 1992).

Technical developments

Based on the studies summarized above, it has been clear for almost a decade that a vaccination approach that could induce high levels of CD8$^+$ T cells might be useful in preventing malaria. Thus malaria researchers, along with those in many other fields, have been interested in identifying vaccine delivery systems that can induce this immune response reliably. Indeed malaria research has been at the forefront of this general area and some of the most important advances for the development of T-cell-inducing vaccines have come from malaria research. Progress has been facilitated by the availability of established rodent challenge models that allow ready assessment of vaccine protective efficacy as well as immunogenicity, the development of new assays of T-cell response, and the possibility of undertaking safely challenge studies with malaria sporozoites in humans (Church *et al.*, 1997), an option available in very few major infectious diseases for which there is no vaccine.

An extensive literature exists on the variety of delivery systems (to use a term embracing adjuvants and all types of vector) that may induce CD8$^+$ T cells by immunization. There is now a general consensus that peptides used alone or with any adjuvant can induce at best only weak CD8$^+$ T-cell responses (Allsopp *et al.*, 1996). However, it is clear that addition of a lipid tail to the CD8$^+$ T-cell epitope improves responses and that a fused CD4$^+$ T-cell epitope, such as the PADRE peptide, may improve responses further (Vitiello *et al.*, 1995). There are reports of induction of CD8$^+$ T-cell responses using bacterial delivery systems such as recombinant *Salmonella* and BCG, but these are less efficient than viral recombinants such as vaccinia and adenovirus. However, the greatest interest in recent years has focused on DNA vaccines as inducers of CD8$^+$ T cells (Hoffman *et al.*, 1995). Although not as immunogenic as recombinant viral vaccines, these can induce CD8$^+$ T-cell responses reliably in mice using either intramuscular or intraepidermal delivery. DNA vaccines are particularly attractive for malaria because of ease of manufacture, lack of expressed vector sequences, thereby avoiding induction of immunity to vector sequences, and good stability, allowing easier use in developing countries. Safety concerns about the possibility of a low rate of insertional mutagenesis have now been addressed substantially (Martin *et al.*, 1999; Parker *et al.*, 1999) so that regulatory authorities have approved the use of these vaccines at least for Phase I studies in healthy volunteers (Wang *et al.*, 1998). However, the main residual concern about DNA vaccines relates to their potency in large animals and humans. It is clear that DNA vaccines are less immunogenic in primates than in rodents when administered intramuscularly and a great variety of approaches are now being evaluated in attempts to improve this modest immunogenicity.

Central to such evaluations of immunogenicity are the specificity and sensitivities of the immunoassays employed. This area has also seen important technical developments over the last few years. Until recently, the standard assay of CD8[+] T-cell function in both rodents and primates has been the chromium release assay. This measures the percentage of target cells, usually either pulsed with the relevant peptide or infected with a virus expressing the relevant target antigen, that can be lysed by a defined number of effector T cells over a 4 h period. However, usually the relative insensitivity of this assay requires that T cells be cultured *in vitro* for at least a week to amplify their frequency, introducing undesired variability because of differences in culture conditions. The introduction of enzyme-linked immunospot (ELISPOT) assays that measure gamma-interferon secretion from single cells has increased the sensitivity of CD8[+] T-cell measurements by at least tenfold (Taguchi *et al.*, 1990; Miyahira *et al.*, 1995; Lalvani & Hill, 1998). Fortunately, measures of gamma-interferon secretion by ELISPOT have been found to correlate well with chromium release assays and allow measurement of CD8[+] T cells without *in vitro* culture. Interestingly, this newer ELISPOT technique has also shown that previous calculations of CD8[+] T-cell precursor frequencies by limiting dilution assays produced marked underestimates (by five- to tenfold) because of the failure of many of the T cells to grow in culture. An alternative approach to the use of ELISPOT assays is to stain intracellular cytokines and measure the number of cells with intracellular gamma-interferon using a fluorescence-activated cell sorting (FACS) machine with and without the addition of the specific peptide. In our hands, this technique gives similar results to ELISPOT but is more laborious when large numbers of peptides are assayed. Another technique introduced recently, involving tetramers of peptide–MHC complexes (Altman *et al.*, 1996; Dunbar *et al.*, 1998), is less useful for most vaccine studies because a different tetramer reagent is required for each peptide assayed and because the sensitivity of tetramer assays is about 100-fold lower than that of ELISPOT assays. However, tetramer reagents are proving useful for the isolation and functional characterization of specific CD8[+] T cells.

Prime–boost approaches

The introduction of gamma-interferon ELISPOT assays has allowed ready and sensitive quantification of CD8[+] T cells induced by various vaccine delivery systems. In general, immunization approaches shown to induce cytotoxic T lymphocyte (CTL) responses with the traditional chromium release assay are found to induce about 50–150 specific CD8[+] T cells per 10^6 splenocytes in *ex vivo* ELISPOT assays of mouse splenocytes. For example, for a rodent malaria CTL epitope, typically Ty-virus-like particles (Ty-VLPs) induce 50 spot-forming cells (SFC) per 10^6 splenocytes, intramuscular plasmid DNA induces about 100 SFC per 10^6

splenocytes and recombinant vaccinia and adenovirus vectors induce 100–200 SFC (Plebanski *et al.*, 1998; Gilbert *et al.*, 1997; Schneider *et al.*, 1998). Interestingly, repeated immunizations with any of these approaches produce little or no increment in peak immune response. More importantly, each of these immunization approaches used alone has produced little or no protection against sporozoite challenge with rodent malaria parasites. Although initially this appeared to contradict the findings from analysis of the protection induced by irradiated sporozoites (i.e. that CD8[+] T cells could be strongly protective), it is now clear that the level of CD8[+] T cells induced is critical for protective immunity to sporozoite challenge.

We have found that by boosting a DNA-primed mouse with a recombinant poxvirus expressing the same malaria antigen, or even a single CD8[+] T-cell epitope, the level of induced response can be increased from 100 to 1000 SFC per 10[6] splenocytes (Schneider *et al.*, 1998). This transformed the level of protection from zero to 100%, and reviews of available data now indicate that greater than a threshold value of about 400 SFC is required for protection in this model. Further analysis revealed that the order of immunization was critical: use of the poxvirus first and the plasmid DNA second did not boost immunogenicity and failed to protect from sporozoite challenge. The nature of the priming immunogen was not critical in that Ty-VLPs (Plebanski *et al.*, 1998), DNA vaccines (Schneider *et al.*, 1998), adjuvanted peptides, recombinant adenoviruses and other priming agents (unpublished results) could all be boosted effectively by the recombinant poxvirus. Several types of poxvirus were effective, including the replicating Western Reserve strain of vaccinia virus, and the replication-impaired strains MVA and NYVAC (Schneider *et al.*, 1998). More recently, we have shown that recombinant fowlpox may also be used to boost a primed CD8[+] T-cell response to malaria epitopes. However, surprisingly, the non-replicating MVA strain was more immunogenic and protective than the replication-competent Western Reserve strain of vaccinia.

This heterologous prime–boost approach using poxvirus boosting appears generalizable to a great variety of epitopes, antigens, pathogens and host species. In various strains of mice, we and others have demonstrated enhanced immunogenicity with epitopes and antigens from three malaria parasites, the simian and human immunodeficiency viruses (Hanke *et al.*, 1998), hepatitis B and C viruses, *Mycobacterium tuberculosis* and melanoma and other tumour epitopes (unpublished data). Data from primates are more limited but enhanced immunogenicity for CD8[+] T cells has been demonstrated for malaria in chimpanzees (J. Schneider and others, unpublished) and for several monkey species using simian immunodeficiency virus (SIV) antigens (Hanke & McMichael, 1999; Kent *et al.*, 1998; Robinson *et al.*, 1999).

To our knowledge, this approach has not yet been tested in humans and therefore we have initiated a Phase I trial in humans to evaluate this prime–boost approach for malaria. This decision was influenced by evidence of the poor to modest immunogenicity of DNA vaccines used alone for malaria in both rodents and primates, and recently in a Phase I clinical trial (Wang *et al.*, 1998). It was also encouraged by the realization that boosting DNA-primed responses with poxvirus recombinants was likely to be a relatively low-cost approach to malaria immunization. Indeed, the current cost of manufacture of poxvirus vaccines is lower dose for dose than for a DNA vaccine.

To the clinic

Several decisions were required to take forward this clinical study. The first was the choice of antigen(s) to be included in the vaccines. We had found that a string of CD8$^+$ T-cell epitopes encoding components of six pre-erythrocytic malaria antigens showed correct intracellular processing and immunogenicity when used in a variety of vectors (Gilbert *et al.*, 1997). This polyepitope string also contains some CD4$^+$ T-cell epitopes such as one from tetanus toxin that all individuals immunized against tetanus will have encountered previously. These CD4$^+$ T-cell epitopes were included in an attempt to enhance further CD8$^+$ T-cell immunogenicity. The polyepitope string encodes 14 CD8$^+$ T-cell epitopes from *P. falciparum* that may be recognizable by over 70% of the population in Europe and West Africa based on the known HLA class I type restriction of these peptides and the population frequencies of the HLA types (Lalvani *et al.*, 1994). To provide coverage of other individuals lacking these HLA types, the epitope string was fused (at the gene level) to the complete sequence of the malaria pre-erythrocytic antigen TRAP (thrombospondin-related adhesion protein) (Robson *et al.*, 1988). This fusion protein has shown good immunogenicity in preclinical studies in mice.

The choice of plasmid DNA vector was fairly straightforward. Most DNA vaccines tested to date in humans have had a similar backbone with a human cytomegalovirus intron A promoter/enhancer and bovine growth hormone terminator and a kanamycin-resistance gene. A similar vector, pSG2, developed in-house was used. Two quite different approaches to DNA vaccine delivery are being compared in this trial. In one group of healthy volunteers, 500 µg DNA is delivered intramuscularly at 3 week intervals; in another, 4 µg DNA is delivered on gold beads intraepidermally, again at 3 week intervals, using a ballistic needle-less delivery device (Powderject Pharmaceuticals, Oxford, UK). The same plasmid is used in each group so this study should provide direct comparative data on the immunogenicity of these routes.

There are several poxvirus vectors that might be used to boost the immune responses, including replication-competent vaccinia virus, the replication-impaired strains MVA and NYVAC, and the avipoxviruses ALVAC (canarypox) and fowlpox. Recombinants of the NYVAC (Ockenhouse *et al.*, 1998), vaccinia and ALVAC viruses have been tested safely in humans previously (Plotkin *et al.*, 1995). We have chosen the MVA strain for several reasons. Firstly, most of the preclinical data on prime–boost regimes in malaria have been generated using this strain. Secondly, MVA is markedly replication-impaired and thus likely to be much safer in humans than replicating vaccinia viruses. Indeed, non-recombinant MVA has been used safely in over 100 000 people for smallpox vaccination (Blanchard *et al.*, 1996). Thirdly, the genome of MVA has been sequenced (Antoine *et al.*, 1998) and the profile of immune defence molecules that have been lost or retained by MVA during its period of extensive passage in cell culture has been evaluated (Blanchard *et al.*, 1998). The loss of the type I and II interferon receptor genes and the retention of the soluble interleukin-1 beta receptor gene indicated a particularly favourable set of such genes for enhanced safety and immunogenicity (Blanchard *et al.*, 1998).

Conclusions

The remarkable advances in molecular and cellular immunology that led to a definition of the molecular and structural basis of T-cell restriction in the late 1980s and early 1990s are now translating into improved assays of specific T-cell immune responses. At the same time, with improved molecular characterization of vaccine delivery systems several new approaches to enhancing T-cell responses by immunization are in development. These are exciting times for those interested in applying these fundamental advances to clinical practice. Vaccine trials now under way should indicate whether the high levels of CD8[+] T cells that have been induced in experimental animals by heterologous prime–boost regimes can be attained in humans. If this is the case, it should open up new avenues for the development of CD8[+] T-cell-inducing vaccines against not just malaria but many viral and some bacterial pathogens, and also enhance the prospects for effective immunotherapy of chronic infectious diseases and some malignancies in the foreseeable future.

Acknowledgements

Numerous colleagues and collaborators in Oxford and elsewhere have contributed to the studies from the Oxford laboratories reviewed here. I particularly wish to acknowledge Tom Blanchard, Sarah Gilbert, Tom Hanke, Carolyn Hannan, Andrew McMichael, Magdalena Plebanski, Joerg Schneider, Robert Sinden and Geoffrey Smith. Most of this work has been funded by the Wellcome Trust. A.V.S.H. is a Wellcome Trust Principal Research Fellow.

References

Aidoo, M., Lalvani, A., Allsopp, C. E. & 13 other authors (1995). Identification of conserved antigenic components for a cytotoxic T lymphocyte-inducing vaccine against malaria. *Lancet* 345, 1003–1007.

Allsopp, C. E., Plebanski, M., Gilbert, S. & 9 other authors (1996). Comparison of numerous delivery systems for the induction of cytotoxic T lymphocytes by immunization. *European Journal of Immunology* 26, 1951–1959.

Altman, J. D., Moss, P. A. H., Goulder, P. J. R., Barouch, D. H., McHeyzer-Williams, M. G., Bell, J. I., McMichael, A. J. & Davis, M. M. (1996). Phenotypic analysis of antigen-specific T lymphocytes. *Science* 274, 94–97.

Antoine, G., Scheiflinger, F., Dorner, F. & Falkner, F. G. (1998). The complete genomic sequence of the modified vaccinia Ankara strain: comparison with other orthopoxviruses. *Virology* 244, 365–396.

Blanchard, T., Rowland-Jones, S., Gotch, F., McMichael, A. & Smith, G. L. (1996). Future vaccines for HIV. *Lancet* 348, 1741.

Blanchard, T. J., Alcamí, A., Andrea, P. & Smith, G. L. (1998). Modified vaccinia virus Ankara undergoes limited replication in human cells and lacks several immunomodulatory proteins: implications for use as a human vaccine. *Journal of General Virology* 79, 1159–1167.

Church, L. W., Le, T. P., Bryan, J. P. & 12 other authors (1997). Clinical manifestations of *Plasmodium falciparum* malaria experimentally induced by mosquito challenge. *Journal of Infectious Diseases* 175, 915–920.

Clyde, D. F. (1975). Immunization of man against falciparum and vivax malaria by use of attenuated sporozoites. *American Journal of Tropical Medicine and Hygiene* 24, 397–401.

Doolan, D. L. & Hoffman, S. L. (1997). Pre-erythrocytic-stage immune effector mechanisms in *Plasmodium* spp. infections. *Philosophical Transactions of the Royal Society of London B Biological Sciences* 352, 1361–1367.

Dunbar, P. R., Ogg, G. S., Chen, J., Rust, N., van der Bruggen, P. & Cerundolo, V. (1998). Direct isolation, phenotyping and cloning of low-frequency antigen-specific cytotoxic T lymphocytes from peripheral blood. *Current Biology* 8, 413–416.

Gilbert, S. C., Plebanski, M., Harris, S. J., Allsopp, C. E., Thomas, R., Layton, G. T. & Hill, A. V. (1997). A protein particle vaccine containing multiple malaria epitopes. *Nature Biotechnology* 15, 1280–1284.

Hanke, T. & McMichael, A. (1999). Pre-clinical development of a multi-CTL epitope-based DNA prime MVA boost vaccine for AIDS. *Immunology Letters* 66, 177–181.

Hanke, T., Blanchard, T. J., Schneider, J., Hannan, C. M., Becker, M., Gilbert, S. C., Hill, A. V., Smith, G. L. & McMichael, A. (1998). Enhancement of MHC class I-restricted peptide-specific T cell induction by a DNA prime/MVA boost vaccination regime. *Vaccine* 16, 439–445.

Hill, A. V., Allsopp, C. E., Kwiatkowski, D. & 7 other authors (1991). Common west African HLA antigens are associated with protection from severe malaria. *Nature* 352, 595–600.

Hill, A. V., Elvin, J., Willis, A. C. & 9 other authors (1992). Molecular analysis of the association of HLA-B53 and resistance to severe malaria. *Nature* 360, 434–439.

Hoffman, S. L., Nussenzweig, V., Sadoff, J. C. & Nussenzweig, R. S. (1991). Progress toward malaria preerythrocytic vaccines. *Science* 252, 520–521.

Hoffman, S. L., Doolan, D. L., Sedegah, M. & 7 other authors (1995). Nucleic acid malaria vaccines. Current status and potential. *Annals of the New York Academy of Sciences* 772, 88–94.

Kent, S. J., Zhao, A., Best, S. J., Chandler, J. D., Boyle, D. B. & Ramshaw, I. A. (1998). Enhanced T-cell immunogenicity and protective efficacy of a human immunodeficiency virus type 1 vaccine regimen consisting of consecutive priming with DNA and boosting with recombinant fowlpox virus. *Journal of Virology* 72, 10180–10188.

Lalvani, A. & Hill, A. V. S. (1998). Cytotoxic T-lymphocytes against malaria and tuberculosis:

from natural immunity to vaccine design. *Clinical Science (Colchester)* **95**, 531–538.

Lalvani, A., Aidoo, M., Allsopp, C. E., Plebanski, M., Whittle, H. C. & Hill, A. V. (1994). An HLA-based approach to the design of a CTL-inducing vaccine against *Plasmodium falciparum*. *Research in Immunology* **145**, 461–468.

Lalvani, A., Hurt, N., Aidoo, M., Kibatala, P., Tanner, M. & Hill, A. V. (1996). Cytotoxic T lymphocytes to *Plasmodium falciparum* epitopes in an area of intense and perennial transmission in Tanzania. *European Journal of Immunology* **26**, 773–779.

Martin, T., Parker, S. E., Hedstrom, R., Le, T., Hoffman, S. L., Norman, J., Hobart, P. & Lew, D. (1999). Plasmid DNA malaria vaccine: the potential for genomic integration after intramuscular injection. *Human Gene Therapy* **10**, 759–768.

Miyahira, Y., Murata, K., Rodriguez, D., Rodriguez, J. R., Esteban, M., Rodrigues, M. M. & Zavala, F. (1995). Quantification of antigen specific CD8+ T cells using an ELISPOT assay. *Journal of Immunological Methods* **181**, 45–54.

Nussenzweig, R. S. & Long, C. A. (1994). Malaria vaccines: multiple targets. *Science* **265**, 1381–1383.

Nussenzweig, R. S., Vanderberg, J., Most, H. & Orton, C. (1967). Protective immunity produced by the injection of x-irradiated sporozoites of *Plasmodium berghei*. *Nature* **216**, 160–162.

Ockenhouse, C. F., Sun, P. F., Lanar, D. E. & 26 other authors (1998). Phase I/IIa safety, immunogenicity, and efficacy trial of NYVAC-Pf7, a pox-vectored, multiantigen, multistage vaccine candidate for *Plasmodium falciparum* malaria. *Journal of Infectious Diseases* **177**, 1664–1673.

Parker, S. E., Borellini, F., Wenk, M. L., Hobart, P., Hoffman, S. L., Hedstrom, R., Le, T. & Norman, J. A. (1999). Plasmid DNA malaria vaccine: tissue distribution and safety studies in mice and rabbits. *Human Gene Therapy* **10**, 741–758.

Plebanski, M., Aidoo, M., Whittle, H. C. & Hill, A. V. (1997). Precursor frequency analysis of cytotoxic T lymphocytes to pre-erythrocytic antigens of *Plasmodium falciparum* in West Africa. *Journal of Immunology* **158**, 2849–2855.

Plebanski, M., Gilbert, S. C., Schneider, J. & 8 other authors (1998). Protection from *Plasmodium berghei* infection by priming and boosting T cells to a single class I-restricted epitope with recombinant carriers suitable for human use. *European Journal of Immunology* **28**, 4345–4355.

Plotkin, S. A., Cadoz, M., Meignier, B. & 7 other authors (1995). The safety and use of canarypox vectored vaccines. *Developments in Biological Standardization* **84**, 165–170.

Robinson, H. L., Montefiori, D. C., Johnson, R. P. & 14 other authors (1999). Neutralizing antibody-independent containment of immunodeficiency virus challenges by DNA priming and recombinant pox virus booster immunizations. *Nature Medicine* **5**, 526–534.

Robson, K. J., Hall, J. R., Jennings, M. W., Harris, T. J., Marsh, K., Newbold, C. I., Tate, V. E. & Weatherall, D. J. (1988). A highly conserved amino-acid sequence in thrombospondin, properdin and in proteins from sporozoites and blood stages of a human malaria parasite. *Nature* **335**, 79–82.

Schneider, J., Gilbert, S. C., Blanchard, T. J. & 7 other authors (1998). Enhanced immunogenicity for CD8+ T cell induction and complete protective efficacy of malaria DNA vaccination by boosting with modified vaccinia virus Ankara. *Nature Medicine* **4**, 397–402.

Schofield, L., Villaquiran, J., Ferreira, A., Schellekens, H., Nussenzweig, R. & Nussenzweig, V. (1987). Gamma interferon, CD8+ T cells and antibodies required for immunity to malaria sporozoites. *Nature* **330**, 664–666.

Taguchi, T., McGhee, J. R., Coffman, R. L., Beagley, K. W., Eldridge, J. H., Takatsu, K. & Kiyono, H. (1990). Detection of individual mouse splenic T cells producing IFN-gamma and IL-5 using the enzyme-linked immunospot (ELISPOT) assay. *Journal of Immunological Methods* **128**, 65–73.

Vitiello, A., Ishioka, G., Grey, H. M. & 9 other authors (1995). Development of a lipopeptide-based therapeutic vaccine to treat chronic HBV infection. I. Induction of a primary cytotoxic T lymphocyte response in humans. *Journal of Clinical Investigation* 95, 341–349.

Wang, R., Doolan, D. L., Le, T. P. & 12 other authors (1998). Induction of antigen-specific cytotoxic T lymphocytes in humans by a malaria DNA vaccine. *Science* 282, 476–480.

Wizel, B., Houghten, R., Church, P., Tine, J. A., Lanar, D. E., Gordon, D. M., Ballou, W. R., Sette, A. & Hoffman, S. L. (1995). HLA-A2-restricted cytotoxic T lymphocyte responses to multiple *Plasmodium falciparum* sporozoite surface protein 2 epitopes in sporozoite-immunized volunteers. *Journal of Immunology* 155, 766–775.

Prospects for new and rediscovered therapies: probiotics and phage

Stig Bengmark

Ideon Research Park, Scheelevägen 18, SE-223 70, Lund, Sweden

'Death sits in the bowels; a bad digestion is the root of all evil'
Hippocrates, ca. 400 BC

Introduction

Health and well-being depend on a complex and dynamic interplay between factors that control vital processes such as appetite, energy balance, metabolic rate, stress response, apoptosis, cell proliferation and repair (Frame, Hart & Leakey, 1998). Systemic homeostasis/cellular mechanisms in balance seem to depend both on internal/gene-mediated processes and external/lifestyle conditions. Lifestyle (eating habits, physical exercise, psychological conditions, stress, social control, etc.) appears to have a profound if not dominant influence on these various processes. It has become increasingly understood that good health and optimal mental and physical performance are not totally compatible with Western lifestyle.

During the last 10–15 years it has become evident that an increasing number of the persons who practise so-called Western lifestyle suffer from a condition usually called *metabolic syndrome X* (MSX), which is known to reduce resistance to disease and to make the individuals vulnerable to conditions such as arteriosclerosis, coronary heart disease, diabetes, rheumatoid arthritis and inflammatory bowel disease. It is suggested that 20–25% of Americans and Europeans suffer from the syndrome (Reaven, 1995), which although increasing still is much less frequent in countries such as Japan and China, and rare in developing countries, especially in rural areas. It is suggested that 50% or more of those seeking medical attendance at hospitals suffer from the condition. It is among these individuals that infections are common and complications frequently encountered when subject to surgical operations.

Common findings in those suffering from the syndrome are: obesity, hypertension, insulin resistance, dyslipidaemia and glucose intolerance. There is much support that an increased exposure to environmental toxic substances and pharmaceutical drugs, use of tobacco and alcohol, lack of physical activity, mental stress and overconsumption of refined food in combination with insufficient intake of plant fibres and probiotic bacteria are among the factors responsible for the condition (Diplock *et al.*, 1998; Hornstra *et al.*, 1998; Salminen *et al.*, 1998; Saris *et al.*, 1998; and others).

What is negative with Western food?

Table 1 summarizes what is generally recognized as negative with Western food. It is especially observed that overconsumption of saturated fatty acids, especially C_{12}, C_{14}, C_{16} and of cholesterol, and also of refined sugar and sodium salts, are contributing to the condition. As important to the development of MSX is, however, a documented underconsumption of fruit and vegetable fibres, antioxidants and of polyunsaturated fatty acids, especially those of the omega-3 type. The World Health Organization (WHO), Commission of the European Communities, most Western governments and various experts consequently recommend that the consumption of saturated fat is drastically reduced to under 10% (presently >40%) of the daily consumed calories (e.g. less than 20 g per day), the use of refined sugar and table salt should be limited and the consumption of fresh fruit and vegetables radically increased (five to eight pieces of fruit and vegetables per day).

The habit of heating food before eating is known to have important negative consequences. Not only does it to a large extent eliminate important nutrients, vitamins (Schroeder, 1971) and antioxidants such as glutathione (Wierzbicka, Hagen & Jones, 1989), it also promotes significant production of mutagenic substances from both animal and plant proteins—for a review see Skog (1993). The production of mutagens increases considerably with cooking/frying temperatures higher than 125 °C, with cooking for more than 10 min and when fat and/or sugar is added. Animal studies demonstrate that the number of intestinal polyps will

Table 1. Summary of what is regarded as characteristic of Western food and in the scientific literature generally regarded as detrimental to health.

1 Too much saturated fat
2 Too little polyunsaturated fat
3 Too much sodium salt
4 Too few fermentable fibres
5 Too much refined sugar
6 Too few antioxidants – cooking, canning and freezing destroy vitamins and antioxidants
7 Too many mutagenic compounds – frying and cooking produce mutagens
8 Too many animal-derived hormones and growth factors, which delay apoptosis and enhance tumour development
9 Too few probiotic micro-organisms

double and the number of cancers triple when half the ingested protein has been heated up to a light brown colour (Zhang *et al.*, 1992). The good news is that the negative consequences most likely are considerably reduced by regular consumption of lactic acid bacteria (LAB). It has been reported that mutagens in urine occurring after eating fried ground beef can be reduced to half with consumption of *Lactobacillus casei* (Hayatsu & Hayatsu, 1993). It is not unlikely that the preventive effect can be expected to be more pronounced with the use of other and more potent LAB.

The Palaeolithic perspective

The human genetic constitution has changed very little since the birth of mankind. The range of diets available to pre-agricultural human beings seems to still determine the nutrition for which human beings in essence are genetically programmed (Eaton & Konnor, 1985). These authors suggested that our genes, which over millions of years adapted to the lifestyle of our prehistoric ancestors, badly tolerate the dramatic changes in lifestyle and especially food habits which occurred following the introduction of agriculture some 10 000 years ago, and further dramatically altered since the industrial revolution, i.e. during the last 100–150 years. The Palaeolithic diet (Table 2) is calculated to have contained much less animal protein, saturated fat, cholesterol and sodium salts, but contained significantly more minerals, plant fibres, antioxidants, polyunsaturated fat (especially omega-3 fatty acids) and particularly LAB. Furthermore, the food eaten by our prehistoric ancestors was raw and uncooked, which is one of the reasons why many sensitive

Table 2. A summary of what is regarded as characteristic of the food of our prehistoric ancestors, often called palaeolithic food. *'Much support that our genes, adapted during millions of years to the lifestyle of our prehistoric ancestors, badly tolerate the dramatic changes, especially in food habits, which have occurred'* (Eaton & Konnor, 1985).

Component	Relative amount
Contained more:	
Minerals	2×
Fibre	4×
Antioxidants	10×
Omega-3 fatty acids	50×
Lactobacilli	$>10^{10}\times$
Contained less:	
Protein	2×
Saturated fatty acids	4×
Sodium	10×

food ingredients, antioxidants and vitamins, and also live bacteria, were supplied in significantly larger quantities. It has been suggested that our forefathers ate more than ten times more vitamins and antioxidants. The Western intake of fibre is today less than 20 g per day (recommended minimum 20–30 g per day), which sharply contrasts to the consumption by rural Chinese (>75 g per day), native Americans living more than 100 years ago (>100 g per day), rural Africans living today (>120 g per day) and other primates such as chimpanzees (>200 g per day) (Eaton *et al.*, 1996). For further information about the palaeolithic diet, see Milton (1999).

LAB consumption drastically reduced

The largest changes have without doubt occurred in the consumption of LAB. Our forefathers preserved their food in the soil, to a large extent, which is why their food was rich in fibre-fermenting LAB such as *Lactobacillus plantarum*, *Lactobacillus pentosus*, *Lactobacillus paracasei* and others commonly found in naturally fermented foods. These LAB are richly supplied with the fermented food consumed in developing countries and also in silage for domestic animals. One can assume that the palaeolithic diet contained billions if not trillion times or more LAB. The consumption of fermented food both in Europe and North America has been drastically reduced especially during the last 100 years. The LAB consumed today, if actively supplied at all, are mainly through dairy products, which contain LAB seeming to have much weaker health effects than fruit and vegetable fermenting LAB.

An adult human is calculated to carry a commensal flora weighing about 1·3 kg (Table 3), of which approximately 1 kg is to be found in the colon, which is suggested to contain at least 10^{12} living bacterial cells. It is calculated that the human body consists of ten times more bacterial cells (10^{14}) than eukaryotic cells (10^{13}). About 400 bacterial species have been detected in the faecal/colonic microflora, but

Table 3. Calculated weight of the commensal flora of the human body, sometimes referred to as 'the microflora organ'. Based on Gustafsson (1985). Values are grams wet weight (appraised weights).

Eyes	1
Nose	10
Mouth	20
Lungs	20
Vagina	20
Skin	200
Intestines	1000

30–40 species seem to constitute 99% of the collection in any one human subject—for further information see Tannock (1997). Among the bacterial genera which are commonly detected as components of the intestinal micro-flora in humans are: *Bacteroides, Bifidobacterium, Clostridium, Enterococcus, Eubacterium, Fusobacterium, Peptostreptococcus, Ruminococcus, Lactobacillus* and *Escherichia.* Each human being has his/her own unique collection of at least *Bifidobacterium* and *Lactobacillus* strains, and could well be identified on the basis of the personal intestinal microflora (Tannock, 1997). A recent study suggests that the most prevalent LAB taxa to be found on the rectal mucosa in healthy humans with so-called Western lifestyle are *L. plantarum, Lactobacillus rhamnosus* and *L. paracasei* subsp. *paracasei,* isolated in 52, 26 and 17%, respectively, of Swedish individuals studied (Ahrné *et al.*, 1998). The colonization rate of commonly used milk-borne probiotic bacteria such as *L. casei, Lactobacillus reuteri* and *Lactobacillus acidophilus* was in the same study only 2, 2 and 0%, respectively.

Palaeolithic and rural foods rich in *L. plantarum*

Fibre-fermenting bacteria such as *L. plantarum* are reported to be much more dominant among rural Asians and Africans, known to consume significantly more plant fibres and fermented foods. *L. plantarum* is also known to be the dominant species in fermented foods such as sourdough, sauerkraut, green olives, natural wines and beers and most Third World staple foods, such as African *ogi, kenkey* and *wara* (Olasupo, Olukoya & Odunfa, 1995), or Asian (Indonesian) *tempeh.* Conse-quently, all those living in rural areas and consuming traditional foods are colo-nized with *L. plantarum.* A study in the US showed that *L. plantarum* is more likely to colonize vegetarians (approx. 2/3 of that population) than omnivores (approx. 1/4 of that population) (Finegold, Sutter & Mathisen, 1983). Other observations support the assumption that LAB do not tolerate well modern, so-called Western lifestyle. It is reported that Swedish children have a different flora than Pakistani children (Adlerberth *et al.*, 1991) and that cosmonauts on return from space flights have lost most of their commensal flora (Table 4), including *Lactobacillus* species such as *L. plantarum* (lost to almost 100%), *L. casei* (lost to almost 100%), *Lactobacillus fermentum* (reduced by 43%), *L. acidophilus* (reduced by 27%), *Lactobacillus salivarius* (reduced by 22%) and *Lactobacillus brevis* (reduced by 12%) (Lencner *et al.*, 1984). These changes could be attributed to poor eating (dried food, no fresh fruit and vegetables) and consequently much reduced supply of fibre and antioxidants, and also to lack of exercise and to mental and physical stress. Many Westerners living on Earth today seem to practise, in a way, an astronaut-like lifestyle.

Table 4. Changes in commensal flora of astronauts before and after flights of various durations, and during rehabilitation after the flight. Based on the results of Lencner *et al.* (1984). Studied changes in *L. acidophilus*, *L. casei*, *L. plantarum* and *Streptococcus* in saliva and stools before and after space flights. Values are the mean number (log ml^{-1}). The effects are attributed to poor eating and stress. '*Many people on Earth live like astronauts and might need supplementation of Lactobacillus*'?

	L. acidophilus	**L. casei**	**L. plantarum**	**Streptococcus**
Saliva				
During preparation	1·8	1·2	0·3	4·2
After standard flight	0·9	0·9	0·0	3·0
After short flight	0·7	0·3	0·0	2·7
After long flight	1·3	1·7	0·0	3·5
During rehabilitation	0·0	1·5	0·0	6·8
Stools				
During preparation	4·0	3·5	2·6	4·9
After standard flight	1·7	0·0	0·5	4·8
After short flight	1·0	0·0	0·8	4·6
After long flight	2·9	0·0	0·0	5·1
During rehabilitation	0·0	4·0	1·7	8·5

Two separate digestion systems

It is not always recognized that we have two rather independent digestive systems, one based on gastric and small intestinal enzymes and another equally important based on enzymes produced by the commensal flora, and mainly active in the large intestine. Supplied fibres (prebiotics) are fermented by the commensal flora and important products (synbiotics) are locally produced and absorbed. Among the products produced in the large intestine are important short-chain fatty acids (SCFAs) and peptides, amino acids and polyamines, and also antioxidants, growth and coagulation factors and messenger molecules such as cytokines and nitric oxide. The concept of food for the large intestine, colonic food, is well recognized today, and it is recommended that at least 10% of the calories or 20% of the food volume should be of the colonic food type. While a daily supply of sufficient quantities of prebiotics/plant fibres is most important to health, there is not yet sufficient evidence to suggest that a daily supply of probiotics (live microbial feed supplements with expected beneficial effects on the host by improving its intestinal microbial balances) should be necessary for a good colonic fermentation function and for optimal health, despite there being accumulating evidence that a daily supply of LAB contributes significantly to the maintenance of good health. But probiotics supplied daily to healthy individuals might exercise their main functions in the parts of the digestive tract, such as the small intestine and eventually also the stomach, which are less favourable to microbial colonization compared to the oral cavity and particularly to the large intestine. It is often discussed why a daily supply of only a few grams of LAB can be expected to have a profound influence on our immune functions and resistance to disease, when the content of commensal bacteria in the colon amounts to about 1 kg, i.e. hundreds of times or more larger than the daily dose of LAB consumed. Schiffrin *et al.* (1997) suggest that the inoculum of LAB mainly targets host cells in the stomach and small intestine, i.e. parts of the digestive tract that are not highly colonized by the commensal flora. Possible targets are Peyer's patches of the gut-associated lymphoid tissue and the epithelium, which are associated with the highest proportion of specialized so-called M cells, known to be more permeable to luminal components than other intestinal cells. Increased antibacterial activity of Peyer's patch lymphocytes has also been reported in mice after administration of yogurts containing live LAB (De Simone *et al.*, 1987).

Are all diseases inflammatory?

The complex interactions between micro-organisms and the human host are far from fully understood. It is not much more than 100 years since Robert Koch (1876) showed the first proof that a particular disease was caused by a specific organism (anthrax)—see Stanier, Doudoroff & Adelberg (1963). During the following 100 years, relatively few diseases could be demonstrated to fulfil what became known as Koch's postulates, the fulfilment of which was regarded as necessary to establish a causal relationship between a specific agent and a specific disease. However, during the last 15–20 years a quiet revolution has been taking place, mainly made possible by modern scientific technology such as PCR. Today there is increasing evidence that infectious agents can be the causes of or precipitating factors for various diseases that were not previously thought to be caused by transmissible agents (Lorber, 1996). Among these various diseases are peptic ulcers, various neurodegenerative diseases, acute renal failure, arthritis, vasculitis, inflammatory bowel disease, diabetes, coronary heart disease and neoplastic disorders. It is not without evidence that Lorber asks the question: 'Are all diseases infectious'? There is, as pointed out by Lorber, every reason to believe that science will in the future show more infectious causes for what until today have been defined as degenerative, inflammatory (non-infectious) or even hereditary diseases. The documented changes in acute phase reactants (cytokines, acute phase proteins, growth and coagulation factors) support the view that inflammation is an important ingredient in most diseases. These molecules are often increased in serum weeks, months or even years before the actual disease is clinically identifiable. Such a process, which seems to be associated with 'environmental disease', should probably better be called 'chronic phase response'. It seems clear that many of the above-mentioned diseases are linked with Western lifestyle, as they do not exist at all or are rare in primitive cultures, and they do not seem to have existed among our Palaeolithic forefathers. Furthermore, there is increasing evidence that consumption of Western food is intimately related to the development of so-called Western diseases (see Bengmark, 1998b), and these diseases should be preventable and also to some extent reversible through radical lifestyle changes, including dramatic diet modifications (Ornish *et al.*, 1998). The main purpose of this chapter is to discuss how and to what extent consumption of pre- and probiotics is or can be involved in the prevention and treatment of these conditions. This is especially burning as we enter the new Millennium and possibly are coming to the end of the antibiotic era, and consequently look forward, with the greatest uncertainty, to our future relationship with infectious diseases—see further Henderson, Poole & Wilson (1996).

Chronic infections in the respiratory tract, stomach and also maybe the prostate and the female genital tract have been associated with significantly increased risk of development of cardiovascular disease (CVD) and also of cancer. Among the bacteria associated with such developments are *Helicobacter pylori* and *Chlamydia pneumoniae*. Several studies, however, have shown that LAB can inhibit the growth of *H. pylori in vitro* and exhibit antagonistic activity against *H. pylori in vivo* (Kabir *et al.*, 1997; Coconnier *et al.*, 1998). It has recently been shown that reduced nitric oxide production is involved in processes associated with CVD, including vaso-constriction, atherosclerosis and thrombosis (for review see Pollard, 1997). A link between endothelial nitric oxide production, insulin sensitivity and protection against CVD has also been suggested (Petrie *et al.*, 1996), and dietary supply of L-arginine has been shown to reduce oxidative stress and preserve endothelial function in hypercholesterolaemic rabbits (Böger *et al.*, 1998). Both chronic *H. pylori* infection (Laurila *et al.*, 1999) and chronic *C. pneumoniae* infection (Laurila *et al.*, 1997) have been associated with a serum lipid profile known to be a risk factor for CVD. Furthermore, *C. pneumoniae* has also been shown to be present in athero-sclerotic lesions (Saikku, 1997), and a marked cell-mediated and humoral immunity to *C. pneumoniae* has been observed in CVD patients (Halme *et al.*, 1997). Although no studies yet exist to prove it, it is tempting to speculate that continuous intake of LAB could at least have a preventive effect against CVD, and also that LAB such as *L. plantarum* will play an important role in such a process.

Nitric oxide – a key player?

It has become increasingly obvious that the composition of the diet and consumption of drugs are important to the gastric pH, the gastric emptying rate, and the killing of pathogens in the upper gastrointestinal (GI) tract. An important entero-salivary circulation of nitrate has been discovered relatively recently (Duncan *et al.*, 1995). Nitrate secreted by the salivary glands, and/or ingested nitrate, is reduced at the surface of the posterior third of the tongue to nitrite by a facultatively anaerobic bacterium. When nitrite is swallowed into an acidified stomach, e.g. if the acidity has not been impaired by supply of H_2-blockers or proton inhibitors, it will immediately lead to a large production of nitric oxide in the lumen of the stomach, and the production via the acidified stomach is said to be significantly larger than can be generated through intrinsic nitric oxide synthase (McKnight *et al.*, 1997). Gastric nitrate concentration rose after nitrate ingestion to 3430 μmol l^{-1} and the nitric oxide produced at a pH of around 2 rose from 18·8 p.p.m. to 89·4 after 60 min. This is regarded as significantly more than required for stimulation of mucosal blood flow, mucus formation, for stimulating motility and for bacteriostasis (Duncan *et al.*, 1995). Antimicrobial action of acidified nitrite has been reported with nitric oxide concentrations as low as 1 p.p.m. Consequently, gastric nitric oxide production is suggested to be crucial for prevention of gastric colonization with pathogenic flora, and acidified nitrite is also shown to effectively eliminate *Candida albicans, Escherichia coli, Shigella, Salmonella, H. pylori* and conditions such as amoebic dysentery and chronic intestinal parasitism (Duncan *et al.*, 1995). A recent study demonstrates that adding 1 mM nitrite *in vitro* to an acidic solution (pH 2) will within 30 min produce a complete kill of *H. pylori*, which is not seen when acid alone is administered ($P < 0·001$) (Dykhuizen *et al.*, 1998). It should especially be observed that gastric nitric oxide production does not occur in germ-free animals and is observed to be significantly less when antibiotics are supplied. Another recent publication (Smith *et al.*, 1999) suggests that, in the oral cavity, organisms such as *Actinomyces* spp. and *Veillonella* spp., but not the most frequent species in the oral cavity, such as *Streptococcus* spp., have a strong capacity to reduce nitrate and to produce nitric oxide. This process is influenced by the pH level in the oral cavity and a sucrose rinse results in a significant increase in intraoral generation of nitric oxide. It is speculated that this might be an important mechanism to inhibit colonization of the oral cavity by other, unwanted, microbial species.

Nitric oxide produced from acidified nitrite might, in the future, be used as a tool to control microbial overgrowth in the upper GI tract, and also to maintain and improve mucosal and splanchnic (digestive tract) blood flow, and to stimulate GI motility. An alternative to oral supply of nitrate/nitrite could be to eat fruit and vegetables known to be rich in nitrates (see Bengmark, 2000b). Support for this opinion comes from a recent publication (Chen & Ran, 1996), where animals with induced pancreatitis were treated with a 10% rhubarb decoction (rhubarb is rich in nitrate). Significant reductions in the rate of microbial translocation to mesenteric lymph nodes and to pancreatic tissue (treated 25% vs controls 100%), in mortality (1/8 vs 5/8 animals) and in serum endotoxin levels (treated $5·41 \pm 3·6$ pg l^{-1} vs controls $61·36 \pm 28·3$ pg l^{-1}, $P < 0·001$) were reported for the rhubarb-treated group. The authors also concluded that a 'remarkable inhibition of gut motility was observed in the control group, but gut motility was significantly improved by administration of rhubarb'.

Studies of acute phase reactants

A lot of information can be obtained from studying the reaction of the body not only when challenged by an acute threat such as microbial invasion, but also when challenged by other types of stress/exterior threats such as burns, physical trauma, surgical operation, tissue ischaemia, tissue infarction, strenuous exercise, child birth and allergic reactions. This reaction, which has been given the name acute phase reaction (APR), is instant and consists of release of a series of mediators such as nitric oxide, cytokines, coagulation and growth factors, prostaglandins and leukotrienes, and also of release of numerous acute phase proteins. Liver transplant patients, who already during the later phase of the operation show a sixfold or more increase in cytokines such as TNF-α and interleukin-6 (IL-6), are likely to develop sepsis during the subsequent post-operative days (Sautner *et al.*, 1995). Accumulating evidence suggests that, during the APR, IL-6 in particular is a key mediator of a pathological hyperinflammation. It is known that an exaggerated IL-6 response (e.g. prolonged and/or extreme elevations of circulating IL-6) is associated with adverse clinical events such as acute respiratory distress and multiple organ failure in patients suffering from conditions such as infection, burns or trauma (Biffl *et al.*, 1996). Among the effects associated with an exaggerated IL-6 response are augmented endothelial adhesion of polymorphonuclear (PMN) cells, increased production of intracellular adhesion molecule-1 (ICAM-1) and priming of the PMNs for an oxidative burst, release of proinflammatory platelet activating factor (PAF), and, associated with this, a delay in PMN apoptosis (Biffl *et al.*, 1996). Increasing evidence supports the contention that our lifestyle, especially our diet, influences the magnitude of the APR and that diet modifications such as reduction in consumption of saturated fat and increase in intake of fibre (prebiotics) and LAB (probiotics) can dramatically modify the APR, prevent overreaction of the APR, and also reduce 'chronic phase response' and prevent diseases associated with Western lifestyle. Synbiotics, i.e. the combination of pre- and probiotics, affect numerous biological processes, some of which are discussed below.

The process of apoptosis

Programmed cell death is one of the important mechanisms by which the body controls both infections, especially of viral origin, and neoplastic transformation. It is important for tissue homeostasis and prevention of disease that aged cells, especially those with premalignant potential, are eliminated. Dairy products (Outwater, Nicholson & Barnard, 1997; Westin & Richter, 1990), rich in saturated fat and various growth factors including insulin growth factor 1 and various cow oestrogens and xenoestrogens (from pesticides), regarded as prominent risk factors for cancer, are known to inhibit or delay apoptosis and to promote malignant cell proliferation, changes associated with increasing luminal concentrations of bile acids (Hague *et al.*, 1993) and modification of the composition of the bowel microflora. It has been suggested that for each 1% intake of saturated fat in the diet the risk of dying from breast cancer will increase by 10% (Jain, Miller & To, 1994), and the risk of failure of treatment of breast cancer will increase by 8% (Holm *et al.*, 1993). There are several reports which demonstrate that restricted food intake (Grasl-Kraupp *et al.*, 1994) and also increased consumption of plant fibres such as pectin, oat, wheat, rye or chicory fibre (inulin) (Hong *et al.*, 1997) will significantly increase the rate of apoptosis and prevent cancer development. Also, cells infected with a virus undergo apoptosis as a defence mechanism to prevent spreading of virus infection (Solary, Dubrez & Eymin, 1996), a mechanism known to be induced by T cells. These processes are most likely enhanced by supply of both fibre and LAB (synbiotics), as the process of apoptosis is known to be promoted by SCFAs (Heerdt, Houston & Augenlicht, 1994; Marchetti *et al.*, 1997). These fatty acids are products of bacterial fermentation of fibre in the large intestine, which is why an oral supply of probiotics (LAB) and prebiotics (various fibres) can be expected to have apoptosis-stimulating effects. It has been observed in experimental animals that feeding beans increases SCFA production sevenfold (Key & Mathers, 1995) and feeding fibres such as oligofructans inhibits induction of colonic preneoplastic lesions (Reddy, Hamid & Rao, 1997).

LAB influence phagocytic functions

The ability of special cells to engulf, kill and eliminate invading micro-organisms and/or defective cells and also to eliminate toxins, mutagens and other poisonous substances is extremely important to health. That enteral and/or parenteral supply of fat has a profound effect on the APR and inhibits immune functions is supported by numerous observations. As an example, it is observed that intravenous infusion of 20% fat emulsions (Intralipid) will, in humans, significantly potentiate endotoxin-induced coagulation activation (Van der Poll *et al.*, 1996). A recent study in mice observed higher IgM and IgG antibody levels, worsened proteinuria and shortened lifespan in mice fed a high-fat diet (200 g fat per kg food) compared to those fed a low-fat diet (50 g fat per kg food) (Lin *et al.*, 1996). *In vitro* LPS stimulation of peritoneal macrophages from the two groups showed a significantly higher release of IL-6 (134 vs 59 ng per 10^6 cells, $P = 0.02$), TNF-α (311 vs 95 pg per 10^6 cells, $P = 0.001$) and PGE_2 (906 vs 449 pg per 10^6 cells, $P = 0.01$) in the group fed a high-fat diet. A diet too rich in polyunsaturated fatty acids can also be negative. Studies in mice with standardized thermal injuries showed a significantly increased mortality on challenge with *Pseudomonas aeruginosa* when 40% of total calories were supplied as fish oil (Peck *et al.*, 1990).

Chemicals and pharmaceuticals to which modern humans are richly exposed also seem to alter the functions of the macrophages, both the bactericidal function and the production and secretion of cytokines. It has been shown that a supply of antibiotics [150 mg Mezlocillin (Bayer) per kg body weight] leads to unwanted effects, such as suppression of various macrophage functions, including chemiluminescence response, chemotactic motility, bactericidal and cytostatic ability and also lymphocyte proliferation (Roszkowski *et al.*, 1988). These effects have not been explored to the extent they should be. Subsequent work by the same group (Pulverer *et al.*, 1990) demonstrated that the reduction in peritoneal macrophage function and in lymphocyte proliferation observed after microbial decontamination of the digestive tract was significantly reconstituted by supply of low-molecular-mass peptides obtained from species of the indigenous GI tract microflora such as *Bacteroides* spp., *Clostridium* spp., *Propionibacterium* spp. and from *Lactobacillus* spp. (Pulverer *et al.*, 1990). The authors conclude that such low-molecular-mass peptides are essential for an adequate immune response of the host. Other studies demonstrate that supply of live or non-viable bacteria or bacterial wall components such as peptidoglucan stimulates macrophage recruitment and function (Kilkullen *et al.*, 1998), and cell-free extracts of both *Bifidobacterium longum* and *L. acidophilus* have been shown to significantly enhance phagocytosis both of inert particles and viable *Salmonella* (Hatcher & Lamprecht, 1993). However, not all LAB are capable of activating macrophages. As an example, Kato, Yokokura & Mutai (1988) observed in mice after intraperitoneal administration of *L. casei* or *Corynebacterium parvum* an increased macrophage activation (increased expression of Ia antigen on the surface), an effect not obtained with *L. fermentum*.

LAB influence cytokine release

Interest has, in more recent years, turned from studying the phagocytosis function to concentrate on studies related to production of cytokine and other signal molecules. There is some support that the cytokine profile seen after oral administration of LAB determines the direction and efficacy of the humoral response, and that this response is modulated differently by different LAB, when supplied. Most of the attention has been given to cytokine production by monocytic cells such as macrophages, but also mononuclear eukaryotic cells are important sources of cytokines. It has thus become increasingly clear that tissues such as intestinal epithelial cells (Eckmann, Kagnoff & Fierer, 1995; Ogle *et al.*, 1997) and prokaryotic cells such as commensal flora and/or supplemented probiotic bacteria (Henderson, Poole & Wilson, 1996; Henderson, Wilson & Wren, 1997), when challenged, will secrete a spectrum of chemoattractants and cytokines or cytokine-like molecules (called bacteriokines by Henderson). As an example, it has been demonstrated in cell cultures that intestinal epithelial cells, on challenge with LPS and PGE_2, produce significant amounts of IL-6, a process which can be blunted by supply of indomethacin (Meyer *et al.*, 1994) and inhibited by nitric oxide (Meyer *et al.*, 1995).

Supplementation with some LAB seems to significantly influence the expression of cytokines by various cells, but the cytokine response varies greatly with the strain of LAB supplied. It is understandable that mainly the effects of milk-borne LAB have so far been studied, as in more recent years it has mainly or only been the dairy industry which has remained interested in the supply of food products containing live LAB. One can expect, however, that as interest in LAB with specific ability to ferment plant fibres is gaining popularity, studies with LAB such as *L. plantarum, L. paracasei* subsp. *paracasei, L. pentosus* and other related lactobacilli will also be undertaken.

The activity of $2'–5'$-synthetase, a marker of interferon-gamma (IFN-γ) expression, in blood mononuclear cells of healthy subjects is found to be significantly increased (approx. 250%) 24h after a LAB-containing meal (Solis-Pereyra, Aattouri & Lemonnier, 1997). Significant increases in cytokine activity compared to controls were also observed when human mononuclear cells were incubated in the presence of the yogurt bacteria *Lactobacillus bulgaricus* (LB) and *Streptococcus thermophilus* (ST), alone or in combination (Yog) (IFN-γ: LB 775%, ST 2100%, Yog 570%; TNF-α, LB 1020%, ST 3180%, Yog 970%; IL-1β, LB 2120%, ST 1540%, Yog 1920%). Another recent study (Marin *et al.*, 1998) compared in both a macrophage model and a T-helper-cell model the *in vitro* ability to induce cytokine production by strains commonly used in yogurt production. Again *S. thermophilus* stimulated macrophage and T-cell cytokine production to a somewhat greater extent than did *L. bulgaricus, Bifidobacterium adolescentis* and *Bifidobacterium bifidum* (Marin *et al.*, 1998), but a significant variability in effect was observed between the four different strains of *S. thermophilus* tried. Also, heat-killed *L. acidophilus* (LA 1) has been shown *in vitro* to increase the production of IL-1α (approx. 300%) and TNF-α (approx. 1000%) by mouse macrophages, an effect said to be of considerably greater magnitude than that produced by other lactobacilli and bifidobacteria (Rangavajhyala *et al.*, 1997).

L. casei, when administered intrapleurally to tumour-bearing mice, induced significant production of cytokines such as IFN-γ, IL-1β and TNF-α, paralleled by inhibited tumour growth and increased animal survival (Matsuzaki, 1998). Supply of this organism also delayed the onset of diabetes in diabetes-prone and alloxan-treated animals (Matsuzaki *et al.*, 1997). It is of special interest to mention that nitric-oxide-donating molecules such as L-arginine and sodium nitroprusside, known to increase nitric oxide production, in experimental animals restored antioxidant status to near normal and prevented alloxan-induced beta-cell damage (Mohan & Das, 1998).

LAB influence immunoglobulin production

Adaptive immunity at mucosal surfaces represents by far the most impressive humoral immune mechanism of the body, providing an immunological barrier to foreign matter, particularly pathogenic micro-organisms, allergenic food proteins and carcinogens. IgA and to some extent IgM are dominant among the immunoglobulins in intestinal secretions. IgA mediates, to a large extent, its function through binding antigens and preventing bacterial and viral colonization and invasion at mucosal surfaces. IgA cooperates with a variety of innate protective mechanisms, but does not participate in the proinflammatory and cytotoxic responses that are readily activated by other immunoglobulin classes, e.g. complement activation and antibody-directed cytotoxic responses (Kagnoff, 1993). Selective IgA deficiency is the most common immunodeficiency in white people, but clinical abnormalities are often difficult to recognize as decreased IgA levels are often compensated for by increased production of IgM. Deficiency in IgA, but not in IgG and IgM, is associated with significantly increased morbidity and mortality in sepsis by opportunistic infections after major surgery, and also with increased rejection after liver transplantation (Van Thiel *et al.*, 1992).

About 80% of all the body's Ig-producing immunocytes are localized in the lamina propria of the gut (Brandtzaeg *et al.*, 1989), and large quantities, especially of IgA, are transferred each day to the gut lumen. The synthesis of IgA is highly dependent on T cells and several cytokines produced by activated lymphocytes influence different steps in the IgA differentiation pathway (Kiyono & McGhee, 1994). Transforming growth factor-β (TGF-β) has been suggested to be a crucial 'switch' factor at least in mice, but cytokines such as IL-2, IL-5 and IL-10 are also known to be involved—for further information see Brandtzaeg (1995). Changes in nutrition, physical activity, sleep, mood, age, gender, circadian rhythm, drug use, medical illness and other innate changes are known to influence lymphocyte function and Ig production and thereby also resistance to disease.

It has been suggested that during fermentation LAB may release components that possess immunomodulatory activity. When the ability of bifidobacteria to induce production of large quantities of IgA by Peyer's patches was studied in tissue culture, only three of 120 strains tested, all isolated from human faeces, had such an ability (Yasui *et al.*, 1992). Two of these were identified as *Bifidobacterium breve* and one as *B. longum*. I am not aware of any study comparing the ability of various *Lactobacillus* spp. to initiate production of large quantities of IgA, but supply of LAB such as *Lactobacillus* GG is reported to significantly increase the IgA immune response in Crohn's disease (Malin *et al.*, 1996), and also to enhance the

IgA response to rotavirus (Kaila *et al.*, 1992). Human intake of *L. acidophilus* is also known to result in a >fourfold increase in IgA response when challenged by *S. typhi* (Solis-Pereyra, Aattouri & Lemonnier, 1997). In a recent experimental study of methotrexate-induced colitis both supplementation of *L. reuteri* (R2LC) and *L. plantarum* (299 V; DSM 9843) significantly increased both small and large intestinal IgA secretion, in both soluble and insoluble fractions, and elevated the numbers of both CD4$^+$ and CD8$^+$ T cells (Mao *et al.*, 1997).

Th1/Th2 balance is essential

It is suggested that a balance between Th1 lymphocytes, primarily associated with cellular immunity, and Th2 lymphocytes, mainly associated with humoral immunity, is essential to health and well-being. Reduced microbial stimulation during early infancy and childhood, especially in developed countries, has been associated with the increasing prevalence of allergy in children and young adults (Björksten, 1994). Reduced microbial stimulation is associated with slower post-natal maturation of the immune system, a delayed development and lack of balance between Th1 and Th2 immunity (Lucey, Clerici & Shearer, 1996). Swedish infants, known to have a high incidence of allergy, have been reported to have a different gut flora than both Pakistani (Adlerberth *et al.*, 1991) and Estonian (Sepp *et al.*, 1997) children.

Production of IgE for dietary protein plays an important role in the mediation of food allergy. Allergic disease is thought to be caused by inappropriate generation and activation of Th2 cells; this process is inhibited by IFN-γ and also by IL-12. Some *Lactobacillus* species stimulate both IFN-γ and IL-12 production (see below), which promotes a Th1-type response and inhibits a Th2-type immune response (Murosaki *et al.*, 1998). Stimulation of human peripheral blood mononuclear cells with various *L. rhamnosus* and *L. bulgaricus* strains leads to induction of Th1-type cytokines IL-12, IL-18 and IFN-γ (Miettinen *et al.*, 1998a). Supply of *L. casei* (Shida *et al.*, 1998) and *L. plantarum* (Murosaki *et al.*, 1998) totally inhibited antigen-induced IgE secretion in ovalbumin- and casein-primed mice, an effect which was not obtained with *Lactobacillus johnsonii*. IL-12 production by peritoneal macrophages was enhanced but IL-4 production of concanavalin-A-stimulated spleen cells was suppressed in the *L. plantarum*-treated animals (Murosaki *et al.*, 1998). Similarly, *L. casei* induced IFN-γ, but suppressed IL-4 and IL-5 (Shida *et al.*, 1998).

L. plantarum and related LAB

As most of the interest in recent years has concentrated on LAB which are useful in yogurts and kefirs, palatability has been among the most important criteria for selection of LAB to be supplied. This is most likely the reason why lactobacilli such

Table 5. Some known characteristics of *L. plantarum*.

1 The most common microbe in naturally fermented foods:
Vegetables, cereals, fish and meat
Sourdough, green olives, natural wines, beers
The food of our ancestors
Foods in Africa and Asia
2 Excellent for preserving nutrients:
n-3 and *n*-6 fatty acids increase 200% in sauerkraut stored up to 12 months
n-3 is 300% higher in silage compared to dried food: hay, barley and oat
Vitamin content is increased in *L. plantarum* fermented vegetables
3 Eliminates nitrate from fermented foods
4 Inhibits pathogens in fermented foods:
Enterobacteriaceae, *Staphylococcus aureus*, enterococci
5 Reduces the number of potential pathogens in the GI tract
6 Metabolizes difficult fibres such as fructooligosaccharides rich in:
Chicory, onions, garlic, artichokes
Wheat, oat, rye
Banana, honey
Bacteria and yeasts
Fodder grass

as *L. acidophilus*, *L. casei* and *Lactobacillus delbrueckii* and bifidobacteria such as *B. adolescentis*, *B. bifidum*, *B. longum* and *Bifidobacterium infantis* so far dominate the studies. Little interest has been directed at studying LAB with the greatest capacity to ferment plant fibres. *L. plantarum* and taxonomically related LAB are the dominant microbes in naturally fermented foods (vegetables, cereals, fish, meat, sourdough, green olives, natural wines, beers, etc.), which are still important ingredients in the food consumed in developing countries (Olasupo, Olukoya & Odunfa, 1995) as they once were in the food of our Palaeolithic ancestors. *L. plantarum* is also the dominant species in fermented food products administered to farm animals (silage). It is thus not surprising that *L. plantarum* is among the dominant LAB in the human GI tract, when consumed (Ahrné *et al.*, 1998). Table 5 summarizes some of its unique features. It seems clear that food products obtained through fermentation with *L. plantarum* are superior in preserving key nutrients such as omega-3 fatty acids, vitamins and antioxidants, and most likely also other sensitive nutrients in food such as glutamine and glutathione, especially when compared to foods stored by modern preservation methods, such as freezing, drying or heating—for a review see Bengmark (1998b). Another interesting ability of *L. plantarum* is to utilize and eliminate nitrates from foods and it has been suggested that this ability should be a decisive criterion in the selection of LAB for bioconservation of vegetables (Andersson, 1985; Hybenová *et al.*, 1995). *L. plantarum* 92H was shown, during a 7-d-long fermentation process, to eliminate 100% nitrate/nitrite; *L. plantarum* 90H to eliminate 83% and *L. delbrueckii* 37H 73%

(Hybenová *et al.*, 1995). Another interesting observation is that *L. plantarum* prevents spoilage when used in *tempeh*, an Indonesian dish based on fermented cooked dehulled beans (Ashenafi & Busse, 1991), an ability also used in the preservation of other food products. The exact mechanisms for the strong ability to biopreserve are not known, but can be expected to be either through direct antioxidant effects of the bacterium, through bacteria-produced molecules such as plantaricins, through nitric oxide production, and/or through the low pH unique to *L. plantarum* ferments.

Oligofructans such as inulin and phleins are receiving increasing attention as nutritional fibres of great importance. These fibres are difficult to ferment and only a few LAB are able to do so. The ability of 712 different LAB to ferment oligofructans was studied in a recent publication (Müller & Lier, 1994). Only 16/712 were able to ferment the phlein- and 8/712 the inulin-type fibre. Apart from *L. plantarum*, only three other LAB species, *L. paracasei* subsp. *paracasei*, *L. brevis* and *Pediococcus pentosaceus*, were able to ferment these relatively resistant fibres.

The survival and ability to induce cytokine production after passage through the stomach and small intestine of four different LAB species, *L. plantarum* (E98), *L. paracasei* (E510), *L. rhamnosus* (E522) and *Bifidobacterium animalis* (E508), were recently studied (Miettinen *et al.*, 1998b). From an originally administered 10^8 cells ml^{-1}, after the passage in the intestinal contents between 10^7 (*L. plantarum*) and 10^2 (*L. rhamnosus*) bacterial cells remained. Most of the strains showed, at this level, a significantly reduced or weak (especially *L. rhamnosus*) ability to induce TNF-α and IL-6. Interestingly, *L. plantarum*, in sharp contrast to the other LAB tested, demonstrated an even greater capacity to induce IL-6 after the passage through the stomach and small intestine. Although the reasons for this special effect have not been studied, it can be speculated that the low pH in the stomach could stress and specifically activate *L. plantarum*.

LAB colonization is important

It is evident that only LAB with the ability to colonize the gut can be expected to express strong clinical effects. Colonization can only be expected by LAB with the ability to adhere directly to mucosal surfaces and/or to mucus. So far, such ability to adhere has only been documented for a few LAB strains but it appears to be clear that yogurt bacteria, and most other LAB used in combination with milk, do not adhere. As examples, most *L. acidophilus*, *L. bulgaricus* and bifidobacteria seem not to have the ability to adhere to intestinal mucosa (Chauvière *et al.*, 1989), at least not when tried *in vitro*, where adhesion most often is studied. A strong colonization ability has been demonstrated for *L. rhamnosus* (GG), which in at least one-third of the individuals persisted in faeces 7 d after the last administration (Goldin *et al.*, 1992). *L. plantarum* (299) has an even stronger ability to colonize and remained in at least two-thirds of the studied individuals for as long as 28 d (Johansson *et al.*, 1993). While the adherence of most LAB is via protease-sensitive mechanisms or via lipid (lactosylceramide) receptors, *L. plantarum* seems to adhere via carbohydrate (mannose) adhesion mechanisms, i.e. to use the same receptors as Gram-negative bacteria such as *E. coli*, *Enterobacter*, *Klebsiella*, *Salmonella*, *Shigella*, *Pseudomonas* and *Vibrio cholerae* (Adlerberth *et al.*, 1996). It has recently been shown *in vitro* that LAB inhibit enteropathogenic *E. coli* adherence by inducing intestinal mucin gene expression and that *L. plantarum* significantly increases the expression of MUC2 and MUC3 mRNA (Mack *et al.*, 1999). These special features make it particularly interesting to try to use *L. plantarum* and taxonomically related LAB as alternatives to antibiotics, and also as most interesting ingredients in functional foods.

Microbial interference treatment (MIT)

It was Pasteur & Joubert (1877) who were the first to observe the antagonistic interaction between some bacterial strains. A few years later, Metchnikoff (1907) suggested consumption of LAB in order to promote health. The use of MIT in prevention and treatment of disease declined with the introduction of antibiotics and chemotherapeutics. Only a small group of scientists continued to stubbornly promote MIT as an alternative method of controlling infection and preventing disease (Jack, Tagg & Ray, 1995). There are several reasons for the renewed interest in MIT seen today:

1 A recognition that antibiotic therapy has not been as successful as was hoped. Although it has solved many medical problems, it has also created important new ones.

2 An increasing awareness of the fact that antibiotic treatment deranges the protective flora, and predisposes to later infections. It has been observed that at least some antibiotics and other drugs suppress immunity.

3 An increasing fear of antibiotic-resistant microbial strains as a result of widespread overprescription and misuse of antibiotics.

4 A fear that the pharmaceutical industry will not be able to develop new and effective antibiotics at a sufficient rate to compete with the development of microbial resistance to old antibiotics.

5 A widespread interest among the public in ecological methods.

Despite dramatic advances in intensive care technology and in the development of new antibiotics, the mortality associated with Gram-negative bacteraemia during the last century continued to remain the same, between 20 and 40% (Wells, Maddaus & Simmons, 1988), and leading causes of bacteraemia are still *E. coli*, *Klebsiella pneumoniae*, other enterobacteria and *P. aeruginosa*. The mortality from such episodes as reported seems today not to differ significantly from what was reported in the preantibiotic era (Felty & Keefer, 1924). In addition, antibiotic-resistant bacteria are emerging as new causes of infectious episodes: bacteraemia with vancomycin-resistant *Enterococcus faecium* is reported to be second to *E. coli* as the most common cause of intensive care bacteraemia episodes (Spera & Farber, 1994), and its mortality (41%) is equal to the overall intensive care unit (ICU) bacteraemic mortality (42%) (Mainous *et al.*, 1997).

WHO experts recommend MIT

The growing threat of antibiotic resistance is today a major and fast-increasing public health problem in both developed and developing countries throughout the world. It was concluded by a WHO Scientific Working Group (1994) that 'the incidence has increased at an alarming pace in recent years and is expected to increase at a similar or even greater rate in the future as antibiotics continue to lose their effectiveness'. It is against this background that global programmes are recommended to dramatically reduce the use of antibiotics 'in animals, plants and fishes, for promotion of life stock growth', to increase efforts to prevent disease 'through increasing immunization coverage with existing vaccines, and through development of newer, more effective and safer vaccines', and also to try older forms of therapy including MIT. There is accumulating evidence that probiotic bacteria can effectively control various enteric pathogens such as *Salmonella typhimurium*, *Shigella*, *Clostridium difficile*, *Campylobacter jejuni* and *E. coli*, and various uropathogens such as *Gardnerella vaginalis*, *Bacteroides bivius*, *Candida albicans* and *Chlamydia trachomatis*—for review and references see Bengmark (1998a).

Nosocomial infection rate remains high

Clinical medicine of today is, in many aspects, not as successful as one would have expected. The use of synthetic drugs to prevent and treat diseases has not been as successful as was originally anticipated, and side effects are unacceptably high. It has been reported from the US that almost 7% of hospital patients develop adverse drug reactions (ADRs), despite the fact that the drugs are prescribed and used as recommended (Lazarou, Pomeranz & Corey, 1998). Apart from the human aspect, ADRs cost society approximately 135 billion US dollars, which is roughly the same as the total cost for care of CVD and diabetes (Bates *et al.*, 1997). It has been calculated that in the US alone 106 000 persons die each year due to ADRs, which is more than the number of persons dying in accidents, of pneumonia or of diabetes (Lazarou, Pomeranz & Corey, 1998). Modern surgery is, despite significant advances in surgical techniques, far from safe. The three leading causes of complications and sequelae, infections, thrombosis and adhesion formation, remain unsolved. It has been calculated that about 2 million Americans (6% of the hospital patients) suffer each year from nosocomial infections; most of the patients have reduced immune functions, and half of the patients are over the age of 65 (Swartz, 1994). Infections are especially common in neutropenic patients (48%), after transplantation (approx. 50%) and after extensive operations such as liver or pancreas resections (approx. 33%), but incidences are also unacceptably high after gastric and colonic resections (approx. 20%). The mortality in acute conditions such as severe pancreatitis is increased four times (approx. 40%) when the pancreatic tissue becomes infected with anaerobic gut bacteria (Isenmann & Büchler, 1994), which occurs in one-third of the patients after 1 week and in as many as two-thirds after 3 weeks of disease.

Although today the clinical manifestations of venous thrombosis can be suppressed, this complication is still found in 40–70% of patients if phlebography is routinely performed. It has recently been suggested that 'one common complication (bacterial infection) facilitates the occurrence of another common complication (venous thrombosis) by synergistic stimulation of the coagulation system' (Van der Poll *et al.*, 1998), both conditions being associated with the common clinical praxis of parenteral nutrition (PN). Furthermore, formation of fibrous adhesions occurs in the peritoneum, pericardium and pleura in more than 90% of surgical patients, and is still the main cause of reoperations, intestinal obstructions and female infertility (Bengmark, 2000a). Western lifestyle is most likely responsible for an exaggerated and prolonged APR in connection with trauma and surgery, and there are reasons to believe that the three conditions, nosocomial infections, thrombosis and adhesion formation, are all associated with our lifestyle, and thus should be possible to control—see Bengmark (2000b, c). It was already observed 30 years ago (Malhotra, 1968) that persons living in rural areas of developing countries, and consuming large quantities of live lactobacilli and vegetable fibres, had significantly longer coagulation times and softer jelly-like clots, compared to those living in urban areas. It has also been observed that persons consuming a high-fibre diet have a low incidence of thrombosis (Frohn, 1976).

The gut = the immune system

It has been suggested that approximately 80% of the human immune system is located in the gut (Brandtzaeg *et al.*, 1989). Furthermore, there are good reasons to associate the above complications and sequelae not only with Western lifestyle but also with the clinical practice of parenteral (lack of enteral) nutrition. If so, early or uninterrupted supply of nutrition via the enteric route should offer opportunities to modulate the APR and reduce the occurrence of complications. Recent studies

Table 6. Changes in clinical outcome and in nutritional and immune parameters in liver resection patients: comparison between enteral nutrition (EN) and parenteral nutrition (PN). Table based on the results of Shirabe *et al.* (1997). Natural killer cell activity (NK cell activity) was assayed against NK-sensitive K-562 cells, provided by the Foundation for the Promotion of Cancer Research, Tokyo, Japan. EN, given from the second post-operative day, contained no fibre but 15% fat; there is no information about antibiotics supplied. There was no difference found in the nutritional parameters [retinol-binding protein (RBP), transferrin, prealbumin and 3-methylhistidine] or in the IgA, IgG and IgM levels. There were significant differences in the percentage of pre-operative values for the immune parameters listed in the table. '*Early EN maintained immunocompetence and . . . . reduced septic complications*'.

	EN	PN	
Lymphocyte count	114	66	$P < 0.05$
Phytohaemagglutinin stimulation response	103	78	$P < 0.05$
NK cell activity	106	49	$P < 0.05$
Complications	8%	31%	

Table 7. Changes in clinical outcome and in various immunity-related parameters in patients with acute pancreatitis: comparison between enteral nutrition (EN) and parenteral nutrition (PN). Table based on the results of Windsor *et al.* (1998). The following parameters were used: sickness score (APACHE II; Knaus, Draper & Wagner, 1985), C-reactive protein (mg l^{-1}), IgM EndoCAb using an ELISA measured as percentage change, and total antioxidant potential quantified using an enhanced chemiluminescence technique (Whitehead, Thorpe & Maxwell, 1992). EN was given within 48 h after onset of disease, contained 36% fat, no fibre and no routine antibiotics. '*Systematic inflammatory response, sepsis, organ failure and ICU stay were globally improved in the EN group*'.

	EN	PN	
APACHE II score	6	8	$P < 0.0001$
C-reactive protein	84	156	$P < 0.005$
IgM anticore endotoxin antibodies (Chromogenix, Mölndal) (%)	−1.1	+29	$P < 0.05$
Total antioxidant potential (chemiluminescence) (%)	−28	+33	$P < 0.05$
Clinical outcome (sepsis, multiple organ failure)	o patients	5 patients	

suggest that enteral nutrition (EN) is more important as a tool to control APR and immune response than to provide calories and nutrients. A recent study compared parenteral hyperalimentation and early EN after major liver resection (Shirabe *et al.*, 1997), and found no differences when studying nutritional parameters, but significant differences when studying immunological parameters such as natural killer cell activity, changes in lymphocyte numbers and response to phytohaemagglutinin (Table 6). Also, most importantly the incidence of infectious complications was 8% in the EN group compared to 31% in the PN group. Similar results (Table 7) have been reported in patients with severe acute pancreatitis (Windsor *et al.*, 1998). The APR, indicated by changes in C-reactive protein and disease severity scores (APACHE II), was significantly improved with EN compared to PN. The IgM anticore endotoxin antibodies (EndoCAb) and total antioxidant potential were both significantly better in the EN group compared to PN. However, most importantly, systemic inflammatory response (SIRS), sepsis, organ failure and stay in intensive care were globally improved in the EN fed patients.

The knowledge that some nutrients and antioxidants have strong immunomodulating functions has led to the production of special immune-enhancing nutrition solutions, based on a mixture of amino acids, polyunsaturated fatty acids and antioxidants. The experience so far has not been what was originally expected—for review see Barton (1997), Dickerson (1997) and Bengmark (2000b). Despite the fact that some compelling data have been presented in the literature, there is much to support the view of these reviewers and others that at present 'routine use of these formulas cannot be recommended'. There seem to be many reasons for the lack of greater success (Bengmark, 1999, 2000b), the most important probably being that so far none of these solutions consider the need of the colon (need of substrate) for the important fermentation and for colonic release of nutrients, antioxidants, and growth and coagulation factors, as none of the commercial immuno-enhancing solutions so far contain fibres (prebiotics). Nor so far has supply of probiotics been considered—see below. Furthermore, EN is often instituted too late to have the ability to significantly affect APR, and is often given together with nutrition solutions rich in fat, which might inhibit the immune functions and counteract the purpose. In addition, EN is often combined with treatment with antibiotics, which might reduce or eliminate the important commensal flora.

Effects of LAB in experimental models

Experimental colitis

Colitis was induced in rats by instillation of a 4% acetic acid solution for 15 s in an exteriorized colonic segment. A uniform colitis was produced with a threefold increase in myeloperoxidase activity of the colonic tissue (index of neutrophil infiltration) and a sixfold increase in plasma exudation into the lumen of the colon. Intracolonic administration of species-specific *L. reuteri* R2LC immediately after instillation of acetic acid, as either a pure bacterial suspension or fermented oat soup, totally, or almost totally, prevented development of colitis (Fabia et al., 1993). Treatment instituted after 24h resulted in a significantly improved healing. In another study, enterocolitis was induced by intraperitoneal injection of 20 mg methotrexate (MTX) per kg body weight, and the animals were pretreated by oral supply of a *L. reuteri* R2LC or *L. plantarum* DSM 9843 (strain 299v) and oat-base-containing diet. Supply of lactobacilli, but not oat base alone, decreased the intestinal myeloperoxidase level, re-established intestinal microecology and reduced bacterial translocation to extraintestinal sites (Mao *et al.*, 1996a), and both treatments, lactobacilli and oat base alone, reduced the levels of plasma endotoxin. The effects of lactobacilli were greater with fermentation than without fermentation or oat base alone, and *L. plantarum* was more effective in reducing intestinal pathogens than *L. reuteri*. Pretreatment with 1% pectin solution for 4 d could also significantly reduce MTX-induced intestinal injury and improve bowel integrity (Mao *et al.*, 1996b).

Experimental intra-abdominal infection

Experimental colitis was induced by caecal ligation and puncture (CLP) following a 5 d pretreatment with an oral supply of either fermented or unfermented oat base or saline. The animals were treated with saline, gentamicin or *Lactobacillus*. No animal in a sham-operated group but 32/36 in an otherwise untreated CLP, so-called control group, demonstrated bacterial growth in blood, compared to 11/24, 8/20 and 12/24, respectively, in the *Lactobacillus*-, gentamicin- and combined gentamicin/*Lactobacillus*-treated groups (S. Nobaek and others, unpublished). The difference was statistically significant between the untreated group and the various treatment groups, but not between the different treatment groups.

Chemical hepatitis

Acute liver injury was induced by intraperitoneal injection of D-galactosamine (1·1 g per kg body weight) following 8 d pretreatment with rectal instillation of one of the following four lactobacilli, with and without simultaneous supply of a 2% arginine solution: *L. rhamnosus* DSM 6594 (= strain 271), *L. plantarum* DSM 9843 (= strain 299v), *L. fermentum* DSM 8704 : 3 (strain 245) and *L. reuteri* (= strain 108). All lactobacilli, with or without added arginine, significantly reduced the extent of liver injury and reduced bacterial translocation, but the most pronounced effect was seen with the combination of *L. plantarum* and arginine, which significantly reduced liver enzymes, hepatocellular necrosis and inflammatory cell infiltration, and bacterial translocation, and reduced the number of *Enterobacteriaceae* in the caecum and colon (Adawi *et al.*, 1997). A subsequent study, recently published, shows that the extent of liver injury and bacterial translocation is increased after supply of *Bacteroides fragilis* and *E. coli*, but significantly inhibited by *L. plantarum* (Adawi *et al.*, 1999). Supply of *B. longum* reduced translocation but had no influence on the extent of liver injury. Oral supply of lactulose in the same model was also highly effective in preventing liver injury and bacterial translocation (Kasravi *et al.*, 1996a). Similar preventive effects could also be obtained by repeat intraperitoneal injection of endotoxin for 3 d before induction of liver injury (Kasravi *et al.*, 1997), while oral supply of *L. reuteri* during the same period showed no beneficial effects. It was also observed that endotoxin pretreatment renders macrophages unresponsive to subsequent stimulation, which might explain its beneficial effects seen in D-galactosamine-induced injury (Kasravi *et al.*, 1996b).

Experimental pancreatitis

Rats were pretreated with *L. plantarum* 299 at a dose of $0.5-1.0 \times 10^9 \, \text{ml}^{-1}$ 4 d before induction of acute pancreatitis (isolation and ligation of the biliopancreatic duct) and the treatment was continued until the end of the study, 4 d after induction. Pathogenic micro-organisms could be cultivated at this time from mesenteric lymph nodes in 14/20 animals and from pancreatic tissue in 10/20 untreated control animals (G. Mangiante and others, unpublished). The dominant flora consisted of *E. coli*, *Enterococcus faecalis*, *Pseudomonas* and *Proteus*. However, when the animals were treated with *L. plantarum* 299, only 4/20 animals demonstrated growth in mesenteric lymph nodes, and 3/20 in pancreatic tissue, the micro-organisms being *E. faecalis* and *E. coli*.

Human experience largely is lacking

Table 8 summarizes some suggested clinical indications for treatment with probiotics/synbiotics. The demand for well-controlled human studies is great. Although several human studies have been undertaken, few have so far been completed, and the studies so far reported have mainly dealt with either diarrhoea or inflammatory bowel syndrome. We have supplied *Lactobacillus* to five consecutive patients suffering from multiple organ failure after GI surgery. The mean APACHE II score fell from 18 before instigation of treatment to 12 and 9 after 5 and 10 d treatment, respectively (Table 9), and all patients could leave the ICU. For further details see Bengmark (2000c).

It is increasingly suggested, as the problems with antibiotic-resistant bacteria escalate (see above), that probiotics are tried at least to partly replace antibiotics. One such indication could be prophylaxis against perioperative infection. Two patients were, in a small pilot study, pretreated over 3 d with an enteral supply of *L. plantarum* 299 and oat fibre and two patients were treated conventionally with antibiotics and lavage before elective colorectal surgery for cancer. During the operation, biopsies were taken from the resected specimen for bacterial cultures. No *Enterobacteriaceae* were found on the mucosa of *L. plantarum*- and oat-fibre-treated patients in contrast to on average 500 colonies per cm^2 in the traditionally treated patients (L. Gianotti, personal communication).

A randomized trial, recently reported, compared selective bowel decontamination (SBD) with EN containing inulin fibre and live or heat-killed *L. plantarum* perioperatively in human liver transplantation (Rayes *et al.*, 1999). Each group

Table 8. Some potential indications for treatment with pre- and probiotics (synbiotics).

Probiotics
Prematures
Infants
Alternative to antibiotics

Synbiotics
'Astronauts'
Antibiotic therapy or chemotherapy
Irradiation
Renal dialysis
Biliary obstruction
Liver cirrhosis, portal hypertension
Cancer

Allergy
Immunodepression
Haematological malignancies
HIV/AIDS
Inflammatory bowel disease
Irritable bowel disease
Rheumatoid arthritis
Hepatitis
Pancreatitis
Stomatitis
Diarrhoea

When infected
After trauma
In major surgery, especially transplantation
In intensive care

Table 9. Changes in sickness score – APACHE II – during treatment with *Lactobacillus* and oat fibre in intensive care patients suffering multiple organ failure.

Patient	Before	After 5 d	After 10 d
1	16	5	4
2	20	12	8
3	21	17	13
4	20	18	16
5	15	9	3
Mean	18	12	9

consisted of 15 patients. In the group treated with SBD, sepsis occurred in 40% (6/15 patients). When inulin fibre and heat-killed *L. plantarum* were supplied, the infection rate fell to 27% (4/15 patients), and in the group supplied with inulin fibre and live *L. plantarum*, the infection rate was only 13% (2/15 patients).

Uncommon cause of bacteraemia

Lactobacillus is rarely a human pathogen, but has been found in connection with various pathological conditions of which infective endocarditis is the most common (Husni *et al.*, 1997). However, the risk is small and only a total of 45 patients have been reported in the English-language literature during the last 50 years. Most, if not all, patients reported had underlying depressed immune functions and suffered from cancer or diabetes. They were almost all on therapy with immunosuppressive drugs or antibiotics or receiving total PN. *Lactobacillus* bacteraemia was reported recently in both a small number of liver transplant recipients (Patel *et al.*, 1994) and a few patients with AIDS (Horwitch *et al.*, 1995). Although 31 of the 45 patients, published in the world literature, died, only one death can so far be attributed to *Lactobacillus* bacteraemia. It is likely that Husni *et al.* (1997) are right in concluding that when 'lactobacillus bacteremia occurs, it serves as a marker of a serious underlying illness and poor long-term prognosis for hospitalised patients'.

Impact of antibiotics

The normal GI microflora is a remarkably stable ecosystem, but disease, hospitalization, drug treatment and irradiation are associated with dysbiotic changes. The most common cause of disturbance in the normal gut flora is the use of antibiotics. Both orally and parenterally applied antibiotics cause disturbances. Of the orally administered drugs, those with poor absorption have the most negative effects. Of parenterally administered antibiotics, mainly those which are secreted in saliva, bile and intestinal secretions will be negative to the flora. However, during an operation, there is usually no salivation or GI secretions, which is why one or two shots with antibiotics can most often be safely used.

Suppression of the normal flora creates a microbiological vacuum, which is readily filled by resistant micro-organisms normally excluded from this site (Cherbut *et al.*, 1991). For more than 15 years valuable information about antibiotics and their effects on flora has been obtained from the table of Van Saene *et al.* (1983) (Table 10). Gismondo (1998) has recently updated the information on the impact of antibiotics on intestinal microflora. Ampicillin is said to be incompletely absorbed and associated with a high incidence of diarrhoea. Piperacillin is excreted with the bile and overgrowth of enterobacteria and *B. fragilis* resistant to

Table 10. Influence of various antimicrobial agents on colonization resistance. Based on Van Saene *et al.*
(1983).

No or small reduction of the normal flora	Significant reduction of the normal flora
Cefaclor	Penicillins
Cefradine	Cephalosporins
Cefotaxime	Aminoglycosides
Ceftazidime	Erythromycin
Tobramycin	Tetracyclines
Co-trimoxazole	Vancomycin
Doxycycline	Chloramphenicol
Amphotericin B	Thioamphenicol
Nystatin	Clindamycin
Polymyxins	Lincomycin
Metronidazoles	
Nalidixic acid	
Trimethoprim	
Cinoxadin	

piperacillin has been observed during piperacillin treatment. However, no over-growth is observed when combined with tazobactam. Faecal excretion of imipenem and meropenem is known to result in no, or only minor, changes in the gut flora of patients receiving these two antibiotics. Significant changes in both the aerobic and anaerobic flora are observed after administration of nitroimidazoles (metronidazole, ornidazole, timidazole), previously regarded as safe for the flora.

Bacteriophages as 'antibiotics'

Bacteriophage therapy, i.e. harnessing of a specific kind of virus to specifically attack and kill pathogenic micro-organisms, is receiving a somewhat renewed interest. This concept was born about 100 years ago but clinical research and implementation of phage therapy have during the last 50 years been carried out mainly, or almost only, in Eastern Europe (for a review see Kutter, 1999). However, the new possibility of sequencing entire microbial genomes and determining the molecular basis for pathogenicity has no doubt created new possibilities for this therapy (Barrow & Soothill, 1997; Kutter, 1999). Viruses are known to have the ability to carry genetic material between susceptible cells and then reproduce those cells. Lytic phages are expected to infect the cell from the outside, reprogramme the host cell and release a burst of phage through breaking open or lysing the cell at relatively regular intervals (Kutter, 1999). Thus far, specific phages for over 100 bacterial genera have been isolated (Ackermann, 1996), but only a few have been well studied (Ackermann *et al.*, 1997). More extensive clinical experience with phage therapy seems only to have been reported from Poland and Georgia. A large multicentre study has been reported from Poland (Slopek *et al.*, 1987) in which phage therapy was tried following unsuccessful therapy with available antibiotics in a mixed group of 518 patients with long-lasting suppurative fistulas, septicaemia, abscesses, chest infections, purulent peritonitis and furunculosis. Full elimination of the suppurative process and healing of local wounds were reported in 84% of patients, and infants seemed to respond better than the elderly and patients with a deteriorated immune system and poor resistance to disease. So far, Western experience is very limited and restricted to studies in animals. Soothill (1992, 1994) investigated the ability of phage treatment to prevent infection in burned experimental animals. He observed an excellent protection against systemic infections with both *P. aeruginosa* and *Actinobacter* when appropriate phages were used, and also prevention of skin-graft rejection by pretreatment with phage against *P. aeruginosa*, concluding that as few as 100 phages protected against infection with 100 million bacteria—five times the LD_{50}.

Phage therapy is more in its infancy than probiotic treatment. However, it cannot be denied that it appears to have significant potential. It is reasonable to assume that new possibilities in molecular biology will open the door for better understanding and use of this interesting mode of treatment. Phage therapy seems to have certain advantages over conventional treatments such as antibiotics. As is pointed out by Kutter (1999), phage numbers have the ability to increase, while the concentrations of antibiotics in the body will quickly decrease from the moment of administration, which gives phage therapy not only the potential of curing but also of preventing disease.

A 'new world' – a new understanding of pathology

I was frustrated during most of my career as a surgeon with the high rate of septic complications globally observed, especially in more extensive surgery. Over more than 30 years I performed extensive surgery of the liver and the pancreas. To prevent infection, I, like most other surgeons at that time, used an antibiotic umbrella perioperatively and during the first post-operative days. On follow-up made some 15 years ago, we observed that for various reasons one-third of liver-resected patients had never received any prophylactic antibiotics. Although the infection rate in the total material was about 33%, no infections had occurred in the group of patients not receiving antibiotics. This made me interested in the role of the commensal flora and created a desire to find methods to refunctionalize the large intestine/to re-establish the 'microflora organ' in sick patients. At that time I had no idea that some 15 years later the use of probiotics should be regarded as a realistic alternative to antibiotics, nor could I foresee that one day probiotics would be considered interesting as a tool to combat environmental diseases such as diabetes, atherosclerosis and cancer. New insights into pathogenesis and recent developments in microbiology and immunology make it likely that some new tools based on pre- and probiotics will soon be available to control infections in clinical medicine and surgery, and also to prevent chronic diseases such as those related to environmental disease/metabolic syndrome.

References

Ackermann, H. W. (1996). Frequency of morphological phage descriptions in 1995. *Archives of Virology* 141, 209–218.

Ackermann, H. W., DuBow, M. S., Gershman, M., Karska-Wysocki, B., Kasatiya, S. S., Loessner, M. J., Mamet-Bratley, M. D. & Regue, M. (1997). Taxonomic changes in tailed phages of enterobacteria. *Archives of Virology* 142, 1381–1390.

Adawi, D., Kasravi, F. B., Molin, G. & Jeppsson, B. (1997). Effect of *Lactobacillus* supplementation with and without arginine on liver damage and bacterial translocation in an acute liver injury model in the rat. *Hepatology* 25, 642–647.

Adawi, D., Molin, G., Ahrné, S. & Jeppsson, B. (1999). Modulation of the colonic bacterial flora affects differently bacterial translocation and liver injury in acute liver injury model. *Microbial Ecology in Health and Disease* 11, 47–54.

Adlerberth, I., Carlsson, B., deMan, P. & 7 other authors (1991). Intestinal colonization with *Enterobacteriaceae* in Pakistani and Swedish hospital-delivered infants. *Acta Paediatrica Scandinavica* 80, 602–610.

Adlerberth, I., Ahrné, S., Johansson, M. L., Molin, G., Hansson, L. Å. & Wold, A. E. (1996). A mannose-specific adhesion mechanism in *Lactobacillus plantarum* conferring binding to the human colonic cell line HT-29. *Applied and Environmental Microbiology* 62, 2244–2251.

Ahrné, S., Nobaek, S., Jeppsson, B., Adlerberth, I., Wold, A. E. & Molin, G. (1998). The normal *Lactobacillus* flora in healthy human rectal and oral mucosa. *Journal of Applied Microbiology* 85, 88–94.

Andersson, R. (1985). Nitrate reduction during

fermentation by Gram-negative bacterial activity in carrots. *International Journal of Food Microbiology* 2, 219–225.

Ashenafi, M. & Busse, M. (1991). Growth of *Bacillus cereus* in fermenting tempeh made from various beans and its inhibition by *Lactobacillus plantarum. Journal of Applied Microbiology* 70, 329–333.

Barrow, P. A. & Soothill, J. S. (1997). Bacteriophage therapy and prophylaxis: rediscovery and renewed assessment of the potential. *Trends in Microbiology* 5, 268–271.

Barton, R. G. (1997). Immune-enhancing enteral formulas: are they beneficial in critically ill patients? *Nutrition in Clinical Practice* 12, 51–62.

Bates, D. W., Spell, N., Cullen, D. J., Burdick, E., Laird, N., Petersen, L. A., Small, S. D., Sweitzer, B. J., Leape, L. L. & the Adverse Drug Events Study Group (1997). The costs of adverse drug events in hospitalized patients. *Journal of the American Medical Association* 277, 307–311.

Bengmark, S. (1998a). Ecological control of the gastrointestinal tract. The role of probiotic bacteria. *Gut* 42, 2–7.

Bengmark, S. (1998b). Ecoimmunonutrition: a challenge for the third millennium. *Nutrition* 14, 563–572.

Bengmark, S. (1999). Gut microenvironment and immune function. *Current Opinion in Clinical Nutrition and Metabolic Care* 2, 83–85.

Bengmark, S. (2000a). Bioadhesive polymers that reduce adhesion formation. In *Peritoneal Surgery*. Edited by G. S. diZerega and others. New York: Springer (in press).

Bengmark, S. (2000b). Nutritional modulation of the acute phase response and immune functions. In *SIRS, MODS and MOF—*

Systemic Inflammatory Response Syndrome, Multiple Organ Dysfunction Syndrome, Multiple Organ Failure—Pathophysiology, Prevention and Therapy. Edited by A. E. Baue, E. Faist & D. Fry. New York: Springer (in press).

Bengmark, S. (2000c). Refunctionalization of the gut. In *SIRS, MODS and MOF—Systemic Inflammatory Response Syndrome, Multiple Organ Dysfunction Syndrome, Multiple Organ Failure—Pathophysiology, Prevention and Therapy.* Edited by A. E. Baue, E. Faist & D. Fry. New York: Springer (in press).

Biffl, W. L., Moore, E. E., Moore, F. A. & Barnett, C. C. (1996). Interleukin-6 delays neutrophil apoptosis via a mechanism involving platelet-activating factor. *Journal of Trauma Injury Infection and Critical Care* **40**, 575–579.

Björksten, B. (1994). Risk factors in early childhood for the development of atopic diseases. *Allergy* **49**, 400–407.

Böger, R. H., Bode-Böger, S. M., Phivthong-ngam, L., Brandes, R. P., Schwedhelm, E., Mügge, A., Böhme, M., Tsikas, D. & Frölich, J. C. (1998). Dietary L-arginine and α-tocopherol reduce vascular oxidative stress and preserve endothelial function in hypercholesterolemic rabbits via different mechanisms. *Arteriosclerosis* **141**, 31–43.

Brandtzaeg, P. (1995). Molecular and cellular aspects of the secretory immunoglobulin system. *APMIS* **103**, 1–19.

Brandtzaeg, P., Halstensen, T. S., Krajci, P., Kvale, D., Rognum, T. O., Scott, H. & Sollid, L. M. (1989). Immunobiology and immunopathology of human gut mucosa: humoral immunity and intraepithelial lymphocytes. *Gastroenterology* **97**, 1562–1584.

Chauvière, G., Barbat, A., Fourniat, J. & Servin, A. L. (1989). Adhesion of *Lactobacillus* onto cultured human enterocyte-like cell lines Caco-2 and HT-29. Comparison of human and non-human strains. In *Les Laits Fermentes. Actualité de la Recherche.* Edited by J. Libby. Montrouge: Eurotext.

Chen, H. & Ran, R. (1996). Rhubarb decoction prevents intestinal bacterial translocation during necrotic pancreatitis. *Journal of West China University of Medical Sciences* **27**, 418–421.

Cherbut, C., Ferre, J. P., Corpet, D. E.,

Ruckebusch, Y. & Deloort-Laval, J. (1991). Alterations in intestinal microflora by antibiotics. Effects on fecal excretion, transit time and colonic motility in rats. *Digestive Diseases and Sciences* **36**, 1729–1734.

Coconnier, M. H., Lievin, V., Hemery, E. & Servin, A. L. (1998). Antagonistic activity against *Helicobacter* infection in vitro and in vivo by the human *Lactobacillus acidophilus* strain LB. *Applied and Environmental Microbiology* **64**, 4573–4580.

De Simone, C., Vesely, R., Negri, R., Biachi-Salvadori, B., Cilli, A. & Lucci, L. (1987). Enhancement of immune response of murine Peyer's patches by a diet supplemented with yogurt. *Immunopharmacology and Immunotoxicology* **9**, 87–100.

Dickerson, R. N. (1997). Immune-enhancing enteral formulas in critically ill patients. *Nutrition in Clinical Practice* **12**, 49–50.

Diplock, A. T., Charleux, J.-I., Crozier-Willi, G., Kok, F. J., Rice-Evans, C., Roberfroid, M., Stahl, W. & Viña-Ribes, J. (1998). Functional food science and defence against oxidative species. *British Journal of Nutrition* **80** (suppl. 1), S77–S112.

Duncan, C., Dougall, H., Johnston, P., Green, S., Brogan, R., Leifert, C., Smith, L., Golden, M. & Benjamin, N. (1995). Chemical generation of nitric oxide in the mouth from enterosalivary circulation of dietary nitrate. *Nature Medicine* **1**, 546–551.

Dykhuizen, R. S., Fraser, A., McKenzie, H., Goilden, M., Leifert, C. & Benjamin, N. (1998). *Helicobacter pylori* is killed by nitrite under acidic conditions. *Gut* **42**, 334–337.

Eaton, S. B. & Konnor, M. (1985). Paleolithic nutrition. A consideration of its nature and current implications. *New England Journal of Medicine* **312**, 283–289.

Eaton, S. B., Eaton, S. B. III, Konnor, M. J. & Shostak, M. (1996). An evolutionary perspective enhances understanding of human nutritional requirements. *Journal of Nutrition* **126**, 1732–1740.

Eckmann, L., Kagnoff, M. F. & Fierer, J. (1995). Intestinal epithelial cells as watchdogs for the natural immune system. *Trends in Microbiology* **3**, 118–120.

Fabia, R., Ar´Rajab, A., Johansson, M. L., Willén,

R., Andersson, R., Molin, G. & Bengmark, S. (1993). The effect of exogenous administration of *Lactobacillus reuteri* R2LC and oat fiber on acetic acid-induced colitis in the rat. *Scandinavian Journal of Gastroenterology* 28, 155–162.

Felty, A. R. & Keefer, C. S. (1924). *Bacillus coli* sepsis: a clinical study of twenty eight cases of blood stream infection by the colon bacillus. *Journal of the American Medical Association* 82, 1430–1433.

Finegold, S. M., Sutter, V. L. & Mathisen, G. E. (1983). Normal indigenous intestinal flora. In *Human Intestinal Microflora in Health and Disease*, pp. 3–31. Edited by D. J. Hentges. London: Academic Press.

Frame, L. T., Hart, R. W. & Leakey, E. A. (1998). Caloric restriction as a mechanism mediating resistance to environmental disease. *Environmental Health Perspectives* 106 (suppl. 1), 313–324.

Frohn, M. J. (1976). Left-leg varicose veins and deep-vein thrombosis. *Lancet* 11, 1019–1020.

Gismondo, M. R. (1998). Antibiotic impact on intestinal microflora. *Gastroenterology International* 11 (suppl.), 29–30.

Goldin, B. R., Gorbach, S. L., Saxelin, M., Bakarat, S., Gualtieri, L. & Salminen, S. (1992). Survival of *Lactobacillus* species (strain GG) in human gastrointestinal tract. *Digestive Diseases and Sciences* 37, 121–128.

Grasl-Kraupp, B., Bursch, W., Ruttkay-Nedecky, B., Wagner, A., Lauer, B. & Schulte-Herman, R. (1994). Food restriction eliminates preneoplastic cells through apoptosis and antagonizes carcinogenesis in rat liver. *Proceedings of the National Academy of Sciences, USA* 91, 9995–9999.

Gustafsson, B. (1985). *The Future of Germ-Free Research*. New York: Alan R. Liss.

Hague, A., Manning, A. M., Hanlon, K. A., Huschscha, L. I., Hart, D. & Paraskeva, C. (1993). Sodium butyrate induces apoptosis in human colonic tumour cell lines in a p53-independent pathway: implications for the possible role of dietary fiber in the prevention of large-bowel cancer. *International Journal of Cancer* 55, 498–505.

Halme, S., Syrjala, H., Bloigu, A., Saikku, P., Leinonen, M., Airaksinen, J. & Sursel, H. M. (1997). Lymphocyte responses to *Chlamydia* antigens in patients with coronary heart disease. *European Heart Journal* 18, 1095–1101.

Hatcher, G. E. & Lamprecht, R. S. (1993). Augmentation of macrophage phagocytic activity by cell-free extracts of selected lactic acid-producing bacteria. *Journal of Dairy Science* 76, 2485–2492.

Hayatsu, H. & Hayatsu, T. (1993). Surprising effect of *Lactobacillus casei* administration on the urinary mutagenicity arising from ingestion of fried ground beef in the human. *Cancer Letters* 73, 173–179.

Heerdt, B. G., Houston, M. A. & Augenlicht, L. H. (1994). Potentiation by specific short-chain fatty acids of differentiation and apoptosis in human colonic carcinoma cell lines. *Cancer Research* 54, 3288–3294.

Henderson, B., Poole, S. & Wilson, M. (1996). Microbial/host interactions in health and disease: who controls the cytokine network? *Immunopharmacology* 35, 1–21.

Henderson, B., Wilson, M. & Wren, B. (1997). Are bacterial exotoxins cytokine network regulators? *Trends in Microbiology* 5, 454–458.

Holm, L. E., Nordevang, E., Hjalmar, M. L., Lidbrink, E., Callmer, E. & Nilsson, B. (1993). Treatment failure and dietary habits in woman with breast cancer. *Journal of the National Cancer Institute* 85, 32–36.

Hong, M. Y., Chang, W. C., Chapkin, R. S. & Lupton, J. R. (1997). Relationship among colonocyte proliferation, differentiation, and apoptosis as a function of diet and carcinogen. *Nutrition and Cancer* 28, 20–29.

Hornstra, G., Barth, C. A., Galli, C. & 7 other authors (1998). Functional food science and cardiovascular system. *British Journal of Nutrition* 80 (suppl. 1), S113–S146.

Horwitch, C. A., Furseth, H. A., Larsson, A. M., Jones, T. L., Oliffe, J. F. & Spach, D. H. (1995). Lactobacillemia in three patients with AIDS. *Clinical Infectious Diseases* 21, 1460–1461.

Husni, R. N., Gordon, S. M., Washington, J. A. & Longworth, D. L. (1997). Lactobacillus bacteremia and endocarditis: review of 45 cases. *Clinical Infectious Diseases* 25, 1048–1055.

Hybenová, E., Drdák, M., Gouth, R. & Gracák, J.

(1995). Utilisation of nitrates — a decisive criterion in the selection of lactobacilli for bioconservation of vegetables. *Zeitschrift für Lebensmittel-Untersuchung und Forschung* 200, 213–216.

Isenmann, R. & Büchler, M. W. (1994). Infection and acute pancreatitis. *British Journal of Surgery* 81, 1707–1708.

Jack, R. W., Tagg, J. R. & Ray, B. (1995). Bacteriocins of Gram-positive bacteria. *Microbiological Reviews* 59, 171–200.

Jain, M., Miller, A. B. & To, T. (1994). Premorbid diet and the prognosis of women with breast cancer. *Journal of the National Cancer Institute* 86, 1390–1397.

Johansson, M. L., Molin, G., Jeppsson, B., Nobaek, S., Ahrné, S. & Bengmark, S. (1993). Administration of different *Lactobacillus* strains in fermented oatmeal soup: in vivo colonization of human intestinal mucosa and effect on the indigenous flora. *Applied and Environmental Microbiology* 59, 15–20.

Kabir, A. M. A., Aiba, Y., Takagi, A., Kamiya, S., Miwa, T. & Koga, Y. (1997). Prevention of *Helicobacter pylori* infection by lactobacilli in a gnotobiotic murine model. *Gut* 41, 49–55.

Kagnoff, M. F. (1993). Immunology of the intestinal tract. *Gastroenterology* 105, 1275–1280.

Kaila, M., Isolauri, E., Soppi, E., Virtanen, E., Laine, S. & Arvilommi, H. (1992). Enhancement of the circulating antibody secreting cell response in human diarrhea by a human *Lactobacillus* strain. *Pediatric Research* 32, 141–144.

Kasravi, F. B., Adawi, D., Hägerstrand, I., Molin, G., Jeppsson, B. & Bengmark, S. (1996a). The effect of pretreatment with endotoxin and *Lactobacillus* on bacterial translocation in acute liver injury. *European Journal of Surgery* 162, 537–544.

Kasravi, F. B., Gebreselassie, D., Adawi, D., Wang, L. Q., Molin, G., Jeppsson, B. & Bengmark, S. (1996b). The effect of endotoxin and *Lactobacillus* pretreatment on peritoneal macrophage behavior in acute liver injury in rat. *Journal of Surgical Research* 62, 63–68.

Kasravi, F. B., Adawi, D., Molin, G., Bengmark, S. & Jeppsson, B. (1997). Effect of oral supplementation of *lactobacilli* on bacterial translocation in acute liver injury induced by D-galactosamine. *Journal of Hepatology* 26, 417–424.

Kato, I., Yokokura, T. & Mutai, M. (1988). Correlation between increase in Ia-bearing macrophages and induction of T-cell-dependent antitumour activity by *Lactobacillus casei* in mice. *Cancer, Immunology and Immunotherapy* 26, 215–221.

Key, F. B. & Mathers, J. C. (1995). Digestive adaptations of rats given white bread and cooked haricot beans (*Phaseolus vulgaris*): large bowel fermentation and digestion of complex carbohydrates. *British Journal of Nutrition* 74, 393–406.

Kilkullen, J. K., Ly, O. P., Chang, T. H., Levenson, S. M. & Steinberg, J. J. (1998). Nonviable *Staphylococcus aureus* and its peptidoglycan stimulate macrophage recruitment, angiogenesis, fibroplasia and collagen accumulation in wounded rats. *Wound Repair and Regeneration* 6, 149–156.

Kiyono, H. & McGhee, J. R. (1994). T helper cells for mucosal immune responses. In *Handbook of Mucosal Immunology*, pp. 263–274. Edited by P. L. Ogra, J. Mestecky, M. E. Lamm, W. Strober, J. R. McGhee & J. Bienenstock. Orlando, FL: Academic Press.

Knaus, W. A., Draper, E. A. & Wagner, D. P. (1985). APACHE II: a severity of disease classification system. *Critical Care Medicine* 13, 818–829.

Kutter, E. (1999). Phage therapy: bacteriophages as antibiotics. http://192.211.16.12/user/T4/PhageTherapy/Phagethea.html

Laurila, A., Bloigu, A., Nayha, S., Hassi, J., Leinonen, M. & Saikku, P. (1997). Chronic *Chlamydia pneumoniae* infection is associated with a serum lipid profile known to be a risk factor for atherosclerosis. *Arteriosclerosis, Thrombosis and Vascular Biology* 17, 2910–2913.

Laurila, A., Bloigu, A., Nayha, S., Hassi, J., Leinonen, M. & Saikku, P. (1999). Association of *Helicobacter pylori* infection with elevated serum lipids. *Atherosclerosis* 142, 207–210.

Lazarou, J., Pomeranz, B. H. & Corey, P. N.

(1998). Incidence of adverse drug reactions in hospitalized patients. *Journal of the American Medical Association* 279, 1200–1205.

Lencner, A. A., Lencner, C. P., Mikelsaar, M. E. & 7 other authors (1984). Die quantitative Zusammensetzung der Lactoflora des Verdauungstrakts vor und nach kosmischen Flügen unterschiedlicher Dauer. *Nahrung* 28, 607–613.

Lin, B. F., Huang, C. C., Chiang, B. L. & Jeng, S. J. (1996). Dietary fat influences Ia antigen expression, cytokines and prostaglandin E_2 production in immune cells in autoimmune-prone NZB × NZW F1 mice. *British Journal of Nutrition* 75, 711–722.

Lorber, B. (1996). Are all diseases infectious? *Annals of Internal Medicine* 125, 844–851.

Lucey, D. R., Clerici, M. & Shearer, G. M. (1996). Type 1 and type 2 cytokine dysregulation in human infectious, neoplastic and inflammatory diseases. *Clinical Microbiology Reviews* 9, 532–562.

Mack, D. R., Michail, S., Wei, S., McDougall, L. & Hollingsworth, M. A. (1999). Probiotics inhibit enteropathogenic *E. coli* adherence in vitro by inducing intestinal mucin gene expression. *American Journal of Physiology (Gastrointestinal and Liver Physiology 39)* 276, G941–G950.

McKnight, G. M., Smith, L. M., Drummond, R. S., Duncan, C. W., Golden, M. & Benjamin, N. (1997). Chemical synthesis of nitric oxide in the stomach from dietary nitrate in humans. *Gut* 40, 241–244.

Mainous, M. R., Lipsett, P. A., O'Brien, M. & the Johns Hopkins SICU Study Group (1997). Enterococcal bacteremia in the surgical intensive care unit. Does vancomycin resistance affect mortality? *Archives of Surgery* 132, 76–81.

Malhotra, S. L. (1968). Studies in blood coagulation, diet and ischemic heart disease in two population groups in India. *British Heart Journal* 30, 303–308.

Malin, M., Suomalainen, H., Saxelin, M. & Isolauri, E. (1996). Promotion of IgA immune response in patients with Crohn's disease by oral bacteriotherapy with *Lactobacillus* GG. *Annals of Nutrition and Metabolism* 40, 137–145.

Mao, Y., Nobaek, S., Kasravi, B., Adawi, D., Stenram, U., Molin, G. & Jeppsson, B. (1996a). The effects of *Lactobacillus* strains and oat fiber on methotrexate-induced enterocolitis in rats. *Gastroenterology* 111, 334–344.

Mao, Y., Kasravi, B., Nobaek, S. & 7 other authors (1996b). Pectin-supplemented enteral diet reduces the severity of methotrexate-induced enterocolitis in rats. *Scandinavian Journal of Gastroenterology* 31, 558–567.

Mao, Y., Yu, J. L., Ljungh, Å., Molin, G. & Jeppsson, B. (1997). Intestinal immune response to oral administration of *Lactobacillus reuteri* R2LC, *Lactobacillus plantarum* DSM 9843, pectin and oatbase on methotrexate-induced enterocolitis in rats. *Microbial Ecology in Health and Disease* 9, 261–270.

Marchetti, M. C., Migliorati, G., Moraca, R., Riccardi, C., Nicoletti, I., Fabiani, R., Mastrandrea, V. & Morozzi, G. (1997). Possible mechanisms involved in apoptosis of colon tumor cell lines induced by deoxycholic acid, short-chain fatty acids, and their mixtures. *Nutrition and Cancer* 28, 74–80.

Marin, M. L., Tejada-Simon, M. V., Lee, J. H., Murtha, J., Ustunol, Z. & Pestka, J. J. (1998). Stimulation of cytokine production in clonal macrophage and T-cell models by *Streptococcus thermophilus*: comparison with *Bifidobacterium* sp. and *Lactobacillus bulgaricus*. *Journal of Food Protection* 61, 859–864.

Matsuzaki, T. (1998). Immunomodulation by treatment with *Lactobacillus casei* Shirota. *International Journal of Food Microbiology* 41, 133–140.

Matsuzaki, T., Nagata, Y., Kado, S., Uchida, K., Hashimoto, S. & Yokokura, T. (1997). Effect of oral administration of *Lactobacillus casei* on alloxan-induced diabetes in mice. *APMIS* 105, 637–642.

Metchnikoff, E. (1907). *The Prolongation of Life. Optimistic Studies*. London: William Heinemann.

Meyer, T. A., Noguchi, Y., Ogle, C. K., Tiao, G., Wang, J. J., Fischer, J. E. & Hasselgren, P. O. (1994). Endotoxin stimulates interleukin-6 production in intestinal epithelial cells. A synergistic effect with prostaglandin E_2. *Archives of Surgery* 129, 1294–1295.

Meyer, T. A., Tiao, G. M., James, J. H., Noguchi, Y., Ogle, C. K., Fischer, J. E. & Hasselgren, P. O. (1995). Nitric oxide inhibits LPS-induced IL-6 production in enterocytes. *Journal of Surgical Research* 58, 570–575.

Miettinen, M., Matikainen, S., Vuopio-Varkila, J., Pirhonen, J., Varkila, K., Kurimoto, M. & Julkunen, I. (1998a). Lactobacilli and streptococci induce interleukin-12 (IL-12), IL-18, and gamma interferon production in human peripheral blood mononuclear cells. *Infection and Immunity* 66, 6058–6060.

Miettinen, M., Alander, M., von Wright, A., Vuopio-Varkila, J., Marteau, P., Huis in't Veld, J. & Mattila-Sandholm, T. (1998b). The survival of and cytokine induction by lactic acid bacteria after passage through a gastrointestinal model. *Microbial Ecology in Health and Disease* 10, 141–147.

Milton, K. (1999). Nutritional characteristics of wild primate foods: do the diets of our closest living relatives have lessons for us? *Nutrition 6*, 488–498.

Mohan, I. K. & Das, U. N. (1998). Effect of L-arginine-nitric oxide system on chemical-induced diabetes mellitus. *Free Radical Biology & Medicine* 8, 757–765.

Müller, M. & Lier, D. (1994). Fermentation of fructans by epiphytic lactic acid bacteria. *Journal of Applied Bacteriology* 76, 406–411.

Murosaki, S., Yamamoto, Y., Ito, K., Inokuchi, T., Kusaka, H., Ikeda, H. & Yoshikai, Y. (1998). Heat-killed *Lactobacillus plantarum* L-137 suppresses naturally fed antigen-specific IgE production by stimulation of IL-12 production in mice. *Journal of Allergy and Clinical Immunology* 102, 57–64.

Ogle, C. K., Guo, X., Hasselgren, P. O., Ogle, J. D. & Alexander, J. W. (1997). The gut as a source of inflammatory cytokines after stimulation with endotoxin. *European Journal of Surgery* 163, 45–51.

Olasupo, N. A., Olukoya, D. K. & Odunfa, S. A. (1995). Studies on bacteriocinogenic *Lactobacillus* isolates from selected Nigerian fermented foods. *Journal of Basic Microbiology* 35, 319–324.

Ornish, D., Scherwitz, L. W., Billings, J. H. & 8 other authors (1998). Intensive lifestyle changes for reversal of coronary heart disease. *Journal of the American Medical Association* 280, 2001–2007.

Outwater, J. L., Nicholson, A. & Barnard, N. (1997). Dairy products and breast cancer: the IGF-1, estrogen and bGH hypothesis. *Medical Hypotheses* 48, 453–461.

Pasteur, L. & Joubert, J. F. (1877). Charbon et septicémie. *C.R. Society Biology Paris* 85, 101–115.

Patel, R., Cockerill, F. R., Porayko, M. K., Osmon, D. R., Ilstrup, D. M. & Keating, M. R. (1994). Lactobacillemia in liver transplant patients. *Clinical Infectious Diseases* 18, 207–212.

Peck, M. D., Alexander, J. W., Ogla, C. K. & Babcock, G. F. (1990). The effect of dietary fatty acids in response to *Pseudomonas* infection in burned mice. *Journal of Trauma* 30, 445–452.

Petrie, J. R., Ueda, S., Webb, D. J., Elliott, H. L. & Cornell, J. M. C. (1996). Endothelial nitric oxide production and insulin sensitivity. *Circulation* 93, 1331–1333.

Pollard, T. M. (1997). Environmental changes and cardiovascular disease. *Yearbook of Physical Anthropology* 40, 1–24.

Pulverer, G., Ko, H. L., Roszkowski, W., Beuth, J., Yassin, A. & Jeljaszewics, J. (1990). Digestive tract microflora liberates low molecular weight peptides with immunotriggering activity. *Zentralblatt für Bakteriologie* 272, 318–327.

Rangavajhyala, N., Shahani, K. M., Sridevi, G. & Srikumaran, S. (1997). Nonlipopolysaccharide component(s) of *Lactobacillus acidophilus* stimulate(s) the production of interleukin 1α and tumour necrosis factor-α by murine macrophages. *Nutrition and Cancer* 28, 130–134.

Rayes, N., Hansen, S., Müller, A. R., Bechstein, W. D., Bengmark, S. & Neuhaus, P. (1999). SBD versus fibre containing enteral nutrition plus *Lactobacillus* or placebo to prevent bacterial infections after liver transplantation. In *Abstracts of the Meeting of the European Society of Transplantation*.

Reaven, G. M. (1995). Pathophysiology of insulin resistance in human disease. *Pathophysiological Reviews* 75, 473–486.

Reddy, B. S., Hamid, R. & Rao, C. V. (1997). Effect of dietary oligofructose and inulin on colonic preneoplastic aberrant crypt foci inhibition. *Carcinogenesis* 18, 1371–1374.

Roszkowski, K., Ko, K. L., Beuth, J., Ohshima, Y., Roszkowski, W., Jeljaszewics, J. & Pulverer, G. (1988). Intestinal microflora of BALB/c-mice and function of local immune cells. *Zentralblatt für Bakteriologie und Hygiene* 270, 270–279.

Saikku, P. (1997). *Chlamydia pneumoniae* and atherosclerosis—an update. *Scandinavian Journal of Infectious Diseases* 104 (suppl.), 53–56.

Salminen, S., Bouley, C., Boutron-Ruault, M.-C. & 7 other authors (1998). Functional food science and gastrointestinal physiology and function. *British Journal of Nutrition* 80 (suppl. 1), S147–S171.

Saris, W. H. M., Asp, N. G. L., Björck, I. & 8 other authors (1998). Functional food science and substrate metabolism. *British Journal of Nutrition* 80 (suppl. 1), S47–S75.

Sautner, T., Függer, R., Götzinger, P., Mittlböck, M., Winkler, S., Roth, E., Steininger, S. & Mühlbacher, F. (1995). Tumour necrosis factor-α and interleukin-6: early indicators of bacterial infection after human orthotopic liver transplantation. *European Journal of Surgery* 161, 97–101.

Schiffrin, E. J., Brassart, D., Servin, A. L., Rochat, F. & Donnet-Hughes, A. (1997). Immune modulation of blood leukocytes in humans by lactic acid bacteria: criteria for strain selection. *American Journal of Clinical Nutrition* 66, 515S–520S.

Schroeder, H. A. (1971). Losses of vitamins and trace minerals resulting from processing and preservation of foods. *American Journal of Clinical Nutrition* 24, 562–573.

Sepp, E., Julge, K., Vasur, M., Naaber, P., Björksten, B. & Mikelsaar, M. (1997). Intestinal microflora of Estonian and Swedish infants. *Acta Paediatrica* 86, 956–961.

Shida, K., Makino, K., Takamizawa, K., Hachimura, S., Ametani, A., Sato, T., Kumagai, Y., Habu, S. & Kaminogawa, S. (1998). *Lactobacillus casei* inhibits antigen-induced IgE secretion through regulation of cytokine production in murine splenocyte cultures.

International Archives of Allergy and Immunology 115, 278–287.

Shirabe, K., Matsumata, T., Shimada, M., Takenaka, K., Kawahara, N., Yamamoto, K., Nishizaki, T. & Sugimachi, K. (1997). A comparison of parenteral hyperalimentation and early enteral feeding regarding systemic immunity after major hepatic resection—the results of a randomized prospective study. *Hepato-Gastroenterology* 44, 205–209.

Skog, K. (1993). Cooking procedures and food mutagens: a literature review. *Food Chemistry and Toxicology* 9, 655–675.

Slopek, S., Weber-Dabrowska, B., Dabrowski, M. & Kucharewica-Krukowska, A. (1987). Results of bacteriophage treatment of suppurative bacterial infections in the years 1981-1986. *Archivum Immunologiae et Therapiae Experimentalis* 35, 563–568.

Smith, A. J., Benjamin, N., Weetman, D. A., Mackenzie, D. & MacFarlane, T. W. (1999). The microbial generation of nitric oxide in the human oral cavity. *Microbial Ecology in Health and Disease* 11, 23–27.

Solary, E., Dubrez, L. & Eymin, B. (1996). The role of apoptosis in the pathogenesis and treatment of diseases. *European Respiratory Journal* 9, 1293–1305.

Solis-Pereyra, B., Aattouri, N. & Lemonnier, D. (1997). Role of food in the stimulation of cytokine production. *American Journal of Clinical Nutrition* 66, 521S–525S.

Soothill, J. S. (1992). Treatment of experimental infections of mice with bacteriophages. *Journal of Medical Microbiology* 37, 258–261.

Soothill, J. S. (1994). Bacteriophage prevents destruction of skin grafts by *Pseudomonas aeruginosa*. *Burns* 20, 209–211.

Spera, R. V., Jr & Farber, B. F. (1994). Multidrug-resistant *Enterococcus faecium*. An untreatable nosocomial pathogen. *Drugs* 48, 678–688.

Stanier, R. Y., Doudoroff, M. & Adelberg, E. A. (1963). The beginnings of microbiology. In *The Microbial World*, 2nd edn, pp. 3–28. Edited by R. Y. Stanier, M. Doudoroff & E. A. Adelberg. New Jersey: Englewood Cliffs.

Swartz, N. N. (1994). Hospital-acquired infections: diseases with increasingly limited therapies. *Proceedings of the National Academy of Sciences, USA* 91, 2420–2427.

Tannock, G. W. (1997). Probiotic properties of lactic-acid bacteria: plenty of scope for fundamental R&D. *Trends in Biotechnology* 15, 270–274.

Van der Poll, T., Coyle, S. M., Levi, M., Boermeester, M. A., Braxton, C. C., Jansen, P. M., Hack, C. E. & Lowry, S. F. (1996). Fat emulsion infusion potentiates coagulation activation during human endotoxemia. *Thrombosis and Haemostasis* 75, 83–86.

Van der Poll, T., Levi, M., Braxton, C. C., Coyle, S. M., ten Cate, J. W. & Lowry, S. F. (1998). Parenteral nutrition facilitates activation of coagulation but not fibrinolysis during human endotoxemia. *Journal of Infectious Diseases* 177, 793–795.

Van Saene, H. K. F., Stoutenbeck, C. P., Miranda, D. R. & Zandstra, D. F. (1983). A novel approach to infection control in the intensive care unit. *Acta Anaesthesiologica Belgica* 3, 193–208.

Van Thiel, D. H., Finkel, R., Friedlander, L., Gaveler, J. S., Wright, H. I. & Gordon, R. (1992). The association of IgA deficiency but not IgG or IgM deficiency with a reduced patient and graft survival following liver transplantation. *Transplantation* 54, 269–273.

Wells, C. L., Maddaus, M. A. & Simmons, R. L. (1988). Proposed mechanisms for translocation of intestinal bacteria. *Reviews of Infectious Diseases* 10, 958–977.

Westin, J. B. & Richter, E. (1990). The Israeli breast-cancer anomaly. *Annals of the New York Academy of Sciences* 609, 269–279.

Whitehead, T. P., Thorpe, G. H. G. & Maxwell, S. R. J. (1992). Enhancement chemiluminescence assay for antioxidant capacity. *Analyses Chemica Acta* 266, 265–277.

Wierzbicka, G. T., Hagen, T. M. & Jones, D. P. (1989). Glutathione in food. *Journal of Food Composition and Analysis* 2, 327–337.

Windsor, A. C. J., Kanwar, S., Li, A. G. K., Barnes, E., Guthrie, J. A., Spark, J. I. & Welsh, F. (1998). Compared with parenteral nutrition, enteral feeding attenuates the acute phase response, and improves disease severity in acute pancreatitis. *Gut* 42, 431–435.

World Health Organization Scientific Working Group on Monitoring and Management of Bacterial Resistance to Antimicrobial Agents (1994). *World Health Organization; Bacterial, Viral Diseases and Immunology.* Geneva WHO/CDS/BVI/95.7.

Yasui, H., Nagaoka, N., Mike, A., Hayakawa, K. & Ohwaki, M. (1992). Detection of *Bifidobacterium* strains that induce large quantities of IgA. *Microbial Ecology in Health and Disease* 5, 155–162.

Zhang, X. M., Stamp, D., Minkin, S., Medline, A., Corpet, D. E., Bruce, W. R. & Archer, M. C. (1992). Promotion of aberrant crypt foci and cancer in rat colon by thermolyzed protein. *Journal of the National Cancer Institute* 84, 1026–1030.

Vaccine production in plants

Garry C. Whitelam

Department of Biology, University of Leicester, University Road, Leicester LE1 7RH, UK

Introduction

Vaccination is established as one of the safest and perhaps most cost-effective medical interventions for disease prevention. Within the last 20 years the application of biotechnological methods has led to the development of several new advances in vaccine technology. In particular, non-replicating subunit vaccines have been developed as alternatives to more conventional vaccines based on killed pathogens. Subunit vaccines may range in complexity from single epitopes fused to carrier proteins through to assembled complexes of several subunits. Since they are based on non-infectious molecules, such as proteins or peptides, subunit vaccines are safe. The ideal subunit vaccine would be effective, inexpensive to produce, easy to store and transport and orally applicable. Although oral subunit vaccines have the advantage of a very simple means of administration, most orally administered vaccines are required in relatively high amounts compared with parenteral vaccines. This is partly because they will be exposed to the acidic and proteolytic environment of the alimentary tract. Therefore, the production of systems that will provide commercially acceptable preparations of oral vaccines is a challenge. This will be especially true in poor countries, where the prohibitive cost of vaccines, together with the lack of a suitable health care infrastructure, have prevented the implementation of appropriate mass immunization programmes. Global vaccination programmes will require changes to be made in the way that vaccines are produced, distributed and administered.

In order to meet some of these challenges, Arntzen and co-workers (see Arntzen, 1998) introduced the concept of transgenic plants as production and delivery systems for subunit vaccines. This concept is based upon the generation of plants that have the vaccine gene of interest stably integrated into the genome. This approach is quite distinct from the use of plant viruses to carry and express vaccine genes in infected plants (see below). There are several attractions to the use of stable transgenic plants for vaccine or therapeutic protein production. Firstly, production could be on an agricultural scale, allowing for the development of large-scale, low-cost production. Since plants are one of the cheapest sources of proteins, they are also potentially a cheap source of recombinant proteins. The generation of transgenic plants is now routine and many millions of acres of transgenic food and non-food crops are grown throughout the world. Secondly, the seeds of stable transgenic lines could provide a simple means of distributing the transgenic crop throughout the world, where it could be grown within the existing agricultural infrastructure. For oral vaccines, the use of edible transgenic crops could greatly simplify administration. An additional advantage of the use of transgenic plants for the production of subunit vaccine antigens is the lack of contamination with animal pathogens or animal products. These advantages are also being exploited in the development of transgenic plants for the large-scale production of recombinant antibodies for use in topical passive immunotherapy.

Production of oral vaccines in transgenic plants

The pathogens that cause diarrhoea were the first targets for edible vaccines produced in transgenic plants. Each year, diarrhoea, caused by either bacterial or viral pathogens, is responsible for many millions of infant deaths, mostly in developing countries. Effective vaccines directed against enterotoxigenic *Escherichia coli* (ETEC) and cholera, the major causes of acute watery diarrhoea, are highly desirable. Both ETEC and *Vibrio cholerae* colonize the small intestine and produce enterotoxins, including *E. coli* heat-labile toxin (LT) and cholera toxin (CT), with related structures and similar functions. To be effective against these enteric pathogens, a vaccine would need to stimulate the mucosal immune system and lead to the production of secretory IgA. These immunoglobulins are found in the mucosal secretions of the intestines, as well as in saliva and respiratory and reproductive tract secretions, and are a major indicator of mucosal immunity. Stimulation of the mucosal immune system is better achieved with an oral vaccine than with a parenteral vaccine. Previously, killed or attenuated vaccines of *V. cholerae* have been shown to be effective oral immunogens (Holmgren *et al.*, 1992) and recombinant CT B subunit (CT-B), produced in bacteria or yeast, has also been used successfully (Clemens *et al.*, 1990; Lebens *et al.*, 1993; Schonberger, Hirst & Pines, 1991).

Haq *et al.* (1995) created plant expression vectors containing the gene for the *E. coli* LT B subunit (LT-B), and used them to transform tobacco and potato plants. A range of individual transformed plants were selected and assayed for accumulation of LT-B. The highest levels of antigen accumulation were observed for plants that had been transformed with a construct that included an endoplasmic reticulum retention signal. To determine whether the plant-produced antigens were orally immunogenic, mice were given crude extracts of transgenic tobacco leaves, containing 12·5 μg antigen, by gavage. The serum and mucosal antibody responses of these animals were compared with a control group of animals that had been given the same dose of the purified LT-B from *E. coli*. Amounts of serum antibodies were similar in the two groups, indicating that the plant-derived antigens were immunogenic. Furthermore, antibodies to LT-B were detected in mucosal samples from mice immunized with the plant extracts. Mucosal IgA, specific for LT-B, was also induced in mice that were fed on a single 5 g dose of tuber from transgenic potato plants. This latter observation establishes the feasibility of using transgenic edible plants for both expression and delivery of oral vaccines.

Subsequently, Mason *et al.* (1998) designed an optimized, synthetic LT-B gene for plant expression. By using an LT-B gene with a codon usage optimized for plants, increased expression of LT-B was achieved. Furthermore, feeding of mice with raw tuber tissue demonstrated immunogenicity of the plant-produced, modified LT-B, as indicated by the presence of serum IgG and faecal IgA responses. When the orally immunized mice were challenged by oral administration of LT, they showed significant protection against toxin challenge. Recently, Tacket *et al.* (1998) have shown in preclinical trials involving 14 healthy adult volunteers that ingestion of a single dose of transgenic potato expressing LT-B led to the presence of gut-derived antibody secretory cells within 7–10 d. In addition, 10 of the 11 volunteers who ingested transgenic potato also showed a rise in anti-LT IgG, and 6 of the 11 volunteers also developed a rise in anti-LT IgA. These results establish the feasibility of using plant-derived oral subunit vaccines for human immunization.

In a similar way, potato has been transformed with gene sequences encoding the CT-B subunit (Arakawa, Chong & Langridge, 1998). Both serum and mucosal antibodies with specificity to CT-B were induced in mice following oral immunization. Furthermore, serum derived from mice fed with the transgenic potato protected Vero cells from CT-B toxicity and immunized mice showed significant reductions in CT-B-induced diarrhoea compared with mice fed on the transformed potato. This study confirms the use of plant-derived subunit vaccines for protection against enteric infections. Also, because CT-B is known to function as a carrier molecule for conjugated peptides for the induction of immunological tolerance, it also opens the possibility for development of food-plant-based therapies for autoimmune diseases.

Transgenic plants have also been used for the production of an oral vaccine against Norwalk virus (NV), which causes epidemic, acute gastroenteritis in humans (Mason *et al.*, 1996). In this instance, transgenic tobacco and potato plants expressing the capsid protein of Norwalk virus (NVCP) were created. The NVCP was found to self-assemble into non-replicating virus-like particles that were orally immunogenic in mice. Extracts of transgenic tobacco leaf tissue, when administered to mice by gavage, led to the development of both serum IgG and secretory IgA with specificity for the recombinant NV. Furthermore, when mice were fed directly with uncooked potato tuber from transgenic plants they also developed a serum IgG response. However, for mice fed with potato tuber the serum titres were significantly lower than those of mice dosed with transgenic tobacco leaf, and only 1 of 11 mice showing a serum response also produced detectable intestinal IgA.

More recently, transgenic alfalfa, a forage plant, has been used for the production of an effective oral and parenteral vaccine against the foot-and-mouth disease virus (FMDV) (Wigdorovitz *et al.*, 1999). FMDV is the causative agent of economically important diseases affecting several meat-producing animals. In this case, plants were transformed with a gene encoding the structural protein VP1 of FMDV. The VP1 protein is known to carry epitopes responsible for the induction of neutralizing antibodies (Carrillo *et al.*, 1998). Transgenic alfalfa plants expressing the VP1 gene were detected by immunoassay using anti-FMDV serum and mice were either immunized intraperitoneally with leaf extracts from transgenic or control plants, or they were fed with freshly harvested transgenic or control leaves. Following inoculation, mice were bled and sera were shown to contain specific anti-FMDV antibodies. Mice that had been fed with leaf tissue from transgenic plants also developed serum antibodies against FMDV. No anti-FMDV antibodies were detected from mice immunized with non-transgenic alfalfa. Orally and intraperitoneally immunized mice were experimentally challenged with FMDV and between 66 and 80% of immunized animals showed protection.

The oral immunogenicity of plant-derived VP1 represents a significant finding. Prior to this study, all of the plant-derived antigens that had been shown to be immunogenic following oral administration were derived from enteric pathogens. These antigens are therefore likely to survive in the harsh environment of the gastrointestinal tract. Furthermore, *E. coli* LT and CT are two of the most powerful oral immunogens identified. These toxins are resistant to proteolysis and denaturation and display strong adjuvant activity. The work with FMDV demonstrates that oral immunization can also be successfully employed using antigens from non-enteric pathogens and for which a systemic immune response is required. This finding is all the more surprising given that it has been considered unlikely that the recombinant VP1 protein would be able to assemble into virus-like particles (Mor, Gomez-Lim & Palmer, 1998).

Production of parenteral vaccines in transgenic plants

The pioneering work of Arntzen and colleagues on the development of transgenic plants for the production of subunit vaccines began with the use of transgenic plants for the synthesis of parenteral vaccines. Mason, Lam & Arntzen (1992) generated transgenic tobacco plants expressing a gene encoding the hepatitis B virus (HBV) surface antigen. HBV is regarded as the most important cause of persistent viraemia in humans and an effective recombinant vaccine had been developed following expression of HBV surface antigen in yeast (Emini *et al.*, 1986). Upon expression in tobacco, the recombinant HBV surface antigen was found to assemble into subviral particles that were indistinguishable from the yeast-derived HBV surface antigen. Subsequently, Thanavala *et al.* (1995) purified HBV surface antigen from transgenic tobacco plants and used it to immunize mice by intraperitoneal injection. The anti-HBV immunoglobulin response of the mice was found to be qualitatively similar to that obtained by immunization of mice with the commercial yeast-derived vaccine.

Transgenic plants have also been developed for the production of parenteral vaccines with veterinary applications. Gomez *et al.* (1998) generated transgenic *Arabidopsis* plants expressing two versions of the glycoprotein S from the swine-transmissible gastroenteritis coronavirus. When mice were immunized with leaf extracts from transgenic plants they subsequently developed a specific serum antibody response.

The development of plant-based systems for the production of parenteral vaccines for use in the clinic will require the development of efficient procedures for antigen isolation and purification. In the experimental examples cited above, crude extracts from transgenic plants have been used to immunize experimental animals. Whilst plant production systems avoid the possibility of vaccine contamination with animal pathogens or other animal-derived agents, there will still be safety issues associated with the production of parenteral vaccines in transgenic plants. In particular, plant secondary metabolites, many of which have biological activity, are a possible cause of concern.

Production of 'plantibodies' for passive immunization

Topical passive immunization for the prevention of infectious diseases is an attractive concept. It is well known that antibodies delivered in milk protect the entire gastrointestinal tract of infants (lactogenic immunity) and that experimental topical application of antibodies can prevent infections of the gastrointestinal, respiratory, vaginal and rectal mucosa (Zeitlin *et al.*, 1998). Plant-derived antibodies, 'plantibodies', with the potential for high-capacity production at relatively low cost, may mean that topical immune therapy is an economically feasible strategy for both human and veterinary health care.

Ma *et al.* (1998) described the use of plant-derived secretory IgA antibodies for the topical immunotherapy of dental caries. Previously, it had been shown that a murine monoclonal IgG directed against the cell surface adhesion protein of *Streptococcus mutans* could specifically prevent bacterial colonization in the oral cavity of humans and non-human primates (Ma, Smith & Lehner, 1987). Ma *et al.* (1998) used transgenic tobacco plants to produce a secretory form of the same monoclonal antibody. A secretory IgA was considered to offer several advantages over IgG, in terms of increased resistance to proteolytic degradation and increased avidity. To create a suitable antibody, α chain domains of an IgA monoclonal antibody were engineered into the constant region of the monoclonal IgG. Then, using a sexual crossing technique, a single transgenic plant was created that expressed four transgenes: the immunoglobulin heavy chain gene, the immunoglobulin light chain gene, a J chain gene and a secretory component gene. It was shown that the chimaeric, secretory IgA/G was able to dimerize with J chain and assemble with secretory component. The affinity constants of the parental monoclonal IgG and the plant-derived secretory IgA/G were found to be very similar. However, in the human oral cavity, the plantibody was found to survive for 3 d compared with only 1 d for the monoclonal IgG. Furthermore, the purified plantibody was shown to afford specific protection in human volunteers against oral streptococcal colonization for a period of at least 4 months. In this study, it was determined that effective topical passive immunotherapy required relatively large amounts of antibody, in the order of 25 mg antibody being used for each course of treatment. The use of a plant expression system was readily able to satisfy this requirement. It is speculated that a future development may be the production of antibodies in edible transgenic plant material, thus eliminating the need for plantibody purification. There are several examples of successful passive immunization through the administration of antibody produced in foodstuffs. For example, Hamada *et al.* (1991) protected against *S. mutans* colonization using antibodies from the yolk of hen eggs.

Zeitlin *et al.* (1998) described another example of the use of a plant-derived antibody for topical immunotherapy. In this study, transgenic tobacco and soybean plants were used to produce a humanized murine monoclonal antibody directed against glycoprotein B of herpes simplex virus (HSV). The plant-derived antibody was found to afford protection against experimental challenge with the virus in a mouse model of the vaginal transmission of HSV. Since HSV infections are incurable, prevention is of particular importance.

Production of vaccines using plant viruses

A large number of recent studies have established that plant viruses can be effective tools for the production and delivery of antigens. In addition to the use of non-replicating virus-like particles as subunit vaccines, virus particles are known to be particularly effective carriers of antigenic determinants. The capsid proteins of several plant viruses are amenable to genetic modification, allowing foreign polypeptides to be inserted in their surface-exposed loops or at their C-termini (Johnson, Lin & Lomonossoff, 1997). Significantly, plant viruses modified in this way often grow as efficiently as wild-type virus and so can provide significant quantities of modified virus particle. The use of plant viruses for this purpose offers several advantages. In addition to the low cost of production, plant viruses are recognized to be non-pathogenic to humans and other animals.

A range of plant viruses have been used for epitope display (Table 1). The first

Table 1. Examples of subunit vaccines displayed on plant viruses.

Plant virus	Epitope displayed	Reference
1 Cowpea mosaic virus	HIV-1 gp41 peptide	McLain *et al.* (1996)
	Mink enteritis virus epitope	Dalsgaard *et al.* (1997)
	Canine parvovirus peptide	Meloen *et al.* (1998)
2 Tobacco mosaic virus	Malarial epitopes	Turpen *et al.* (1995)
	Murine zona pellucida peptide	Fitchen, Beachy & Hein (1995)
3 Alfalfa mosaic virus	Rabies virus and HIV peptide	Yusibov *et al.* (1997)

reported example involves exploiting the self-assembling properties of the tobacco mosaic virus (TMV) coat protein (Haynes *et al.*, 1986). A poliovirus epitope was fused to the surface of the TMV coat protein, which was produced in the form of virus-like polymers in *E. coli* cells. The polymers were found to be immunogenic in rats.

The majority of examples involve the purification of the virus-like particles from infected plants and their use as parenteral vaccines. However, recently a few examples of the use of plant viruses for the production of oral or intra-nasal vaccines have been described. For example, Modelska *et al.* (1998) used alfalfa mosaic virus to display peptides representing rabies virus (RV) epitopes. Mice immunized intraperitoneally or orally with the modified virus-like particles isolated from infected tobacco or spinach plants are able to mount a local and systemic immune response against RV. Significantly, following oral administration of the antigen, in the absence of adjuvants, the synthesis of both IgG and IgA was stimulated and the immunized mice were protected against experimental challenge with RV. Durrani *et al.* (1998) described the use of cowpea mosaic virus to express a 22-amino-acid peptide from the transmembrane protein gp41 of human immunodeficiency virus 1 (HIV-1). Purified virus particles, given in conjunction with the widely used mucosal adjuvant CT, were shown to induce an HIV-specific mucosal and serum antibody response following intra-nasal immunization.

Recently, a plant virus has been used in a novel way to produce a specific vaccine for lymphoma. Since each clone of malignant B cells expresses a unique cell surface immunoglobulin, this can be used as a tumour-specific marker for the development of patient-specific cancer vaccines. It has been shown that the tumour surface immunoglobulin, if rendered immunogenic by conjugation to a carrier, can be used to effectively vaccinate patients in chemotherapy-induced remission. Vaccines based upon a single chain Fv (scFv) protein are also capable of eliciting anti-idiotype-specific responses and blocking tumour progression in mouse models of lymphoma (Hakim, Levy & Levy, 1996). McCormick *et al.* (1999) used a viral vector to transiently express an scFv protein, derived from a mouse B cell lymphoma, in plants. This rapid plant expression system utilizes a hybrid TMV to introduce specific gene sequences into whole plants, leading to systemic production of recombinant protein. There is no requirement for tissue culture and the entire process takes only a few days. McCormick *et al.* (1999) were able to show that infected tobacco plants were capable of producing a soluble scFv protein which generated an antibody response in vaccinated animals and which was an effective vaccine in a murine tumour challenge model of lymphoma.

Conclusions

Vaccine production and delivery systems based upon transgenic plants employ simple and well-established procedures and are particularly suited to the delivery of oral vaccines. The numerous experimental examples of the use of plant-based vaccines for application in both human and animal health care suggest that such vaccines will be widely available in the near future. A particular potential benefit of plant-based vaccine production and delivery systems is that they may overcome one of the main drawbacks associated with the development of oral vaccines, namely the requirement for a commercially acceptable production procedure. In order to overcome the inherent non-responsiveness of the gastrointestinal tract, the harshness of that environment and the effects of dilution, orally administered vaccines need to be used in much higher doses than parenteral vaccines. The use of edible transgenic plants for vaccine production and delivery could provide bulk quantities of low-cost vaccine. However, there are still some very basic problems that will need to be addressed. One of the greatest challenges is the generally low level of accumulation of foreign antigens in transgenic plant tissues. The optimization of vaccine genes for plant expression, together with new developments in the identification of plant promoters, may go some way to solving this problem. Of course, the accumulation of some antigens may be limited by adverse effects on the host plant. For example, Mason *et al.* (1998) observed that transgenic potato plants expressing the highest levels of *E. coli* LT-B displayed a stunting of growth and reduction in tuber yield, presumably due to a toxic effect of the antigen.

Additional challenges associated with the development of plant-derived vaccines and therapeutic proteins relate to the post-translational processing of proteins in plants. In particular, it is known that glycosylation patterns in plants differ from those in mammals and some plant glycoproteins are extremely antigenic in mammals (Chrispeels & Faye, 1996). Whilst this will not be a major concern in the development of oral subunit vaccines, which are not glycoprotein, it will be an issue in applications involving plant-derived antibodies (Ma & Hein, 1995). Transgenic plants do have the potential for large-scale, low-cost production of recombinant antibodies for topical passive immunization and tremendous scope for further developments exists in this area.

References

Arakawa, T., Chong, D. K. X. & Langridge, W. H. R. (1998). Efficacy of a food plant-based oral cholera toxin B subunit vaccine. *Nature Biotechnology* **16**, 292–297.

Arntzen, C. J. (1998). Pharmaceutical foodstuffs — oral immunization with transgenic plants. *Nature Medicine* **4**, 502–503.

Carrillo, C., Wigdorovitz, A., Oliveros, J. C., Zamorano, P. I., Sadir, A. M., Gomez, N., Salinas, J., Escribano, J. M. & Borca, M. V. (1998). Protective immune response to foot-and-mouth disease virus with VP1 expressed in transgenic plants. *Journal of Virology* **72**, 1688–1690.

Chrispeels, M. J. & Faye, L. (1996). The production of glycoproteins with defined non-immunogenic glycans. In *Transgenic Plants: a Production System for Industrial and Pharmaceutical Proteins*, pp. 99–113. Edited by M. R. L. Owen & J. Pen. Chichester: Wiley.

Clemens, J. D., van Loon, F., Sack, D. A. & 10 other authors (1990). Field trial of oral cholera vaccines in Bangladesh — serum vibriocidal and antitoxic antibodies as markers of the risk of cholera. *Journal of Infectious Diseases* **163**, 1235–1242.

Dalsgaard, K., Uttenthal, A., Jones, T. D. & 12 other authors (1997). Plant-derived vaccine protects target animals against a viral disease. *Nature Biotechnology* **15**, 248–252.

Durrani, Z., McInerney, T. L., McLain, L., Jones, T., Bellaby, T., Brennan, F. R. & Dimmock, N. J. (1998). Intranasal immunization with a plant virus expressing a peptide from HIV-1 gp41 stimulates better mucosal and systemic HIV-1-specific IgA and IgG than oral immunization. *Journal of Immunological Methods* **220**, 93–103.

Emini, E. A., Ellis, R. W., Miller, W. J., McAleer, W. J., Scolnick, E. M. & Gerety, R. J. (1986). Production and immunological analysis of recombinant hepatitis-B vaccine. *Journal of Infection* **13** (Suppl. A), 3–9.

Fitchen, J., Beachy, R. N. & Hein, M. B. (1995). Plant virus expressing hybrid coat protein with added murine epitope elicits autoantibody response. *Vaccine* **13**, 1051–1057.

Gomez, N., Carrillo, C., Salinas, J., Parra, F.,

Borca, M. V. & Escribano, J. M. (1998). Expression of immunogenic glycoprotein S polypeptides from transmissible gastroenteritis coronavirus in transgenic plants. *Virology* **249**, 352–358.

Hakim, I., Levy, S. & Levy, R. (1996). Protein and DNA vaccination with scFv fused to IL-1 beta peptide induce anti B cell tumour immunity. *FASEB Journal* **10**, 2031.

Hamada, S., Horikoshi, T., Minami, T., Kawabata, S., Hiraoka, J., Fujiwara, T. & Ooshima, T. (1991). Oral passive immunisation against dental caries in rats by use of hen egg yolk antibodies specific for cell associated glucosyltransferase of *Streptococcus mutans*. *Infection and Immunity* **59**, 4161–4167.

Haq, T. A., Mason, H. S., Clements, J. D. & Arntzen, C. J. (1995). Oral immunization with a recombinant bacterial-antigen produced in transgenic plants. *Science* **268**, 714–716.

Haynes, J. R., Cunningham, J., Von Seefried, A., Lennick, M., Garvin, R. T. & Shen, S. H. (1986). Development of a genetically engineered, candidate polio vaccine employing the self-assembling properties of the tobacco mosaic virus coat protein. *Bio/Technology* **4**, 637–641.

Holmgren, J., Svennerholm, A. M., Jertborn, M., Clemens, J., Sack, D. A., Salenstedt, R. & Wigzell, H. (1992). An oral B subunit whole cell vaccine against cholera. *Vaccine* **10**, 911–914.

Johnson, J., Lin, T. & Lomonossoff, G. (1997). Presentation of heterologous peptides on plant viruses: genetics, structure and function. *Annual Review of Phytopathology* **35**, 67–86.

Lebens, M., Johansson, S., Osek, J., Lindblad, M. & Holmgren, J. (1993). Large-scale production of *Vibrio cholerae* toxin B subunit for use in oral vaccines. *Bio/Technology* **11**, 1574–1578.

Ma, J. K. C. & Hein, M. B. (1995). Plant antibodies for immunotherapy. *Plant Physiology* **109**, 341–346.

Ma, J. K. C., Smith, R. & Lehner, T. (1987). Use of monoclonal antibodies in local passive immunization to prevent colonization of human teeth by *Streptococcus mutans*. *Infection and Immunity* **55**, 1274–1278.

Ma, J. K. C., Hikmat, B. Y., Wycoff, K., Vine, N. D., Chargelegue, D., Yu, L., Hein, M. B. & Lehner, T. (1998). Characterization of a recombinant plant monoclonal secretory antibody and preventive immunotherapy in humans. *Nature Medicine* 4, 601–606.

McCormick, A. A., Kumagai, M. H., Hanley, K., Turpen, T. H., Hakim, I., Grill, L. K., Tuse, D., Levy, S. & Levy, R. (1999). Rapid production of specific vaccines for lymphoma by expression of the tumour-derived single-chain Fv epitopes in tobacco plants. *Proceedings of the National Academy of Sciences, USA* 96, 703–708.

McLain, L., Durrani, Z., Wisniewski, L. A., Porta, C., Lomonossoff, G. P. & Dimmock, N. J. (1996). Stimulation of neutralizing antibodies to human immunodeficiency virus type 1 in three strains of mice immunized with a 22 amino acid peptide of gp41 expressed on the surface of a plant virus. *Vaccine* 14, 799–810.

Mason, H. S., Lam, D. M. K. & Arntzen, C. J. (1992). Expression of hepatitis-B surface-antigen in transgenic plants. *Proceedings of the National Academy of Sciences, USA* 89, 11745–11749.

Mason, H. S., Ball, J. M., Shi, J. J., Jiang, X., Estes, M. K. & Arntzen, C. J. (1996). Expression of Norwalk virus capsid protein in transgenic tobacco and potato and its oral immunogenicity in mice. *Proceedings of the National Academy of Sciences, USA* 93, 5335–5340.

Mason, H. S., Haq, T. A., Clements, J. D. & Arntzen, C. J. (1998). Edible vaccine protects mice against *Escherichia coli* heat-labile enterotoxin (LT): potatoes expressing a synthetic LT-B gene. *Vaccine* 16, 1336–1343.

Meloen, R. H., Hamilton, W. D. O., Casal, J. I., Dalsgaard, K. & Langeveld, J. P. M. (1998). Edible vaccines. *Veterinary Quarterly* 20, S92–S95.

Modelska, A., Dietzschold, B., Sleysh, N., Fu, Z. F., Steplewski, K., Hooper, D. C., Koprowski, H. & Yusibov, V. (1998). Immunization against rabies with plant-derived antigen. *Proceedings*

of the National Academy of Sciences, USA 95, 2481–2485.

Mor, T. S., Gomez-Lim, M. A. & Palmer, K. E. (1998). Perspective: edible vaccines—a concept coming of age. *Trends in Microbiology* 6, 449–453.

Schonberger, O., Hirst, T. R. & Pines, O. (1991). Targeting and assembly of an oligomeric bacterial enterotoxoid in the endoplasmic reticulum of *Saccharomyces cerevisiae*. *Molecular Microbiology* 5, 2663–2671.

Tacket, C. O., Mason, H. S., Losonsky, G., Clements, J. D., Levine, M. M. & Arntzen, C. J. (1998). Immunogenicity in humans of a recombinant bacterial antigen delivered in a transgenic potato. *Nature Medicine* 4, 607–609.

Thanavala, Y., Yang, Y.-F., Lyons, P., Mason, H. S. & Arntzen, C. J. (1995). Immunogenicity of transgenic plant-derived hepatitis B surface antigen. *Proceedings of the National Academy of Sciences, USA* 92, 3358–3361.

Turpen, T. H., Reinl, S. J., Charoenvit, Y., Hoffman, S. L., Fallarme, V. & Grill, L. K. (1995). Malarial epitopes expressed on the surface of recombinant tobacco mosaic virus. *Bio/Technology* 13, 53–57.

Wigdorovitz, A., Carrillo, C., Santos, M. J. D. & 8 other authors (1999). Induction of a protective antibody response to foot and mouth disease virus in mice following oral or parenteral immunization with alfalfa transgenic plants expressing the viral structural protein VP1. *Virology* 255, 347–353.

Yusibov, V., Modelska, A., Steplewski, K., Agadjanyan, M., Weiner, D., Hooper, D. C. & Koprowski, H. (1997). Antigens produced in plants by infection with chimeric plant viruses immunize against rabies virus and HIV-1. *Proceedings of the National Academy of Sciences, USA* 94, 5784–5788.

Zeitlin, L., Olmsted, S. S., Moench, T. R. & 7 other authors (1998). A humanized monoclonal antibody produced in transgenic plants for immunoprotection of the vagina against genital herpes. *Nature Biotechnology* 13, 1361–1364.

Conserved epitopes in bacterial lipopolysaccharides and cross-reactive antibodies in the treatment of endotoxicosis and Gram-negative septic shock

Sven Müller-Loennies,[1] Franco di Padova,[2] Lore Brade,[1] Didier Heumann[3] and Ernst Theodor Rietschel[1]

[1]*Research Center Borstel, Center for Medicine and Biosciences, Department for Immunochemistry and Biochemical Microbiology, Parkallee 22, D-23845 Borstel, Germany*
[2]*Preclinical Research Novartis, CH-4002 Basel, Switzerland*
[3]*Division of Infectious Diseases, CHUV, CH-1011 Lausanne, Switzerland*

Introduction

It has been known for more than a century that many symptoms of severe infections with Gram-negative bacteria, such as fever, tachycardia, leucopenia and hypotension, are due to the action of a heat-stable toxin associated with the bacterial cell wall, which is termed endotoxin (Pfeiffer, 1892; Westphal & Lüderitz, 1954). Chemical characterization of endotoxins revealed that they were amphiphilic macromolecules composed of a polysaccharide portion and a lipid anchor, called lipid A (Westphal & Lüderitz, 1954). Therefore, endotoxins are nowadays synonymously referred to as lipopolysaccharides (LPSs); however, it was subsequently realized that not all LPSs are endotoxically active (Rietschel *et al.*, 1996a). Many of the symptoms seen during severe infection and septic shock can be elicited by isolated endotoxins, as well as natural and synthetic lipid A (Galanos *et al.*, 1972, 1985), whereas the isolated polysaccharide portion is not toxic (Westphal & Lüderitz, 1954). Therefore, lipid A represents the endotoxic component of LPS. The polysaccharide region constitutes the main target of the humoral immune response in mammals due to its exposed location on the bacterial cell surface. It shields bacteria from the attack of the immune defence system and prevents phagocytosis, the deposition of complement factors and the formation of the membrane attack complex.

LPSs, in general, are highly active endotoxins in mammals (Rietschel & Brade,

Dedicated to our friend and esteemed colleague Professor Dr med. Max Schlaak on the occasion of his 65th birthday (May 24th, 1999).

1992), and powerful stimulatory molecules capable of locally or systemically activating the host's immune defence. The essential role of LPS and its lipid A component in causing fever, hypotension, leucopenia, tachycardia, tachypnoea, disseminated intravascular coagulation (DIC) accompanied by acute respiratory distress syndrome and multi-organ failure (MOF) in the final stages of septic shock (Vincent, 1996) was not always accepted. These effects were rather believed to be the result of destructive effects of the invading micro-organism. The role of LPS in pathogenicity was also in dispute for decades because similar host responses were seen after infections with Gram-positive microbes (Stegmayr *et al.*, 1992) or could be induced even by events other than infection (Parillo, 1993). Likewise, positive blood cultures (bacteraemia) or endotoxin in the blood (endotoxaemia) were not a prerequisite for the diagnosis of septic shock (Marchant *et al.*, 1995). In order to shed light on the role of endotoxins during infection, bacterial mutants were constructed which express lipid A of very low toxicity (Nichols *et al.*, 1997; Sunshine *et al.*, 1997; Khan *et al.*, 1998). These mutants multiply *in vivo* without causing lethality, thus proving an essential role for LPS in lethal infections.

LPSs are released in small quantities during cell fission, but in larger amounts during cell death, in particular upon antibiotic treatment or complement-mediated lysis. Bacteria in the gut may be a source of LPS, although in healthy persons LPS and bacteria do not penetrate the intact gut mucosa at high rates (Brearly *et al.*, 1985). However, after invasive treatment or severe burns, larger amounts of LPS and/or bacteria may enter the bloodstream (Bahrami *et al.*, 1996), presumably due to a malfunctioning of the gut mucosal barrier. Nevertheless, some invasive strains of, for example, *Escherichia coli*, *Shigella* and *Yersinia*, are able to adhere to epithelial cells and penetrate the intact gut wall (Levine, 1987). The most common cause of endotoxic shock is a generalized infection with a Gram-negative pathogen that spreads from a local focus into different organs. In former days, natural infection as the main source of endotoxins occurred more frequently due to lower standards of hygiene and limited knowledge on the sources and routes of infection. Identification of pathogens and development of diagnostic tools as well as the discovery of antibiotics and their introduction into medicine concomitant with the improvement of hygiene conditions led to major improvements in the treatment and prevention of such infections. However, by absolute numbers, the mortality due to Gram-negative bacterial sepsis remained almost unaltered (Nogare, 1991). Due to the introduction of invasive procedures, including surgical treatment and mechanical ventilation in intensive care units, nosocomially contracted infections represent nowadays the main cause of septic shock. It is estimated that approximately 500 000 patients each year are affected in the US with an associated lethality of about 25%. Although the contribution of endotoxins to virulence during infections with different pathogens with mixed infections and at different locations of the body remains to be elucidated, it is evident that LPS represents one of the essential pathogenic factors of Gram-negative bacteria. Endotoxin, therefore, also represents an important target for the diagnosis and therapy of Gram-negative sepsis.

Chemical structure and conformation of endotoxins

Enterobacterial LPS consists of three domains, i.e. the lipid A component, the core region and the O-specific chain. Of these domains, the lipid A component is the most conserved structure among different Gram-negative bacteria (Zähringer, Lindner & Rietschel, 1994). Furthermore, it represents the endotoxic component of LPS (Galanos *et al.*, 1985).

O-specific chain

Bacteria with a smooth colony appearance attach a polysaccharide (*O-specific chain*) to the core region, which is therefore called S-form LPS. In many cases it represents a heteropolymer made up from repeating units (Fig. 1). These are smaller oligosaccharides possessing from two to eight monomeric sugar residues (Jann & Jann, 1984; Knirel & Kochetkov, 1994). In enterobacterial LPS, these polysaccharide chains may contain up to 50 units (Jann & Jann, 1984). The O-chain region of LPS may penetrate capsular and slime layers which are frequently found on bacterial cells, rendering them accessible to binding proteins such as serum factors and antibodies. Due to the large number of different sugars involved in the biosynthesis of O-specific chains, which are subject to further modification by, for example, acetylation, methylation, phosphorylation, amidation and substitution with amino acids, bacterial cells produce a very high number of different chemical structures, which are targets of specific antibodies (Galanos *et al.*, 1977). The recognized epitopes represent the O-antigenic determinants (called O factors), which are of great diagnostic value and which represent the structural basis of the Kauffmann–White scheme for serotyping of *Salmonella* (Kauffmann, 1972).

Core region

Chemical analysis of LPS obtained from *Salmonella enterica* revealed that the same subset of carbohydrates was present in similar amounts in different serovars, whereas other carbohydrates were present in variable amounts. This led to the assumption that *Salmonella* LPS possessed a common structure which carried the O-specific chain (Galanos *et al.*, 1977). This LPS domain, which connects the lipid A domain to the O-specific chain, was termed *core region*. Chemical analysis of mutant bacteria defective in LPS biosynthetic pathways [rough (R) mutants] obtained from *Salmonella* led to the differentiation of several chemotypes (R chemotypes) which could be distinguished serologically (R factors; Galanos *et al.*,

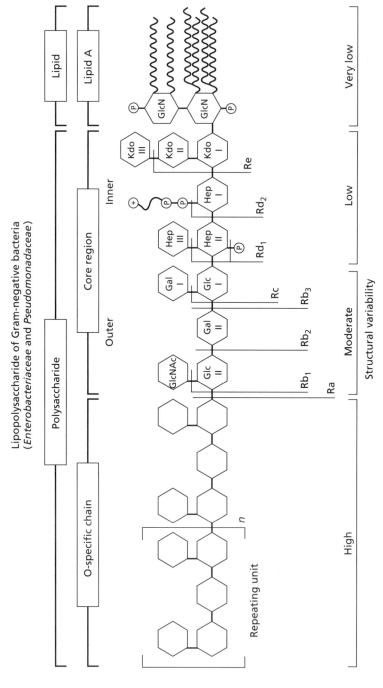

Fig. 1. Schematic representation of the structure of wild-type (S) and rough (R) mutant LPS as present in *Enterobacteriaceae* and *Pseudomonadaceae*. According to chemical, biosynthetic, biological and genetic criteria, LPS can be divided into three regions: O-specific chain, core oligosaccharide and lipid A. The O-specific chain represents a polymer of repeating units characteristic for each bacterial strain. The terms Ra–Re refer to structures of LPS from rough mutants which, due to genetic defects, synthesize a truncated core oligosaccharide and, therefore, lack an O-specific chain. The smallest LPS structure which can be found in still viable enterobacteria consists of lipid A and two Kdo residues (Re mutant). Saccharide groups are depicted by hexagons. GlcN, D-glucosamine; Kdo, 3-deoxy-D-*manno*-oct-2-ulosonic acid (2-keto-3-deoxy-D-*manno*–octonic acid); Hep, L-*glycero*-D-*manno*-heptose; Glc, D-glucose; Gal, D-galactose; GlcNAc, *N*-acetyl-D-glucosamine; P, phosphate, –⊕, 2-aminoethanol. The depicted structure of the outer core is found most frequently in *S.enterica* but may be different in other bacteria. Modified from Brabetz & Rietschel (1999).

1977). In *Enterobacteriaceae*, such as *E. coli* and *S. enterica*, these chemotypes are referred to as Ra–Re. Ra describes the largest core structure and Re was assigned to the smallest core structure (Fig. 1), which is devoid of all core sugars except for a 3-deoxy-D-*manno*-oct-2-ulosonic acid (Kdo)-disaccharide.

The core region of LPS consists of a complex oligosaccharide and, as compared to O-specific chains, shows limited structural variability (Holst & Brade, 1992). In the genus *Salmonella*, only two core types have been found so far (Holst & Brade, 1992; Olsthoorn *et al.*, 1998), and in *E. coli*, five different core types are expressed (Fig. 2; Holst & Brade, 1992). A typical structure present in LPS core structures of *Enterobacteriaceae* is a tetrasaccharide, α-D-Glc-(1→3)-α-L-α-D-Hep-(1→3)-L-α-D-Hep-(1→5)-Kdo, which is also found in LPS of *Yersinia*, *S. enterica*, *E. coli*, *Shigella* and *Citrobacter* (Holst & Brade, 1992).

In enterobacteria and some other families, an *inner core region* and an *outer core region* can be differentiated. The outer core contains predominantly pyranosidic hexoses, such as D-glucose (Glc), D-galactose (Gal), 2-amino-2-deoxy-D-glucose (GlcN) or 2-amino-2-deoxy-D-galactose (GalN) (Figs 1 and 2). Characteristic carbohydrate components of the inner core region are octulosonic acids and heptopyranoses (Hep*p*). These heptoses mainly possess the L-*glycero*-D-*manno* configuration (Holst & Brade, 1992).

The vast majority of analysed LPS structures contain Kdo in their inner core region (Holst & Brade, 1992), which is indispensable for the viability of LPS-containing Gram-negative bacteria. The smallest LPS structure sufficient for bacterial growth and multiplication was isolated from a deep rough mutant of *Haemophilus influenzae*. In this LPS, only one Kdo residue phosphorylated in positions 4 or 5 is attached to lipid A (Helander *et al.*, 1988; Whitfield & Valvano, 1993; Mamat *et al.*, 1999).

Lipid A

The lipid anchor of the LPS molecule is termed *lipid A* (Zähringer, Lindner & Rietschel, 1994). Its architecture is, to a large degree, structurally conserved in different bacterial groups, probably reflecting its essential role for the functioning of the outer membrane (Rietschel *et al.*, 1991). The linkage between the polysaccharide part and lipid A is formed by Kdo (Figs 1 and 2). Because of the extreme acid lability of this ketosidic linkage, lipid A can be obtained in free form after treatment of LPS with weak acid (Westphal & Lüderitz, 1954) and thus becomes accessible for a detailed structural analysis (Zähringer, Lindner & Rietschel, 1994).

Despite the presence of a common structural architecture, represented by the substitution of a hydrophilic carbohydrate backbone by fatty acids and phosphate groups, lipid A preparations extracted from bacteria show some variability, i.e. biosynthetically mature and immature molecular species are present in the same

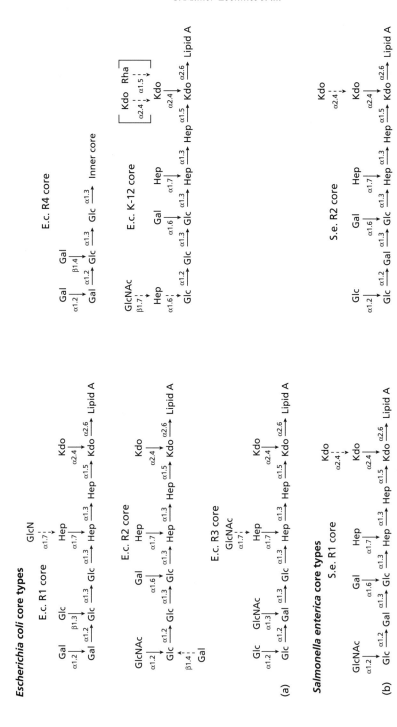

Fig. 2. Chemical structure of core oligosaccharides present in *E. coli* and *S. enterica*. Kdo, 3-deoxy-D-*manno*-oct-2-ulosonic acid; Hep, L-*glycero*-D-*manno*-heptose; broken arrows indicate non-stoichiometric substitutions. For other abbreviations see Fig. 1. The inner core structure of *E. coli* R4 has not been determined. *S. enterica* core structures have been named S. e. R1 and S. e. R2 (O. Holst, personal communication).

culture. Identification of natural heterogeneity with respect to the lipid A structure is of considerable biomedical and pharmaceutical interest because of its important role in mediating the toxic effects exerted by LPS even at low concentration (see below).

An exhaustive and detailed review on the chemical structure of lipid A of different origin was published by Zähringer, Lindner & Rietschel (1994). A lipid A structure which is widely distributed in nature is present in *E. coli* LPS (Fig. 3). It consists of a 1,4′-bisphosphorylated β1,6-linked GlcpN-disaccharide (GlcpN I-GlcpN II, lipid A backbone) which is substituted by (primary) *R*-3-hydroxytetradecanoic acid [14 : 0(3-OH)] residues in positions 2, 3, 2′ and 3′. The hydroxyl group of 14 : 0(3-OH) in positions 2′ (tetradecanoic acid, 14 : 0) and 3′ (dodecanoic acid, 12 : 0) is substituted by secondary acyl groups, leading to an asymmetric distribution of fatty acids. This type of lipid A constitutes as well the main species found in *Salmonella* and other enterobacterial and non-enterobacterial genera (Zähringer, Lindner & Rietschel, 1994). In *Salmonella*, in addition a heptaacylated lipid A species is present in non-stoichiometric amounts which harbours hexadecanoic acid (16 : 0) as secondary fatty acid at position 2 of GlcpN I. The attachment site of the core region is the hydroxyl group at position 6 of GlcpN II (Holst & Brade, 1992). Therefore, in LPS-associated lipid A only one hydroxyl group of the carbohydrate

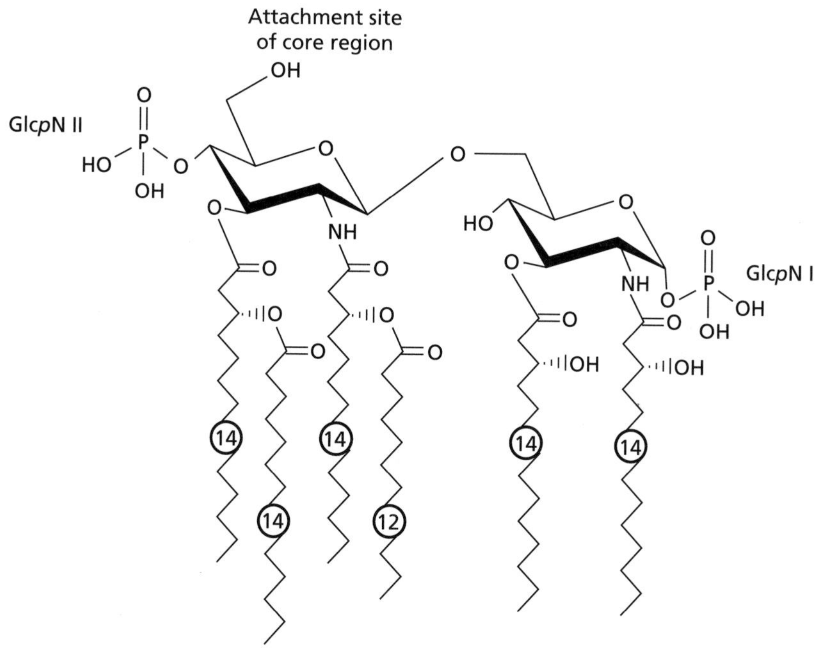

Fig. 3. Chemical structure of *E. coli* lipid A. The fully protonated form is shown. The hydroxyl group in position 6 of the non-reducing GlcN represents the attachment site of the glycosyl region in LPS.

backbone disaccharide is free, which is located at position 4 of GlcpN I. It is not known whether it is of biological significance that this hydroxyl group is not substituted. Different types of lipid A are found in other non-enterobacterial LPS. Structural variability results from chemical differences in the hydrophilic headgroup and in the hydrophobic acylation pattern (Zähringer, Lindner & Rietschel, 1994).

Structure–activity studies of lipid A molecules have revealed that modifications in the chemical structure of *E. coli* lipid A lead to dramatic losses of endotoxic activity (Rietschel *et al.*, 1996a). Thus, aiming at the definition of 'toxophore groups' which are responsible for the induction of different effects, we investigated different lipid A structures, of natural or synthetic origin, as well as partial structures thereof and compared their ability to induce cytokine release from human monocytes or murine macrophages to *E. coli* lipid A, so far the most potent stimulus. Molecules which lacked one constituent or had a different distribution of constituents were endotoxically less active or inactive with the exception of an α-2-phosphonooxyethyl derivative of synthetic *E. coli* lipid A (Rietschel *et al.*, 1996a) and the replacement of GlcpN II by GlcpN3N (2,3-diamino-2,3-dideoxy-D-glucose). Molecules with five or seven fatty acids and monophosphorylated derivatives were 100–1000 times less active than molecules in which only one fatty acid was dislocated. In contrast to the α-derivative, the β-2-phosphonooxyethyl derivative of synthetic *E. coli* lipid A had reduced biological activity. Monosaccharide partial structures were the least active and structures with four fatty acids lacked endotoxic activity in human test systems. In contrast, the attachment of Kdo to lipid A slightly increased biological activity (Rietschel *et al.*, 1990), which indicated a modulating effect on biological activity. All endotoxically active lipid A studied so far contains D-gluco-configured pyranosidic GlcpN or GlcpN3N residues which are present as a β1,6-linked homo- or heterodimer as hydrophilic backbone. Full endotoxicity was only observed for molecules phosphorylated in position 1 and 4′ and acylated with six fatty acids in an asymmetrical (*E. coli*) or symmetrical (*Neisseria meningitidis*) manner. Lipid A containing ketoacids is endotoxically inactive (Müller-Loennies *et al.*, 1998).

Of prime importance with respect to bioactivity of endotoxins are the number and distribution of fatty acids, as well as their chain length (Loppnow *et al.*, 1989; Ulmer *et al.*, 1992; Flad *et al.*, 1993). The tetraacylated lipid A precursor Ia (compound 406, lipid IVa) of *E. coli* was completely inactive in cytokine release assays from human mononuclear cells (Rietschel *et al.*, 1996a). The attachment of secondary 14 : 0 to position 3′ and 12 : 0 to position 2′ (*E. coli* lipid A) rendered the same molecule highly active. However, longer chain lengths, as found in lipid A of *Rhizobium leguminosarum*, *Chlamydia psittaci* and *Bacteroides fragilis*, resulted in reduced activity, whereas replacement of the *R*-configured 14 : 0(3-OH) by the *S*-configured isomer did not influence biological activity. In summary, since there is no single chemical group responsible for toxicity, it appears that the combination of structural features leads to an endotoxic conformation of the molecule which may be adopted by different chemical structures. This conformation has been established by low angle X-ray diffraction experiments. It was demonstrated that endotoxically active lipid A (like that of *E. coli*) adopts a conical–concave conformation in which the cross-section of the hydrophobic region is larger than that of the hydrophilic (GlcN-disaccharide) domain (Seydel, Brandenburg & Rietschel, 1994). In contrast, individual molecules of endotoxically inactive structures (such as the tetraacyl precursor Ia) possess a cylindrical conformation (Schromm *et al.*, 1998).

Biological activity of endotoxins

The activation of target cells involves different mechanisms depending on the concentration of LPS. LPS is known to interact with a variety of different serum proteins and several membrane-bound receptors. At low endotoxin concentrations an important event is complex formation with LPS-binding protein (LBP; Schumann *et al.*, 1990) which catalyses the interaction with membrane-bound CD14 (mCD14) necessary for target cell activation (Wright *et al.*, 1990; Lee *et al.*, 1992, 1993). In vascular cells which do not express mCD14, activation proceeds via complexation with soluble CD14 (sCD14). Since mCD14 is anchored in the membrane via glycosylphosphatidylinositol, which lacks a cytoplasmic domain (Haziot *et al.*, 1988), the interaction with a signal transducing molecule appeared to be necessary for cell activation. Functional signal transducing proteins were identified as members of the Toll-like receptor family [interleukin-1 (IL-1) receptor, TLR1–5] (Kirschning *et al.*, 1998; Yang *et al.*, 1998; Poltorak *et al.*, 1998; Hoshino *et al.*, 1999; Qureshi *et al.*, 1999). Members of this family signal through the same signalling pathway as IL-1, which differs in upstream events from the cascade employed by the tumour necrosis factor (TNF)–receptor complex (Chow *et al.*, 1999). In particular, a point mutation in TLR4 has been identified as the reason for LPS hyporesponsiveness in C3H/HeJ mice. A previous finding (Kirschning *et al.*, 1998; Yang *et al.*, 1998) that co-expression of TLR2 and mCD14 in human embryonic kidney cells 293 was sufficient to confer LPS responsiveness was not confirmed in these studies (Poltorak *et al.*, 1998; Qureshi *et al.*, 1999). However, construction of a chromosomal deletion mutant mouse that lacked the TLR4 gene in both alleles responded to injection of LPS with TNF production (Vogel *et al.*, 1999). When backcrossed with C3H/HeJ mice, F$_1$ mice showed intermediate TNF production. This observation has been explained to result from a dominant negative effect of the mutated TLR4 as present in C3H/HeJ mice. Thus other events such as LPS binding to CD55 or other mechanisms of activation may play a role in cell activation by LPS. Despite considerable progress in recent years, the events which lead to cell activation upon LPS stimulation are not yet fully understood.

It is known that LPS-mediated activation results in the rapid activation of *protein tyrosine kinases* (PTKs), although the PTK specifically involved in LPS signalling is unknown. A range of PTK inhibitors was shown to affect LPS-induced production of TNF-α and IL-1β (Weinstein *et al.*, 1992; Dong, O'Brian & Fidler, 1993; Shapira *et al.*, 1994), indicating that PTKs play a central role. A protein kinase that is tyrosine-phosphorylated in response to LPS in mammalian cells is p38 (Han *et al.*, 1994). This enzyme is a member of the mitogen-activated protein (MAP) kinase or extracellular signal-regulated protein kinase (ERK) family which is involved in many cell surface receptor-mediated signalling pathways in eukaryotic cells (Robinson & Cobb, 1997). These cascades consist of a three-kinase module which comprises a MAP kinase, a MAP kinase/ERK kinase (MEK) kinase that activates the MAP kinase, and a MEK kinase (MEKK) that activates the MEK. Among the most prominent tyrosine-phosphorylated proteins in LPS-stimulated macrophages are the MAP kinases ERK1/ERK2 (Weinstein *et al.*, 1992; Dong, Qi & Fidler, 1993). However, because of the fact that the activation of MAP kinases is not sufficient to induce all the biological responses triggered by LPS (Hambleton, McMahon & DeFranco, 1995), the MAP kinase pathway is not the only mechanism mediating cellular responses to LPS. Several studies have suggested that treatment of macrophages with LPS results in the activation of *phospholipase C*. Apart from the Ca^{2+}- and phospholipid-dependent *protein kinase C* (PKC), involved in LPS signal transduction (Shapira *et al.*, 1994) in macrophages (Dong, O'Brian & Fidler, 1993a), the 1,2-diacylglycerol-independent PKC isoform zeta has also been implicated in LPS responsiveness (Fujihara *et al.*, 1994). However, since LPS-mediated responses without PKC activation have been reported (Dong, Lu & Zhang, 1989; Glaser, Asmis & Dennis, 1990), the role of PKC in LPS-induced signal transduction remains controversial. Upon activation of cells with LPS, pertussis-toxin (PT)-sensitive guanine nucleotide-binding (G) protein is activated (Jakway & DeFranco, 1986; Daniel-Issakani, Spiegel & Strulovici, 1989) and macrophages express high levels of mRNA encoding the PT-sensitive G-protein subunit $G_i\alpha_2$ (Xie *et al.*, 1993).

Transcription factors that appear to be involved in the activation of LPS-inducible genes include *NF-κB, AP-1*, the *Ets* family, *Erg-1/2* and *NF-IL6*. NF-κB in the cytoplasm forms homo- or heterodimers with structurally related proteins (May & Ghosh, 1998) in the cytosol. These dimeric complexes are present in an inactive form bound to an inhibitor protein, I-κB. Phosphorylation of I-κB induces dissociation of NF-κB from I-κB, followed by translocation of NF-κB to the nucleus in an active form that binds to DNA (May & Ghosh, 1998).

Activation of NF-κB plays an important role in the induction of various genes by LPS (Müller, Ziegler-Heitbrock & Bäuerle, 1993; Tebo *et al.*, 1994). As LPS-mediated activation of NF-κB also occurred in macrophages from C3H/HeJ (LPS-hyporesponsive) mice, although at higher LPS concentration, it was suggested that NF-κB activity alone is not sufficient for LPS activation (Ding *et al.*,

1995). In the murine macrophage cell line J774, an increase in *junB* and *c-jun* mRNA was observed in response to LPS (Fujihara *et al.*, 1993) and expression of *c-fos* was induced by LPS in murine macrophages (Collart *et al.*, 1987). Since these are part of the AP-1 complex DNA-binding protein composed of hetero- or homodimers of several proto-oncogenes (Newell, Deisseroth & Lopez-Berestein, 1994), AP-1 appears to be involved in transcriptional regulation of LPS responses.

Of the Ets transcription factor family, the two members Ets and Elk-1 were demonstrated to be LPS-inducible in macrophages (Boulukos *et al.*, 1990), and PU.1, another Ets family member, has been implicated in LPS signalling (Kominato *et al.*, 1995; Lodie *et al.*, 1997). In addition to Erg-2, a member of the early growth response (*egr*) gene family (Coleman *et al.*, 1992), NF-IL6 also appears to be involved in transcriptional activation of LPS-inducible genes in macrophages (Dendorfer, Oettgen & Libermann, 1994).

The importance of endogenously produced mediators for the induction of endotoxic symptoms became evident from the observation that typical endotoxic effects can be elicited by recombinant cytokines such as *TNF-α, IL-1β* and *IL-6* (Vogel, 1990). Additionally, endotoxic symptoms were correlated with serum levels of these mediators (Zabel *et al.*, 1989) and antibodies against them suppressed endotoxic activities of LPS (Beutler & Cerami, 1988). Finally gene-deficient mice (Ko mice) unable to produce such mediators or respond to them proved to be resistant to LPS challenge. The cellular source of these mediators was identified by cell transfer experiments from endotoxin-sensitive to -resistant mice (Freudenberg, Keppler & Galanos, 1986). Although a large number of different cell types are able to respond to LPS challenge, only transfer of macrophages but not granulocytes or lymphocytes sensitized a mouse strain (C3H/HeJ) which is naturally resistant to LPS. Thus stimulation of immune-competent cells leads to the formation of proinflammatory cytokines which are endowed with biological activity and which play a central role in the development of endotoxic effects. The key step involved in endotoxin action appears to be the activation of monocytes/macrophages to produce cytokines such as TNF-α, and IL-1α and β, IL-6, IL-10, IL-12 and IL-18, which themselves elicit pleiotropic biological effects. Activation of the clotting cascade and increased levels of tissue-plasmin activator inhibitor lead to reduced hydrolysis of fibrin by plasmin and to DIC, which in concert with released reactive oxygen species is among the main factors leading to MOF.

Treatment of Gram-negative infection and septic shock

Successful strategies for the treatment of Gram-negative infection, in general, and the development of septic shock, in particular, primarily aim at the elimination of the infection by either surgery or antibiotic treatment. At a time, however, when resistance of bacteria to classical antibiotics becomes a significant problem of worldwide concern, new strategies must be envisaged. Since LPS is essential for the viability of Gram-negative bacteria, it should be considered a potential target for the development of new pharmaceuticals designed to inhibit LPS biosynthesis. Since Kdo is indispensable for the generation of functional LPS, inhibitors of Kdo biosynthesis have been regarded as a potent new class of drugs (Unger, 1981). Analogues have been synthesized (Goldman *et al.*, 1987; Claesson *et al.*, 1987; Sarabia-Garcia, Lopez-Herrera & Pino-Gonzalez, 1994) which inhibited the activation of Kdo by CMP-Kdo synthetase prior to incorporation into LPS and, therefore, caused cessation of LPS biosynthesis. These compounds were reported to be bacteriostatic *in vitro*. Other synthetic compounds directed against further enzymes involved in Kdo biosynthesis, such as arabinose-5-phosphate isomerase, inhibited the enzyme but did not show antibacterial activity (Bigham *et al.*, 1984).

Likewise, inhibitors of the very early steps of LPS biosynthesis, such as one that inhibits enzyme LpxC which catalyses the *N*-deacetylation of 3-*O*-acylated UDP-GlcNAc during lipid A biosynthesis, were developed and exhibited excellent antibacterial activity against *E. coli* (Onishi *et al.*, 1996). Unfortunately, these inhibitors were inactive against *Pseudomonas* strains, major pathogens causing nosocomial infections and septic shock. Recently, the crystal structure of UDP-GlcNAc acyltransferase (LpxA) involved in the first step of lipid A biosynthesis was elucidated, providing a basis for the design of inhibitors of this enzyme (Raetz, 1996).

A related conceivable strategy employs inhibitors which should interfere with later steps of the early LPS core biosynthesis, yielding viable bacteria which, however, are unable to attach their protective O antigen. Such bacteria would be cleared more efficiently by phagocytic cells of the reticuloendothelial system and would be more susceptible to the action of complement. In addition, in such microbes cross-reactive epitopes may be better accessible for the recognition by cross-protective antibodies. An example of defined interference with LPS biosynthesis was reported (Mamat, Rietschel & Schmidt, 1995) for the *trans*-acting RNA derived from the *Acetobacter methanolicus* phage Acm1 which blocked LPS biosynthesis by several members of *Enterobacteriaceae*, such as *E. coli*, *S. enterica* and *Klebsiella pneumoniae*, resulting in the down-regulation of O side chain biosynthesis and a concomitant reduction of pathogenicity (Mamat, Rietschel & Schmidt, 1995).

The use of antibiotics carries the risk of LPS release from bacteria and thus may worsen the clinical situation with an important impact on the activation of the immune response, reviewed by Morrison (1998). The antibiotics moxolactam, gentamicin or chloramphenicol differ in their capacity to release LPS from bacteria. Similarly, treatment of *Pseudomonas aeruginosa* with imipenem resulted in the release of less endotoxin than treatment with ceftazidime, an effect attributed to the interaction of the antibiotics with different penicillin-binding proteins. These initial observations have gained support from more recent studies and, more importantly, could be transferred to the *in vivo* situation, where imipenem treatment proved to be superior over ceftazidime in *Pseudomonas*-induced infections in rats. Imipenem treatment was accompanied by lower circulating TNF-α levels, probably due to lower amounts of released endotoxin. A recently conducted double-blind clinical trial confirmed these observations (Prins *et al.*, 1995).

Apart from killing bacteria by antibiotics, it would be desirable during sepsis to be able to interfere at all steps of the immunological activation cascade. In this context, several approaches have been developed to positively influence the outcome of septic episodes. Among these, the application of therapeutics which interfere with the release and activity of lipid and peptide mediators was investigated (Strieter *et al.*, 1988; Kunkel *et al.*, 1988; Zabel *et al.*, 1989; Ferguson-Chanowitz *et al.*, 1990; Izbicki *et al.*, 1991). As a one-sided influence on the sensitive host equilibrium of the released mediators bears some risk, it seems that a more general approach such as the elimination of endotoxins from the circulation represents a promising strategy. This strategy interferes with the sepsis cascade at a very early step and may be at the same time an anti-infective and an anti-endotoxic treatment.

Neutralization of endotoxin

Once released from the bacterial cell, potentially dangerous LPS is confronted with several host mechanisms aiming at its elimination and neutralization. Natural detoxification of endotoxins is achieved by hepatocytes of the liver. The main route of endotoxins that enter the blood from the gut leads through the portal vein to the liver, where they may be taken up by Kupffer cells, released into the bloodstream, and subsequently redistributed into hepatocytes which finally secrete them together with bile (Van Deventer, Ten Cate & Tytgat, 1988; Freudenberg *et al.*, 1992). Alternatively, endotoxins may enter the bloodstream via the thoracic duct, thereby circumventing the liver. Degradation of endotoxins leading to less toxic lipid A partial structures or even antagonists was achieved by macrophage- and neutrophil-derived enzymes. Thus, dephosphorylation of the lipid A backbone in eukaryotic cells and serum was found (Peterson & Munford, 1987; Hampton & Raetz, 1991). This phosphatase activity was identified in cell lysates and intact cells (peritoneal macrophages), where it was localized in lysosomes. Another natural defence mechanism may be the cleavage of secondary fatty acids by a neutrophil-derived acyloxyacyl hydrolase (Munford & Hall, 1986, 1989). The *in vivo* fate of endotoxins is determined as well by serum factors that complex LPS. Thus, complexation with bactericidal/permeability-increasing protein (BPI) and high-density lipoprotein (HDL) (Flegel *et al.*, 1993; Wurfel *et al.*, 1994; Horwitz, Williams & Nowakowski, 1995; Yu & Wright, 1996) leads to transport to the liver and uptake by hepatocytes. One of the natural functions of BPI is the intracellular killing of invading Gram-negative bacteria (Schlag *et al.*, 1999). However, BPI may also act as a natural feedback inhibitor, preventing an overshoot of a self-destructive inflammatory response of an infected host. BPI is found in the azurophilic granules of neutrophils and constitutes a cationic protein with a molecular mass of about 55 kDa. Because of a strong affinity for LPS, it is able to attack the outer membrane of Gram-negative bacteria (Wiese *et al.*, 1997a, b, 1998). Its lysine-rich N-terminus harbours all antibacterial properties (Mannion *et al.*, 1989). Thus, LPS is neutralized by BPI (Elsbach & Weiss, 1993; Beamer, Carroll & Eisenberg, 1997).

It is evident from the high mortality and morbidity of Gram-negative sepsis that the natural defence mechanisms do not confer sufficient protection at increased endotoxin levels. Successful strategies for the treatment of Gram-negative sepsis, in general, and the development of LPS-induced septic shock, in particular, should ideally interfere at the very start of the LPS recognition pathway. The selective blockade of an overwhelming activation of different target cells would represent an important approach to control septic shock. In principle, each step of the endo-toxic activation cascade could serve as a basis for the development of new therapy concepts (see Levin *et al.*, 1995 for a review). However, the early phases concerning the released bacterial toxin itself as the initial stimulus and its interaction with enhancing and inhibiting humoral factors (LBP, sCD14, BPI), as well as cell-bound receptors (CD14) or signal transducing molecules (TLR4, possibly other Toll molecules), are of special interest. This concept is supported by the consideration that focusing on the early events of the septic cascade would lead to an increased selectivity and, therefore, to less side effects caused by an antiseptic drug. In addition, an early therapeutic blockade or even prophylactic medication seems to be required from the clinician's point of view since, in practice, the development of septic shock is often difficult to predict (Wenzel *et al.*, 1996). In this respect, non-toxic LPS, LPS partial structures and lipid A analogues (reviewed by Rietschel *et al.*, 1996b) which possess antagonizing activity appear to be promising candidates.

Amplification of the natural defence mechanisms has been successfully employed in the fight against endotoxicosis. BPI can be used to reduce endotoxaemia-caused mortality in mice and rats (Evans *et al.*, 1995). The ability of BPI to limit LPS-mediated reactions is particularly important in clinical settings. Thus clinical trials are currently underway to analyse in meningococcal sepsis the therapeutic potential of a recombinant N-terminal fragment of BPI which harbours all anti-endotoxic activities (Giroir *et al.*, 1997).

Apart from BPI, sCD14 has also been described to suppress LPS-mediated reactions *in vitro* and *in vivo* (Schütt *et al.*, 1992). This effect is, however, only seen at high serum concentrations of sCD14, whereas low levels rather augment the endotoxic activity of LPS. The stimulatory activity was readily explained by the activation of mCD14-negative cells by sCD14–LPS complexes. The mechanisms underlying the inhibitory effects of sCD14 at high concentrations, however, remain to be elucidated. On the other hand, in a model system of bacterial meningitis (Cauwels *et al.*, 1999), increased levels of sCD14 which derived from invading neutrophils were found in the cerebrospinal fluid (CSF). Intracerebral coinjection of recombinant soluble (rs) CD14 together with *Streptococcus pneumoniae* led to increased TNF and IL-6 levels in the CSF. In addition, in this experiment, rsCD14-dependent progressive bacterial growth was observed. These effects were seen over a range of physiological concentrations of rsCD14 and thus in the brain sCD14 seems to exert proinflammatory activity. However, it may be speculated that the *in vivo* function of sCD14 is different in tissues and in the circulation, where it may prevent the association of LPS with mCD14 leading to cell activation. A similar observation was made for LBP, of which low amounts act in proinflammatory fashion (Heumann *et al.*, 1998), whereas large amounts are protective in endotoxicosis (Lamping *et al.*, 1998). Recently, LBP detoxified LPS by transferring LPS into HDL particles *in vitro* (Lamping *et al.*, 1998). Furthermore, LBP-deficient mice are dramatically more susceptible to an intraperitoneal *Salmonella* infection in comparison to wild-type mice (Jack *et al.*, 1997). However, in the latter study increased transfer of LPS into HDL in LBP-deficient mice was not observed. Since numerous studies suggest a detrimental role of LBP in endotoxaemia (Heumann *et al.*, 1998), the protective role of LBP in bacterial infection is not understood. Infections of LBP-deficient mice with other bacterial strains and other sites of infection should provide a better understanding of LBP function in bacterial infection and in shock induced by Gram-negative bacteria.

Lipid A antagonists

Certain natural and synthetic partial structures as well as analogues of lipid A are endotoxically inactive in human assay systems (Loppnow *et al.*, 1989). At the same time, such preparations exhibited antagonistic properties against the biological activity of LPS and lipid A (Flad *et al.*, 1993; Ulmer *et al.*, 1992; Lynn & Golenbock, 1992). One example of an endotoxically inactive yet antagonistically very potent compound is the tetraacyl lipid A precursor Ia (compound 406 or lipid IVa), which suppresses the LPS-induced mRNA formation of TNF-α and IL-1 in human monocytes (Feist *et al.*, 1992). At higher doses, precursor Ia competes with LPS for binding to CD14 (Heine *et al.*, 1994). At low concentrations, the antagonistic activity appears to be CD14-independent (Kitchens, Ulevitch & Munford, 1992). In contrast to the human system, precursor Ia exhibits agonistic activity in mice (Galanos *et al.*, 1984). Therefore, it is not possible to use this compound in murine sepsis models.

Other antagonists were identified which lacked endotoxicity in the murine system, including non-toxic lipid A of *Rhodobacter sphaeroides* and *Rhodobacter capsulatus.* An analogue (E5531) of the lipid A structure of *R. capsulatus* was chemically synthesized which inhibited LPS-induced monokine production in human monocytes and whole blood (Christ *et al.*, 1995). In contrast to precursor Ia, E5531 does not exhibit any agonistic activity in murine cells, thus allowing murine *in vivo* models to be established. When administered simultaneously with LPS in mice, E5531 prevented an increase of TNF-α plasma levels and protected mice from LPS lethality. Even more important was the observation that E5531, when injected in combination with β-lactam antibiotics, protected mice from lethal peritonitis in a model situation which correlates in various aspects to human peritonitis and sepsis. Compound E5531 has been investigated in a Phase I clinical trial, where it suppressed in a dose-dependent manner LPS-induced effects such as the production of bioactive cytokines.

Neutralizing antibodies

Passive immunization with polyclonal antisera or mAbs directed against endo-toxins represents another strategy to eliminate or neutralize LPS. Structural analyses of LPS obtained from different pathogenic as well as non-pathogenic bacteria revealed a common architecture. A promising approach for the immunotherapy of septic shock, therefore, appears to be passive immunization with antibodies directed against structurally conserved regions of LPS. Such antibodies are supposedly cross-reactive with different Gram-negative pathogens and hopefully cross-protective. The induction of such antibodies was attempted by immunization with polysaccharide-deprived free lipid A presented in an appropriate form as well as with rough mutants displaying the core region on their surface (Galanos, Lüderitz & Westphal, 1971; Galanos *et al.*, 1972; Braude & Douglas, 1972; Rietschel & Galanos, 1977; Bruins *et al.*, 1977; Johns, Bruins & McCabe, 1977; Brade *et al.*, 1987a; Baumgartner *et al.*, 1991; Raponi *et al.*, 1991; McCabe, 1992). In an attempt to induce broadly cross-reactive and cross-protective antibodies, polyclonal antisera and several mAbs have been prepared (see below), which were reported to be cross-reactive *in vitro* and cross-protective *in vivo*.

In principle, all regions of the LPS molecule are immunogenic. In the course of a natural infection, antibodies directed mainly against the O-specific chain are formed and the O antisera were highly protective against homologous pathogenic bacteria and their LPS. Unfortunately, this approach is of limited use in the clinical situation due to the high structural variability of O antigens of different serotypes and the rapid establishment of shock, leaving little or no time for the determination of the causative serotype. Nevertheless, these investigations showed that protection by passive immunization represents a potentially successful strategy. Even more encouraging was the observation that larger amounts of antisera against R-type LPS conferred protection against different serotypes and were thus cross-protective. Passive immunization of rabbits with an antiserum obtained after immunization with *E. coli* O111 suppressed the local Shwartzman reaction provoked by homologous and heterologous LPS (Tate *et al.*, 1966).

Based on these initial studies, an Re mutant strain of *S. enterica* sv. Minnesota (strain R595) as well as the Rc mutant of *E. coli* O111 : B4 (strain J-5) have been used in the vast majority of the follow-up studies. Passive immunization with these polyclonal antisera was reported to provide cross-protection against *K. pneumoniae, E. coli* O17 and O4 (Ziegler *et al.*, 1973), *P. aeruginosa* (Ziegler *et al.*, 1975), several serovars of *Neisseria* spp. (Davis, Ziegler & Arnold, 1978), as well as *H. influenzae* type b (Marks *et al.*, 1982). These promising results encouraged Ziegler *et al.* (1982) to investigate whether human antiserum against *E. coli* J-5 possessed protective activity in patients with established Gram-negative bacteraemia and endotoxic shock. A randomized clinical trial revealed that the mortality in bacteraemic cases was lowered by 50% in comparison to controls. The most frequent isolates were *E. coli* (38%), *P. aeruginosa* (26%), *Klebsiella* (14%) and *Enterobacter* (8%). At the same time, Dunn & Ferguson (1982) reported that *E. coli* J-5 rabbit antiserum reduced mortality in a sepsis model in guinea pigs. The serum contained significant amounts of IgG antibodies which cross-reacted with *E. coli* O111 : B4, *K. pneumoniae, P. aeruginosa* and *Enterobacter aerogenes.* However, studies by other groups failed to confirm the cross-reactivity of antibodies elicited by *E. coli* J-5 LPS (Peter *et al.*, 1982; Siber, Kania & Warren, 1985; Miner *et al.*, 1986; Greisman & Johnston, 1988). Possible reasons for these conflicting results have been summarized (Greisman & Johnston, 1988) and include LPS contamination of antisera, the induction of tolerance and the presence of O antibodies among other factors.

In order to identify the antibody specificity which was responsible for the protective effect, several *E. coli* J-5-induced mAbs were prepared and subsequently investigated for their protective capacity. Among these, a human monoclonal IgM antibody (HA-1A, Centoxin) which was obtained after immunizing a volunteer with *E. coli* J-5 was investigated in detail (Ziegler *et al.*, 1991). First results appeared promising, and several studies reported cross-protection against a vast number of different pathogens. The results indicated that this antibody bound to lipid A as present in S- as well as R-type LPS and intact bacteria. However, a more critical analysis of the data revealed that differences in HA-1A-treated and placebo-treated groups were not statistically significant (Baumgartner & Glauser, 1993). A second trial was conducted, which did not confirm the protective activity of HA-1A (McCloskey *et al.*, 1994). Introduction of an improved ELISA with antigens complexed to HDL thereby increasing sensitivity and specificity (Heumann *et al.*, 1991) revealed that with large amounts of antibody, non-specific cross-reactivity was observed. The molecular basis for this cross-reactivity was due to reactivity with a carbohydrate epitope expressed on isolated and purified lipid A, the i antigen present on cord red blood cells, on human B lymphocytes and on certain autoantigens (Bhat *et al.*, 1993). The ligand bound by the antibody consisted of an acylated disaccharide and HA-1A; therefore, it had the characteristics of a polyreactive cold agglutinin.

Apart from HA-1A, a murine mAb (E5) was analysed in clinical trials. E5 was lipid-A-specific (Wood *et al.*, 1992) and cross-reacted with S- and R-type LPS of different clinically relevant bacteria (Parent *et al.*, 1992). It inhibited vasodilation in rats after LPS administration (Schwartz *et al.*, 1993) and reduced mortality caused by intravenous injection of *P. aeruginosa* (Romulo, Palardy & Opal, 1993). A first clinical trial showed protective effects in patients with Gram-negative infection which, however, were not confirmed in a second clinical trial (Bone *et al.*, 1995).

Two murine mAbs were obtained after immunization with a deep rough mutant of *S. enterica* sv. Typhimurium (Re chemotype) (mAb 26-20) and free lipid A (mAb 8-2) (Cornelissen *et al.*, 1993). Both antibodies reacted in ELISA with Re LPS, inhibited TNF-α release from macrophages, the activation of polymorphonuclear cells, and reduced mortality of mice challenged intravenously with S-type LPS of *E. coli* O111 : B4 and Re LPS from *S. enterica* sv. Minnesota.

Antibodies reactive with *E. coli* J-5 LPS were identified in human non-immune sera (Law & Marks, 1985). Such an antibody (mAb T88) was isolated from a B-cell line, after fusion from human spleen cells with a human–murine tumour cell line. The selection was achieved by monitoring reactivity with *E. coli* J-5, *S. enterica* sv. Minnesota R595 and S-form LPS, as well as the ability to opsonize bacteria and lyse them in the presence of complement (Winkelhake *et al.*, 1992).

The antibodies obtained after immunization with Re *S. enterica* sv. Minnesota (strain R595) or Rc (strain *E. coli* J-5) LPS are believed to be directed against the lipid A component. If these antibodies were of biological significance, it could be expected that immunization with lipid A leads to the formation of an antibody response with higher titres of cross-reactive antibodies presumed to be cross-protective. Lipid A was immunogenic if presented in an appropriate form by complexation with BSA (Galanos *et al.*, 1972), HDL (Heumann *et al.*, 1991) or liposomes (Brade *et al.*, 1986, 1987a), or when loaded onto erythrocytes and heat-killed bacteria (Galanos, Lüderitz & Westphal, 1971; Brade *et al.*, 1987a). Antibodies induced in this way recognized free lipid A from different pathogenic bacteria. The epitope of these antibodies was located in the common lipid A backbone structure of these LPS molecules; fatty acids are not part of this determinant (Brade *et al.*, 1987a; Kuhn *et al.*, 1992; Brade, Holst & Brade, 1993). However, all polyclonal antisera and mAbs prepared with free lipid A did not afford protection against endotoxic effects of LPS.

A molecular explanation for the lack of protection offered by anti-lipid A antibodies was provided by the observation that the primary hydroxy group of the non-reducing GlcN of the lipid A backbone is important for the interaction with antibodies (Brade *et al.*, 1997). Derivatization of this hydroxy group by a methyl group abolished binding by antibodies. Since in native LPS this position serves as the attachment site of the core and thus is occupied, this epitope is cryptic in native LPS. Therefore, lipid A antibodies are not expected to bind to LPS and to offer protection during Gram-negative infections (Kuhn, 1993).

A possible explanation for the protective capacity of LPS antisera was provided by comparing anti-LPS immunoglobulin and standard immunoglobulin in a multi-centred, placebo-controlled clinical trial (Intravenous Immunoglobulin Study Group, 1992). When immunoglobulin against LPS was prepared by enrichment of anti-Re LPS antibodies from serum, surprisingly, standard immunoglobulin had a higher protective activity. One explanation could be that protective antibody specificities were eliminated during enrichment. Thus these epitope specificities must lie in further distal parts of the LPS molecule, that is, in the outer region of the inner core (heptose region) or the outer core itself.

Anti-core antibodies

With knowledge of the structure of the inner core region of LPS and availability of synthetic compounds, it was possible to determine the epitope specificities of several mAbs against the inner core (Kdo and heptose region) on a molecular level. Four mAbs obtained after immunization with Re mutant LPS, which is made up only of lipid A and Kdo, were characterized in detail (Brade *et al.*, 1987b; Rozalski *et al.*, 1989). Two of them (clones 20 and 25) only required the Kdo region for binding and did not require the presence of lipid A constituents. The other two mAbs (clones 17 and 22) required parts of both lipid A and the Kdo region for binding. These antibody specificities were also detected in polyclonal rabbit antisera in which the majority of antibodies were antibodies of the latter type (clones 17 and 22), whereas true core antibodies (type 20 and 25) were present only in small amounts or were absent. These mAbs were not cross-reactive with LPS of the wild-type as it occurs in most clinical isolates. mAbs elicited against the heptose region (e.g. mAb S36-20 and mAb S32-2) are also not cross-reactive (Swierzko *et al.*, 1994). They recognize a more complex epitope formed by the complete inner core (Kdo and heptose region) and lipid A. They are highly specific for either the Rd2 (containing one heptose) or the Rd1 (containing two heptoses) chemotypes, since they do not react with LPS of the Ra, Rb, Rc or Re chemotype.

A novel immunization scheme using bacteria with a complete LPS core structure was employed by us, which led to the generation of truly anti-core antibodies which exhibited cross-reactivity and cross-protectivity against LPS of different bacterial origin (Di Padova *et al.*, 1993). It was thus for the first time convincingly demonstrated that cross-reactive antibodies do exist and that they possess neutralizing and cross-protective properties in different model systems. Five mAbs were obtained which were cross-reactive for all known *E. coli* and

S. enterica serotypes. These antibodies were unique in recognizing their epitopes in R-form as well as in S-form LPS, demonstrating that the outer as well as the inner core region are accessible for immune recognition despite the presence of the O-specific chain (Fig. 4).

The fine specificities of these antibodies differed slightly since differences in binding to various rough mutants were observed. This may be due to the recognition of different epitopes, but may as well result from a different presentation of the same epitope in rough mutants or different core types. If so, the antibodies may differ in the topology of their combining sites, leading to a difference in sensitivity for steric hindrance by neighbouring residues. Thus mAb H5 13-23 (IgG3κ) and H5 415-6 (IgG2aκ) are less reactive with the *E. coli* R3 core. Further differences were observed in the binding to LPS derived from *S. enterica* rough mutants. Whereas all antibodies recognized *S. enterica* sv. Minnesota Ra LPS (complete core, strain R60) and also the Rb chemotype (Rb2, strain R345) with the exception of H1 61-2 (IgG1κ), only the antibodies WN1 222-5 (IgG2aκ) and WN1 58-9 (IgG2bκ) recognized epitopes in the inner core region. This follows from the observation that some residual binding to deep rough LPS derived from *S. enterica* sv. Minnesota mutant strains RcP⁻ and Rd1P⁻ was demonstrated. However, since the reactivity with R-form LPS of *S. enterica* sv. Typhimurium and with other *Salmonella* strains was lower for antibodies H5 12-23 and H5 415-6, the recognition of different epitopes seems likely. None of these antibodies bind to Re LPS of *E. coli* or *S. enterica* mutants or to free lipid A. It is interesting to note that the minimal *E. coli* core structure recognized by these antibodies is the RcP⁺ chemotype as expressed by *E. coli* J-5, although these antibodies are not induced by immunization of mice with *E. coli* J-5.

This is explained by the fact that a different type of antibody is induced by immunization with R-type LPS, where the epitope consists of terminal sugar residues which are able to protrude into a binding pocket created by the antibody complementarity determining regions. However, when the inner core sugars are capped by the outer core or in addition with the O polysaccharide, this type of interaction is not possible. In this case, it is likely that the interaction resembles that observed for O-specific antibodies which are directed mainly against immunodominant side chain deoxy-sugars, which decorate the polysaccharide chain. As an example, the Fab fragment of antibody Se155-4, which is specific for the O polysaccharide of *Salmonella* serogroup B, was shown to accommodate a trisaccharide epitope in its combining site with the immunodominant side chain sugar buried in a deep pocket. A tight interaction is achieved by a predominance of aromatic amino acids involved in stacking interactions and hydrogen bonds mediated in part by a structured water molecule trapped in the binding site (Cygler, Rose & Bundle, 1991). By analogy, it may be that the side chain heptose which is α1,7-linked to the second heptose of the inner core in LPS (Fig. 2) is important for a tight

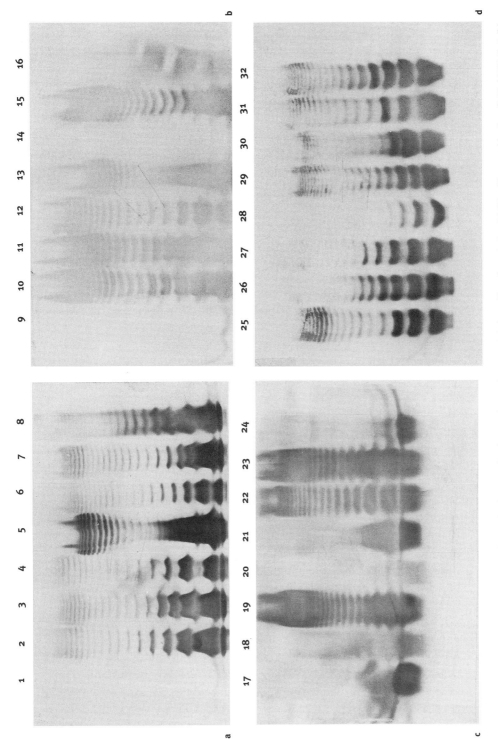

Fig. 4. WN1 222-5 recognition pattern of S-form and R-form *E. coli* and *S. enterica* sv. Minnesota LPS after sodium deoxycholate-PAGE and immunoblotting of LPS. (a) H161-1; (b) H5 13-23; (c) H5 415-6; (d) WN1 222-5. Lanes 1–8 and 9–16, *S. enterica* sv. Minnesota Re (R595), *E. coli* O4, O6, O12, O15, O86, *S. enterica* sv. Abortus equi. Lanes 17–24, *E. coli* O18K⁻, *E. coli* O111, O86, O18, O16, O15, O12, O6, O4, O18rf. Lanes 25–32, *E. coli* R2, O86, O16, O15, O12, O6, O4. Reprinted from Di Padova *et al.* (1996) by courtesy of Marcel Dekker Inc.

163

interaction with antibodies such as WN1 222-5. Since this structural element is widespread among different bacteria, this may form the molecular basis for the observed cross-reactivity. The conformational similarity of regions of the core structure in *Salmonella* and *E. coli* core types (Fig. 2) has been predicted by semi-empirical calculations for the outer core (Jansson *et al.*, 1989). The availability of purified phosphorylated and dephosphorylated oligosaccharides representing the complete core structures of *E. coli* J-5 (Müller-Loennies, Holst & Brade, 1994; Müller-Loennies *et al.*, 1999) will allow the future determination of the minimal *E. coli* LPS structure which is necessary and sufficient for recognition by core antibodies. The oligosaccharides may also allow a detailed conformational investigation of complexes with mAb.

Because it has not been possible so far to determine the fine specificity of these antibodies in detail, several different assay systems, such as ELISA with purified LPS, HDL–LPS complexes, competitive ELISA, PAGE followed by immunoblotting as well as passive haemolysis, have been utilized to confirm cross-reactivity and exclude non-specific interactions and the induction of conformational changes by immobilization.

Cross-reactive and cross-protective anti-core antibody WN1 222-5

Of the available antibodies directed against the core region, one mAb (WN1 222-5) was investigated in more detail. This mAb was selected because preliminary affinity determinations by surface plasmon resonance had indicated high affinity.

In the *in vivo* situation, LPS, released from the bacteria, is recognized by mAb in the presence of serum proteins. It was, therefore, first determined whether WN1 222-5 bound LPS in the presence of HDL, which was shown to be involved in LPS complexation *in vivo* (Freudenberg *et al.*, 1980; Munford, Hall & Dietschy, 1981; Baumberger, Ulevitch & Dayer, 1991; Massamiri, Tobias & Curtiss, 1997). HDL did not inhibit antibody reactivity (Table 1). Antibody reactivity was also observed when LPS was intercalated into plasma membranes or coated onto erythrocytes.

Since antibodies enhance killing and clearance of bacteria by opsonization and the activation of complement, tests were done to determine whether mAb WN1 222-5 was able to bind intact heat-killed or viable bacteria. In ELISA performed with immobilized heat-killed bacteria from a large collection of clinical isolates, mAb WN1 222-5 bound strongly to all blood, faecal and urinary isolates of *E. coli* and *S. enterica*, to some *Citrobacter* and weakly with some *Enterobacter* and *Klebsiella* strains (Di Padova *et al.*, 1993). mAb WN1 222-5 bound living *E. coli* O18 : K1 bacteria coated onto microtitre plates, and this reactivity was enhanced upon antibiotic treatment. However, in flow cytometry, WN1 222-5 failed to bind to viable *E. coli* O18 : K1, whereas *E. coli* O111 : B4 was efficiently recognized even in the absence of antibiotics. Therefore, the core epitope involved

Table 1. Comparison of coating with free LPS and HDL–LPS complexes for measuring the binding of WN1 222-5 to S-form LPS. Optical densities obtained in an ELISA at WN1 222-5 antibody concentration of 500 ng ml^{-1} and after 6 min incubation time with a 1/1000 dilution of horseradish-peroxidase-conjugated secondary antibody.

LPS	LPS	HDL–LPS
E. coli O4	1·16	>2·00
E. coli O6	1·14	0·79
E. coli O8	0·81	0·40
E. coli O111	0·53	0·46
E. coli O127	0·86	0·60
S. enterica sv. Minnesota	0·62	0·61
S. enterica sv. Typhimurium	0·78	1·07

in binding by mAb WN1 222-5 appears to be accessible in some but not all bacterial strains.

Next, the ability of mAb WN1 222-5 to bind to free LPS was investigated. It was shown to react with all serotypes of *E. coli* and *S. enterica* independent of the presence and chain length of the O-specific chain, i.e. the number of repeating units. It also recognized all R-type LPS belonging to the Ra to Rc chemotype. Further, the ability of mAb WN1 222-5 to neutralize (Edmond, Poxton & Di Padova, 1993) the endotoxic activity of LPS *in vitro* and *in vivo* was investigated. As mentioned above, a reliable determination of the neutralization of endotoxic effects requires that the antibody preparations are not contaminated by endotoxin because of a possible induction of endotoxin tolerance. All preparations of mAb WN1 222-5 proved negative in the *Limulus* amoebocyte lysate (LAL) assay and to be non-pyrogenic in rabbits. Moreover, heat treatment abolished protective properties and mAb WN1 222-5 was not active *in vivo* against LPS to which it did not bind *in vitro*.

mAb WN1 222-5 was demonstrated to inhibit the release of proinflammatory cytokines TNF-α and IL-6 from peritoneal cells. Anti-endotoxic activity was observed for S- and R-form LPS and correlated well with binding capacity. Furthermore, mAb WN1 222-5 inhibited the activity of these LPSs in the LAL assay.

In vivo effects like pyrogenicity (in rabbits) and lethality (in GalN-sensitized mice) can be elicited by intravenous injection of purified endotoxins. mAb WN1 222-5 suppressed endotoxin-induced pyrogenicity and lethality (Table 2; Di Padova *et al.*, 1993) which was associated with a decrease in plasma levels of TNF-α (Bailat *et al.*, 1997). This finding correlates with the thesis that, in the GalN sensitization model, lethality is directly correlated with TNF-α levels (Lehmann, Freudenberg & Galanos, 1987). Lethality due to intravenous injection of bacteria, however, is independent of TNF-α levels but is rather correlated with efficient

Table 2. Effects of anti-LPS antibodies on lethality and TNF levels in blood of D-GalN-sensitized C57BL/6J mice injected with LPS of *E. coli* O111 (experiment I), O127 (experiment II) and O18 (experiment III).

Treatment	Mouse group [peak TNF level (ng ml⁻¹)]*	
	Survivors	Non-survivors
Experiment I		
Saline	2 (0)	6 (5·1)
Type-specific mAb	8 (0)†	0
WN1 222-5	7 (0)	1 (0·5)
SDZ 219-800	8 (0)†	0
Experiment II		
Isotype-matched antibody	3 (0)	9 (2·1)
WN1 222-5	12 (0)†	0
Experiment III		
Saline	0	5 (1·4)
Type-specific mAb	5 (0)†	0
WN1 222-5	4 (0·1)	1 (5·1)
SDZ 219-800	5 (0)†	0

*Values in parentheses are the median measured in plasma 1·5 h after LPS challenge.
†$P <0.01$ using the Mann–Whitney–Wilcoxon test comparing non-survivors in the control group with survivors among anti-LPS-treated mice.

clearance of bacteria. In an infection model, mAb WN1 222-5 was able to confer significant protection (six of eight mice survived after injection of *E. coli* O111, whereas all control mice died; Bailat *et al.*, 1997).

However, after i.p. challenge of mice with the same bacterial strain, mAb WN1 222-5 failed to protect mice at 10^4 c.f.u., whereas a type-specific antibody directed against the O-specific chain was able to guarantee survival up to a challenge with 10^5 c.f.u. This discrepancy may be explained by differences in tissue distribution and growth conditions of the bacteria and by altered accessibility of the epitope recognized by mAb WN1 222-5.

Finally, a model of gut ischaemia was chosen to analyse the possible protective activity of mAb WN1 222-5 in hypovolaemic shock (Bahrami *et al.*, 1997).

Shock-induced gut ischaemia is characterized by a systemic inflammatory response, leading to lung injury and death. In rats subjected to hypovolaemic shock, mAb WN1 222-5 reduced endotoxin and TNF-α levels in the blood, reduced lung damage, and most importantly, lowered lethality significantly (28·6% in the treatment group vs 78·6% in the control group) (Bahrami *et al.*, 1997).

The mechanisms by which mAb WN1 222-5 exerts its protective effect may not be necessarily the same in the various models investigated. Protection against viable bacteria is likely to depend on the ability to activate complement and opsonize bacteria leading to an enhanced clearance by phagocytic cells. However, the activity against free LPS *in vivo* leading to reduction of plasma cytokine levels, pyrogenicity and lethality in GalN-sensitized mice, or the *in vitro* reduction of cytokine release by macrophages and neutralizing effects in the LAL assay are somewhat more difficult to explain, since the antibody does not bind to lipid A. Lipid A represents the bioactive region of the LPS molecule and is essential for the induction of the LPS effects *in vitro* and *in vivo*. As mAb WN1 222-5 inhibits the recognition and uptake of S- and R-form LPS by cells expressing mCD14, it seems likely that the protective effect *in vivo* is due to the interference of complex formation between LPS, LBP and mCD14, thus preventing binding, which is a crucial event in the activation of target cells at low concentrations of LPS. Its ability to inhibit the LAL assay *in vitro* may be explained by steric hindrance, conformational changes which may render the LPS molecule inactive or which may lead to different biologically inactive supramolecular structures.

For a conceivable application in patients, a humanized chimeric IgG1 isotype (SDZ 219-800) was created from WN1 222-5 to achieve better tolerance and half-life of the antibody. All cross-reactive and cross-protective activities of mAb WN1 222-5 were retained in mAb SDZ 219-800 (Bailat *et al.*, 1997).

In another study aimed at the induction of cross-reactive antibodies, Nnalue (1998) investigated differences in antisera obtained after immunization with *S. enterica* S- and R-type LPS of Ra and Re chemotypes. Anti-Ra antisera were cross-reactive with different *S. enterica* strains. However, reactivity against *E. coli* was only observed after cell lysis and proteinase K treatment. Since the same core structure is widely distributed in different *Salmonella* spp., this result was not surprising. The data confirmed, however, that outer core regions of LPS were accessible to antibodies even in the presence of the O-specific chain of various serotypes. The failure to cross-react with viable *E. coli* to a significant extent is likely due to different structural properties of the outermost portion of the outer core region against which these antibodies are directed.

mAb WN1 222-5 and SDZ 219-800 fundamentally differ from other core-reactive antibodies. Their epitope comprises the outer and inner core and is not located at the non-reducing glycosyl terminus. Their protective capacity is not linked to lipid A reactivity, to a repetitive determinant or to the IgM isotype. Their anti-endotoxic, i.e. protective, properties are likely to be associated with high affinity.

Conclusions

The search for cross-reactive antibodies against LPS of different bacterial origin to be used for prevention and treatment of severe infections began half a century ago with the discovery of bacterial rough mutants, with a defect in LPS biosynthesis, and the realization that similar core structures are found in different pathogenic Gram-negative bacteria. First results with polyclonal antisera appeared to be promising, and initiated a large number of studies, but ultimately failed to provide convincing evidence that cross-reactive and cross-protective antibodies could be generated. It was only recently that by use of a new immunization scheme it was demonstrated that it is, in principle, possible to obtain cross-reactive mAbs which offer protection in various *in vitro* and *in vivo* model systems. Increasing knowledge of LPS core structures of pathogenic bacteria other than *E. coli* and *S. enterica*, in particular such as *Pseudomonas* and *Klebsiella*, will allow attempts to generate antibodies with even broader specificities. Worldwide concern over severe infections caused by Gram-negative bacteria is increasing as is interest in one of the principal inducers of septic shock, i.e. endotoxin. Both this concern and growing interest represent important driving forces in the development of new strategies against bacterial sepsis. In this context, the development of cross-protective antibodies remains a powerful option which must be and will be explored in great detail by scientists and clinicians in the years to come.

References

Bahrami, S., Redl, H., Yao, Y. M. & Schlag, G. (1996). Involvement of bacteria/endotoxin translocation in the development of multiple organ failure. *Current Topics in Microbiology and Immunology* 216, 239–258.

Bahrami, S., Yao, Y. M., Leichtfried, G., Redl, H., Schlag, G. & Di Padova, F. E. (1997). Monoclonal antibody to endotoxin attenuates hemorrhage-induced lung injury and mortality in rats. *Critical Care Medicine* 25, 1030–1036.

Bailat, S., Neumann, D., Le Roy, D., Baumgartner, J. D., Rietschel, E. Th., Glauser, M. P. & Di Padova, F. E. (1997). Similarities and disparities between core-specific and O-chain-specific antilipopolysaccharide monoclonal antibodies in models of endotoxemia and bacteremia in mice. *Infection and Immunity* 65, 811–814.

Baumberger, C., Ulevitch, R. J. & Dayer, J. M. (1991). Modulation of endotoxic activity of lipopolysaccharide by high-density lipoprotein. *Pathobiology* 59, 378–383.

Baumgartner, J. D. & Glauser, M. P. (1993). Immunotherapy of endotoxemia and septicemia. *Immunobiology* 187, 464–477.

Baumgartner, J. D., Heumann, D., Calandra, T. & Glauser, M. P. (1991). Antibodies to lipopolysaccharides after immunization of humans with the rough mutant *Escherichia coli* J5. *Journal of Infectious Diseases* 163, 769–772.

Beamer, L. J., Carroll, S. F. & Eisenberg, D. (1997). Crystal structure of human BPI and two bound phospholipids at 2.4 angstrom resolution. *Science* 276, 1861–1864.

Beutler, B. & Cerami, A. (1988). Tumor necrosis, cachexia shock, and inflammation: a common mediator. *Annual Reviews of Biochemistry* 57, 505–518.

Bhat, N. M., Bieber, M. M., Chapman, C. J., Stevenson, F. K. & Teng, N. N. (1993). Human antilipid A monoclonal antibodies bind to human B cells and the i antigen on cord red blood cells. *Journal of Immunology* 151, 5011–5021.

Bigham, E. C., Gragg, C. E., Hall, W. R., Kelsey, J. E., Mallory, W. R., Richardson, D. C., Benedict, C. & Ray, P. H. (1984). Inhibition of arabinose 5-phosphate isomerase. An approach to the inhibition of bacterial lipopolysaccharide biosynthesis. *Journal of Medicinal Chemistry* 27, 717–726.

Bone, R. C., Balk, R. A., Fein, A. M. & 7 other authors (1995). A second large controlled clinical study of E5, a monoclonal antibody to endotoxin: results of a prospective, multicenter, randomized, controlled trial. *Critical Care Medicine* 23, 994–1006.

Boulukos, K. E., Pognonec, P., Sariban, E., Bailly, M., Lagrou, C. & Ghysdael, J. (1990). Rapid and transient expression of Ets2 in mature macrophages following stimulation with cMGF, LPS, and PKC activators. *Genes & Development* 4, 401–409.

Brabetz, W. & Rietschel, E. Th. (1999). *Bakterielle Endotoxine: Chemische Konstitution und Biologische Wirkung.* Nordrhein-Westfälische Akademie der Wissenschaften. Vorträge, N 440. Westdeutscher Verlag.

Brade, L., Rietschel, E. Th., Kusumoto, S., Shiba, T. & Brade, H. (1986). Immunogenicity and antigenicity of synthetic *Escherichia coli* lipid A. *Infection and Immunity* 51, 110–114.

Brade, L., Brandenburg, K., Kuhn, H. M., Kusumoto, S., Macher, I., Rietschel, E. T. & Brade, H. (1987a). The immunogenicity and antigenicity of lipid A are influenced by its physicochemical state and environment. *Infection and Immunity* 55, 2636–2644.

Brade, L., Kosma, P., Appelmelk, B. J., Paulsen, H. & Brade, H. (1987b). Use of synthetic antigens to determine the epitope specificities of monoclonal antibodies against the 3-deoxy-D-*manno*-octulosonate region of bacterial lipopolysaccharide. *Infection and Immunity* 55, 462–466.

Brade, L., Holst, O. & Brade, H. (1993). An artificial glycoconjugate containing the bisphosphorylated glucosamine disaccharide backbone of lipid A binds lipid A monoclonal antibodies. *Infection and Immunity* 61, 4514–4517.

Brade, L., Engel, R., William, J. C. & Rietschel, E. Th. (1997). A nonsubstituted primary hydroxyl group in the 6′ position of free lipid A is required for binding of lipid A monoclonal antibodies. *Infection and Immunity* 65, 3961–3965.

Braude, A. I. & Douglas, H. (1972). Passive immunization against the local Shwartzman reaction. *Journal of Immunology* 108, 505–512.

Brearly, S., Harris, R. I., Stone, P. C. W. & Keighley, M. R. B. (1985). Endotoxin levels in portal and systemic blood. *Digestive Surgery* 2, 70–72.

Bruins, S. C., Stumacher, R., Johns, M. A. & McCabe, W. R. (1977). Immunization with R mutants of *Salmonella minnesota. Infection and Immunity* 17, 16–20.

Cauwels, A., Frei, K., Sansano, S., Fearns, C., Ulevitch, R., Zimmerli, W. & Landmann, R. (1999). The origin and function of soluble CD14 in experimental bacterial meningitis. *Journal of Immunology* 162, 4762–4772.

Chow, J. C., Young, D. W., Golenbock, D. T., Christ, W. J. & Gusovsky, F. (1999). Toll-like receptor-4 mediates lipopolysaccharide-induced signal transduction. *Journal of Biological Chemistry* 274, 10689–10692.

Christ, W. J., Asano, O., Robidoux, A. L. & 19 other authors (1995). E5531, a pure endotoxin antagonist of high potency. *Science* 268, 80–83.

Claesson, A., Luthman, K., Gustafsson, K. & Bondesson, G. (1987). A 2-deoxy analogue of KDO as the first inhibitor of the enzyme CMP-KDO synthetase. *Biochemical and Biophysical Research Communications* 143, 1063–1068.

Coleman, D. L., Bartiss, A. H., Sukhatme, V. P., Liu, J. & Rupprecht, H. D. (1992). Lipopolysaccharide induces Egr-1 mRNA and protein in murine peritoneal macrophages. *Journal of Immunology* 149, 3045–3051.

Collart, M. A., Belin, D., Vassalli, J. D. & Vassalli, P. (1987). Modulations of functional activity in differentiated macrophages are accompanied by early and transient increase or decrease in c-fos gene transcription. *Journal of Immunology* 139, 949–955.

Cornelissen, J. J., Mäkel, I., Algra, A., Benaissa-Trouw, B. J., Schellekens, J. F. P., Kraaijeveld, C. A. & Verhoef, J. (1993). Protection against lethal endotoxemia by anti-lipid A murine monoclonal antibodies: comparison of efficacy with that of human anti-lipid A monoclonal antibody HA-1A. *Journal of Infectious Diseases* 167, 876–881.

Cygler, M., Rose, D. R. & Bundle, D. R. (1991). Recognition of a cell-surface oligosaccharide of pathogenic *Salmonella* by an antibody Fab fragment. *Science* 253, 442–445.

Daniel-Issakani, S., Spiegel, A. M. & Strulovici, B. (1989). Lipopolysaccharide response is linked to the GTP binding protein, Gi2, in the promonocytic cell line U937. *Journal of Biological Chemistry* 264, 20240–20247.

Davis, C. E., Ziegler, E. J. & Arnold, K. F. (1978). Neutralization of meningococcal endotoxin by antibody to core glycolipid. *Journal of Experimental Medicine* 147, 1007–1017.

Dendorfer, U., Oettgen, P. & Libermann, T. A. (1994). Multiple regulatory elements in the interleukin-6 gene mediate induction by prostaglandins, cyclic AMP, and lipopolysaccharide. *Molecular and Cellular Biology* 14, 4443–4454.

Ding, A., Hwang, S., Lander, H. M. & Xie, Q. W. (1995). Macrophages derived from C3H/HeJ (Lpsd) mice respond to bacterial lipopolysaccharide by activating NF-kappa B. *Journal of Leukocyte Biology* 57, 174–179.

Di Padova, F. E., Brade, H., Barclay, R. & 8 other authors (1993). A broadly cross-protective monoclonal antibody binding to *Escherichia coli* and *Salmonella* lipopolysaccharides. *Infection and Immunity* 61, 3863–3872.

Di Padova, F., Gram, H., Barclay, G. R., Poxton, I. R., Liehl, E. & Rietschel, E. Th. (1996).

Monoclonal antibodies to endotoxin core as a new approach in endotoxemia therapy. In *Novel Therapeutic Strategies in the Treatment of Sepsis*, pp. 13–31. Edited by D. C. Morrison & J. L. Ryan. New York: Marcel Dekker.

Dong, Z. Y., Lu, S. & Zhang, Y. H. (1989). Effects of pretreatment with protein kinase C activators on macrophage activation for tumor cytotoxicity, secretion of tumor necrosis factor, and its mRNA expression. *Immunobiology* 179, 382–394.

Dong, Z., O'Brian, C. A. & Fidler, I. J. (1993). Activation of tumoricidal properties in macrophages by lipopolysaccharide requires protein-tyrosine kinase activity. *Journal of Leukocyte Biology* 53, 53–60.

Dong, Z., Qi, X. & Fidler, I. J. (1993). Tyrosine phosphorylation of mitogen-activated protein kinases is necessary for activation of murine macrophages by natural and synthetic bacterial products. *Journal of Experimental Medicine* 177, 1071–1077.

Dunn, D. L. & Ferguson, R. M. (1982). Immunotherapy of gram-negative bacterial sepsis: enhanced survival in a guinea pig model by use of rabbit antiserum to *Escherichia coli* J5. *Surgery* 92, 212–219.

Edmond, D. M., Poxton, I. R. & Di Padova, F. (1993). The accessibility of cross-reactive anti-lipopolysaccharide-core monoclonal antibodies to *Escherichia coli* grown in sub-MICs of temocillin and other antibiotics. *Journal of Antimicrobial Chemotherapy* 31, 673–680.

Elsbach, P. & Weiss, J. (1993). The bactericidal/permeability-increasing protein (BPI), a potent element in host-defense against gram-negative bacteria and lipopolysaccharide. *Immunobiology* 187, 417–429.

Evans, T. J., Carpenter, A., Moyes, D., Martin, R. & Cohen, J. (1995). Protective effects of a recombinant amino-terminal fragment of human bactericidal/permeability-increasing protein in an animal model of gram-negative sepsis. *Journal of Infectious Diseases* 171, 153–160.

Feist, W., Ulmer, A. J., Wang, M. H. & 9 other authors (1992). Modulation of lipopolysaccharide-induced production of tumor necrosis factor, interleukin 1 and interleukin 6 by synthetic precursor Ia of lipid A. *FEMS Microbiology Immunology* **89**, 73–90.

Ferguson-Chanowitz, K. M., Katocs, A. S., Jr, Pickett, W. C., Kaplan, J. B., Sass, P. M., Oronsky, A. L. & Kerwar, S. S. (1990). Platelet-activating factor or a platelet-activating factor antagonist decreases tumor necrosis factor-α in the plasma of mice treated with endotoxin. *Journal of Infectious Diseases* **162**, 1081–1086.

Flad, H.-D., Loppnow, H., Rietschel, E. Th. & Ulmer, A. J. (1993). Agonists and antagonists for lipopolysaccharide-induced cytokines. *Immunobiology* **187**, 303–316.

Flegel, W. A., Baumstark, M. A., Weinstock, C., Berg, A. & Northoff, H. (1993). Prevention of endotoxin-induced monokine release by human low- and high-density lipoproteins and by apolipoprotein A. *Infection and Immunity* **61**, 5140–5146.

Freudenberg, M. A., Bog-Hansen, T. C., Back, U. & Galanos, C. (1980). Interaction of lipopolysaccharides with plasma high-density lipoprotein in rats. *Infection and Immunity* **28**, 373–380.

Freudenberg, M. A., Keppler, D. & Galanos, C. (1986). Requirement for lipopolysaccharide-responsive macrophages in galactosamine-induced sensitization to endotoxin. *Infection and Immunity* **51**, 891–895.

Freudenberg, N., Piotraschke, J., Galanos, C., Sorg, C., Askaryar, F. A., Klosa, B., Usener, H. U. & Freudenberg, M. A. (1992). The role of macrophages in the uptake of endotoxin by the mouse liver. *Virchows Archiv B Cell Pathology* **61**, 343–349.

Fujihara, M., Muroi, M., Muroi, Y., Ito, N. & Suzuki, T. (1993). Mechanism of lipopolysaccharide-triggered junB activation in a mouse macrophage-like cell line (J774). *Journal of Biological Chemistry* **268**, 14898–14905.

Fujihara, M., Connolly, N., Ito, N. & Suzuki, T. (1994). Properties of protein kinase C isoforms (beta II, epsilon, and zeta) in a macrophage cell line (J774) and their roles in LPS-induced nitric oxide production. *Journal of Immunology* **152**, 1898–1906.

Galanos, C., Lüderitz, O. & Westphal, O. (1971). Preparation and properties of antisera against the lipid-A component of bacterial lipopolysaccharides. *European Journal of Biochemistry* **24**, 116–122.

Galanos, C., Rietschel, E. Th., Lüderitz, O. & Westphal, O. (1972). Biological activities of lipid A complexed with bovine-serum albumin. *European Journal of Biochemistry* **31**, 230–233.

Galanos, C., Lüderitz, O., Rietschel, E. Th. & Westphal, O. (1977). Newer aspects of the chemistry and biology of bacterial lipopolysaccharides with special reference to their lipid A component. In *Biochemistry of Lipids II*, pp. 239–371. Edited by T. W. Goodwin. Baltimore: University Park Press.

Galanos, C., Lehmann, V., Lüderitz, O. & 7 other authors (1984). Endotoxic properties of chemically synthesized lipid A part structures. Comparison of synthetic lipid A precursor and synthetic analogues with biosynthetic lipid A precursor and free lipid A. *European Journal of Biochemistry* **140**, 221–227.

Galanos, C., Lüderitz, O., Rietschel, E. Th. & 9 other authors (1985). Synthetic and natural *Escherichia coli* lipid A express identical endotoxic activities. *European Journal of Biochemistry* **148**, 1–5.

Giroir, B. P., Quint, P. A., Barton, P. & 7 other authors (1997). Preliminary evaluation of recombinant amino-terminal fragment of human bactericidal/permeability-increasing protein in children with severe meningococcal sepsis. *Lancet* **350**, 1439–1443.

Glaser, K. B., Asmis, R. & Dennis, E. A. (1990). Bacterial lipopolysaccharide priming of P388D1 macrophage-like cells for enhanced arachidonic acid metabolism. Platelet-activating factor receptor activation and regulation of phospholipase A2. *Journal of Biological Chemistry* **265**, 8658–8664.

Goldman, R., Kohlbrenner, W., Lartey, P. & Pernet, A. (1987). Antibacterial agents specifically inhibiting lipopolysaccharide synthesis. *Nature* **329**, 162–164.

Greisman, S. E. & Johnston, C. A. (1988). Failure of antisera to J5 and R595 rough mutants to reduce endotoxemic lethality. *Journal of Infectious Diseases* **157**, 54–64.

Hambleton, J., McMahon, M. & DeFranco, A. L. (1995). Activation of Raf-1 and mitogen-

Kirschning, C. J., Wesche, H., Merrill, A. T. & Rothe, M. (1998). Human toll-like receptor 2 confers responsiveness to bacterial lipopolysaccharide. *Journal of Experimental Medicine* **188**, 2091–2097.

Kitchens, R. L., Ulevitch, R. J. & Munford, R. S. (1992). Lipopolysaccharide (LPS) partial structures inhibit responses to LPS in a human macrophage cell line without inhibiting LPS uptake by a CD14-mediated pathway. *Journal of Experimental Medicine* **176**, 485–494.

Knirel, Y. A. & Kochetkov, N. K. (1994). The structure of lipopolysaccharides in Gram-negative bacteria. III. The structure of O-antigens. *Biochemistry (Moscow)* **59**, 1325–1383.

Kominato, Y., Galson, D., Waterman, W. R., Webb, A. C. & Auron, P. E. (1995). Monocyte expression of the human prointerleukin 1 beta gene (IL1B) is dependent on promoter sequences which bind the hematopoietic transcription factor Spi-1/PU.1. *Molecular and Cellular Biology* **15**, 59–68.

Kuhn, H.-M. (1993). Cross-reactivity of monoclonal antibodies and sera directed against lipid A and lipopolysaccharides. *Infection* **21**, 179–186.

Kuhn, H.-M., Brade, L., Appelmelk, B. J., Kusumoto, S., Rietschel, E. Th. & Brade, H. (1992). Characterization of the epitope specificity of murine monoclonal antibodies directed against lipid A. *Infection and Immunity* **60**, 2201–2210.

Kunkel, S. L., Spengler, M., May, M. A., Spengler, R., Larrick, J. & Remick, D. (1988). Prostaglandin E2 regulates macrophage-derived tumor necrosis factor gene expression. *Journal of Biological Chemistry* **263**, 5380–5384.

Lamping, N., Dettmer, R., Schroeder, N. W., Pfeil, D., Hallatschek, W., Burger, R. & Schuhmann, R. R. (1998). LPS-binding protein (LBP) protects mice from septic shock caused by LPS or Gram-negative bacteria. *Journal of Clinical Investigation* **101**, 2065–2071.

Law, B. J. & Marks, M. I. (1985). Age-related prevalence of human serum IgG and IgM antibody to the core glycolipid of *Escherichia coli* J5, as measured by ELISA. *Journal of Infectious Diseases* **151**, 988–994.

Lee, J. D., Dato, K., Tobias, P. S., Kirkland, T. N. & Ulevitch, R. J. (1992). Transfection of CD14 into 70Z/3 cells dramatically enhances the sensitivity to complexes of lipopolysaccharide (LPS) and LPS binding protein. *Journal of Experimental Medicine* **175**, 1697–1705.

Lee, J. D., Kravchenko, V., Kirkland, T. N., Han, J., Mackman, N., Moriarty, A., Leturcq, D., Tobias, P. S. & Ulevitch, R. J. (1993). Glycosyl-phosphatidylinositol-anchored or integral membrane forms of CD14 mediate identical cellular responses to endotoxin. *Proceedings of the National Academy of Sciences, USA* **90**, 9930–9934.

Lehmann, V., Freudenberg, M. A. & Galanos, C. (1987). Lethal toxicity of lipopolysaccharide and tumor necrosis factor in normal and D-galactosamine-treated mice. *Journal of Experimental Medicine* **165**, 657–663.

Levin, J., Alving, C. R., Munford, R. S. & Redl, H. (editors) (1995). *Bacterial Endotoxins. Lipopolysaccharides from Genes to Therapy. Progress in Clinical and Biological Research*, vol. 392. New York: Wiley-Liss.

Levine, M. (1987). *Escherichia coli* that cause diarrhea: enterotoxigenic, enteropathogenic, enteroinvasive, enterohemorrhagic, and enteroadherent. *Journal of Infectious Diseases* **155**, 377–389.

Lodie, T. A., Savedra, R. J., Golenbock, D. T., Van Beveren, C. P., Maki, R. A. & Fenton, M. J. (1997). Stimulation of macrophages by lipopolysaccharide alters the phosphorylation state, conformation, and function of PU.1 via activation of casein kinase II. *Journal of Immunology* **158**, 1848–1856.

Loppnow, H., Brade, H., Dürrbaum, I., Dinarello, C. A., Kusumoto, S., Rietschel, E. Th. & Flad, H.-D. (1989). Interleukin 1 induction capacity of defined lipopolysaccharide partial structures. *Journal of Immunology* **142**, 3229–3238.

Lynn, W. A. & Golenbock, D. T. (1992). Lipopolysaccharide antagonists. *Immunology Today* **13**, 271–276.

McCabe, W. R. (1992). Antibody to endotoxin in the treatment of gram-negative sepsis. *Journal of the American Medical Association* **267**, 2325.

McCloskey, R. V., Straube, R. C., Sanders, C., Smith, S. M. & Smith, C. R. (1994). Treatment activated protein kinase in murine macrophages partially mimics lipopolysaccharide-induced signaling events. *Journal of Experimental Medicine* 182, 147–154.

Hampton, R. Y. & Raetz, C. R. (1991). Macrophage catabolism of lipid A is regulated by endotoxin stimulation. *Journal of Biological Chemistry* 266, 19499–19509.

Han, J., Lee, J. D., Bibbs, L. & Ulevitch, R. J. (1994). A MAP kinase targeted by endotoxin and hyperosmolarity in mammalian cells. *Science* 265, 808–811.

Haziot, A., Chen, E., Ferrero, E., Low, M. G., Siver, R. & Goyert, S. M. (1988). The monocyte differentiation antigen, CD14, is anchored to the cell membrane by a phosphatidylinositol linkage. *Journal of Immunology* 141, 547–552.

Heine, H., Brade, H., Kusumoto, S., Kusama, T., Rietschel, E. Th., Flad, H. D. & Ulmer, A. J. (1994). Inhibition of LPS binding on human monocytes by phosphonooxyethyl analogs of lipid A. *Journal of Endotoxin Research* 1, 14–20.

Helander, I., Lindner, B., Brade, H., Altmann, K., Lindberg, A. A., Rietschel, E. Th. & Zähringer, U. (1988). Chemical structure of the lipopolysaccharide of *Haemophilus influenzae* strain I-69 Rd$^-$/b$^+$. *European Journal of Biochemistry* 177, 483–492.

Heumann, D., Baumgartner, J. D., Jacot-Guillarmod, H. & Glauser, M. P. (1991). Antibodies to core lipopolysaccharide determinants: absence of cross-reactivity with heterologous lipopolysaccharides. *Journal of Infectious Diseases* 163, 762–768.

Heumann, D., Lengacher, S., Le Roy, D., Jongeneel, C. V. & Glauser, M. P. (1998). The pathogenic role of LBP in gram-negative sepsis and septic shock. *Progress in Clinical and Biological Research* 397, 379–386.

Holst, O. & Brade, H. (1992). Chemical structure of the core region of lipopolysaccharides. In *Molecular Biochemistry and Cellular Biology*, pp. 135–170. Edited by D. C. Morrison & J. L. Ryan. Boca Raton, FL: CRC Press.

Horwitz, A. H., Williams, R. E. & Nowakowski, G. (1995). Human lipopolysaccharide-binding protein potentiates bactericidal activity of human bactericidal permeability-increasing protein. *Infection and Immunity* 63, 522–527.

Hoshino, K., Takeuchi, O., Kawai, T., Sanjo, H., Ogawa, T., Takeda, Y., Takeda, K. & Akira, S. (1999). Cutting edge: Toll-like receptor 4 (TLR4)-deficient mice are hyporesponsive to lipopolysaccharide: evidence for TLR4 as the Lps gene product. *Journal of Immunology* 162, 3749–3752.

Intravenous Immunoglobulin Study Group (1992). Prophylactic intravenous administration of standard immune globulin as compared with core-lipopolysaccharide immune globulin in patients at high risk of postsurgical infection. *New England Journal of Medicine* 327, 234–240.

Izbicki, J. R., Raedler, C., Anke, A. & 8 other authors (1991). Beneficial effect of liposome-encapsulated muramyl tripeptide in experimental septicemia in a porcine model. *Infection and Immunity* 59, 126–130.

Jack, R. S., Fan, X., Bernheiden, M. & 10 other authors (1997). Lipopolysaccharide-binding protein is required to combat a murine gram-negative bacterial infection. *Nature* 389, 742–745.

Jakway, J. P. & DeFranco, A. L. (1986). Pertussis toxin inhibition of B cell and macrophage responses to bacterial lipopolysaccharide. *Science* 234, 743–746.

Jann, K. & Jann, B. (1984). Structure and biosynthesis of O-antigens. In *Chemistry of Endotoxin*, pp. 138–186. Edited by E. Th. Rietschel. Amsterdam: Elsevier.

Jansson, P. E., Wollin, R., Bruse, G. W. & Lindberg, A. A. (1989). The conformation of core oligosaccharides from *Escherichia coli* and *Salmonella typhimurium* lipopolysaccharides as predicted by semi-empirical calculations. *Journal of Molecular Recognition* 2, 25–36.

Johns, M. A., Bruins, S. C. & McCabe, W. R. (1977). Immunization with R mutants of *Salmonella minnesota*. *Infection and Immunity* 17, 9–15.

Kauffmann, F. (1972). *Serological Diagnosis of Salmonella Species. Kauffmann–White Scheme*. Copenhagen: Munksgaard International Publishers.

Khan, S. A., Everest, P., Servos, S. & 7 other authors (1998). A lethal role for lipid A in *Salmonella* infections. *Molecular Microbiology* 29, 571–579.

of septic shock with human monoclonal antibody HA-1A. A randomized, double-blind, placebo-controlled trial. *Annals of Internal Medicine* 121, 1–5.

Mamat, U., Rietschel, E. Th. & Schmidt, G. (1995). Repression of lipopolysaccharide biosynthesis in *Escherichia coli* by an antisense RNA of *Acetobacter methanolicus* phage Acm1. *Molecular Microbiology* 15, 1115–1125.

Mamat, U., Seydel, U., Grimmecke, D., Holst, O. & Rietschel, E. T. (1999). Lipopolysaccharides. In *Comprehensive Natural Products Chemistry*. Edited by Sir D. Barton, K. Nakanishi & O. Meth-Cohn. Vol. III, *Carbohydrates and Their Derivatives Including Tannins, Cellulose, and Related Lignins*, pp. 179–239. Edited by B. M. Pinto. Oxford: Elsevier.

Mannion, B., Kalatzis, A., Weiss, J. & Elsbach, P. (1989). Preferential binding of the neutrophil cytoplasmic granule-derived bactericidal/permeability-increasing protein to target bacteria. Implications and use as a means of purification. *Journal of Immunology* 142, 2807–2812.

Marchant, A., Alegre, M. L., Hakim, A. & 7 other authors (1995). Clinical and biological significance of interleukin-10 plasma levels in patients with septic shock. *Journal of Clinical Immunology* 15, 266–273.

Marks, M. I., Ziegler, E. J., Douglas, H., Corbeil, L. B. & Braude, A. I. (1982). Induction of immunity against lethal *Haemophilus influenzae* type b infection by *Escherichia coli* core lipopolysaccharide. *Journal of Clinical Investigation* 69, 742–749.

Massamiri, T., Tobias, P. S. & Curtiss, L. K. (1997). Structural determinants for the interaction of lipopolysaccharide binding protein with purified high density lipoprotein: role of apolipoprotein A-I. *Journal of Lipid Research* 38, 516–525.

May, M. J. & Ghosh, S. (1998). Signal transduction through NF-kappa B. *Immunology Today* 19, 80–88.

Miner, K. M., Manyak, C. L., Williams, E. & 8 other authors (1986). Characterization of murine monoclonal antibodies to *Escherichia coli* J5. *Infection and Immunity* 52, 56–62.

Morrison, D. C. (1998). Antibiotic-mediated release of endotoxin and the pathogenesis of Gram-negative sepsis. In *Endotoxin and Sepsis*, pp. 199–207. Edited by J. Levin, M. Pollack, T. Yokochi & M. Nakano. New York: Wiley-Liss.

Müller, J. M., Ziegler-Heitbrock, H. W. & Bäuerle, P. A. (1993). Nuclear factor kappa B, a mediator of lipopolysaccharide effects. *Immunobiology* 187, 233–256.

Müller-Loennies, S., Holst, O. & Brade, H. (1994). Chemical structure of the core region of *Escherichia coli* J-5 lipopolysaccharide. *European Journal of Biochemistry* 224, 751–760.

Müller-Loennies, S., Zähringer, U., Seydel, U., Kusumoto, S., Ulmer, A. J. & Rietschel, E. T. (1998). What we know and what we don't know about the chemical structure of lipopolysaccharide in relation to bioactivity. *Progress in Clinical and Biological Research* 397, 51–72.

Müller-Loennies, S., Holst, O., Lindner, B. & Brade, H. (1999). Isolation and structural analysis of phosphorylated oligosaccharides obtained from *Escherichia coli* J-5 lipopolysaccharide. *European Journal of Biochemistry* 260, 235–249.

Munford, R. S. & Hall, C. L. (1986). Detoxification of bacterial lipopolysaccharides (endotoxins) by a human neutrophil enzyme. *Science* 234, 203–205.

Munford, R. S. & Hall, C. L. (1989). Purification of acyloxyacyl hydrolase, a leukocyte enzyme that removes secondary acyl chains from bacterial lipopolysaccharides. *Journal of Biological Chemistry* 264, 15613–15619.

Munford, R. S., Hall, C. L. & Dietschy, J. M. (1981). Binding of *Salmonella typhimurium* lipopolysaccharides to rat high-density lipoproteins. *Infection and Immunity* 34, 835–843.

Newell, C. L., Deisseroth, A. B. & Lopez-Berestein, G. (1994). Interaction of nuclear proteins with an AP-1/CRE-like promoter sequence in the human TNF-alpha gene. *Journal of Leukocyte Biology* 56, 27–35.

Nichols, W. A., Raetz, C. R. H., Clementz, T., Smith, A. L., Hanson, J. A., Ketterer, M. R., Sunshine, M. & Apicella, M. A. (1997). *htrb* of *Haemophilus influenzae*. Determination of biochemical activity and effects on virulence and lipopolysaccharide toxicity. *Journal of Endotoxin Research* 4, 163–172.

Nnalue, N. A. (1998). α-GlcNAc-1→2-α-Glc, the Salmonella homologue of a conserved lipopolysaccharide motif in the Enterobacteriaceae, elicits broadly cross-reactive antibodies. *Infection and Immunity* **66**, 4389–4396.

Nogare, D. (1991). Southwestern internal medicine conference: septic shock. *American Journal of Medical Science* **302**, 50–65.

Olsthoorn, M. M., Petersen, B. O., Schlecht, S., Haverkamp, J., Bock, K., Thomas-Oates, J. E. & Holst, O. (1998). Identification of a novel core type in *Salmonella* lipopolysaccharide. Complete structural analysis of the core region of the lipopolysaccharide from *Salmonella enterica* sv. Arizonae O62. *Journal of Biological Chemistry* **273**, 3817–3829.

Onishi, H. R., Pelak, B. A., Gerckens, L. S. & 8 other authors (1996). Antibacterial agents that inhibit lipid A biosynthesis. *Science* **274**, 980–982.

Parent, J. B., Gazzano-Santoro, H., Wood, D. M., Lim, E., Pruyne, P. T., Trown, P. W. & Conlon, P. J. (1992). Reactivity of monoclonal antibody E5 with endotoxin. II. Binding to short- and long-chain smooth lipopolysaccharides. *Circulatory Shock* **38**, 63–73.

Parillo, J. E. (1993). Pathogenic mechanism of septic shock. *New England Journal of Medicine* **328**, 1471–1477.

Peter, G., Chernow, M., Keating, M. H., Ryff, J. C. & Zinner, S. H. (1982). Limited protective effect of rough mutant antisera in murine *Escherichia coli* bacteremia. *Infection* **10**, 228–232.

Peterson, A. A. & Munford, R. S. (1987). Dephosphorylation of the lipid A moiety of *Escherichia coli* lipopolysaccharide by mouse macrophages. *Infection and Immunity* **55**, 974–978.

Pfeiffer, R. (1892). Untersuchungen über das Choleragift. *Zeitschrift für Hygiene* **11**, 393–412.

Poltorak, A., He, X., Smirnova, I. & 11 other authors (1998). Defective LPS signaling in C3H/HeJ and C57BL/10ScCr mice: mutations in Tlr4 gene. *Science* **282**, 2085–2088.

Prins, J. M., van Agtmael, M. A., Kuijper, E. J., van Deventer, S. J. & Speelman, P. (1995). Antibiotic-induced endotoxin release in patients with gram-negative urosepsis: a double-blind study comparing imipenem and ceftazidime. *Journal of Infectious Diseases* **172**, 886–891.

Qureshi, S. T., Lariviere, L., Leveque, G., Clermont, S., Moore, K. J., Gros, P. & Malo, D. (1999). Endotoxin-tolerant mice have mutations in Toll-like receptor 4 (Tlr4). *Journal of Experimental Medicine* **189**, 615–625.

Raetz, C. R. H. (1996). Bacterial lipopolysaccharides: a remarkable family of bioactive macroamphiphiles. In *Escherichia coli and Salmonella: Cellular and Molecular Biology*, pp. 1035–1063. Edited by F. C. Neidhardt and others. Washington, DC: American Society for Microbiology.

Raponi, G., Keller, N., Overbeek, B. P., Rozenberg-Arska, M., Torensma, R. & Verhoef, J. (1991). Immunization of mice with antibiotic-treated *Escherichia coli* results in enhanced protection against challenge with homologous and heterologous bacteria. *Journal of Infectious Diseases* **163**, 122–127.

Rietschel, E. Th. & Brade, H. (1992). Bacterial endotoxins. *Scientific American* **267**, 54–61.

Rietschel, E. Th. & Galanos, C. (1977). Lipid A antiserum-mediated protection against lipopolysaccharide- and lipid A-induced fever and skin necrosis. *Infection and Immunity* **15**, 34–49.

Rietschel, E. Th., Brade, L., Holst, O., Kulshin, V. A., Lindner, B., Moran, A. P., Schade, U., Zähringer, U. & Brade, H. (1990). Molecular structure of bacterial endotoxin in relation to bioactivity. In *Cellular and Molecular Aspects of Endotoxin Reactions*, pp. 15–32. Edited by A. Nowotny, J. J. Spitzer & E. J. Ziegler. Amsterdam: Elsevier.

Rietschel, E. Th., Seydel, U., Zähringer, U. & 7 other authors (1991). Bacterial endotoxin: molecular relationships between structure and activity. *Infectious Disease Clinics of North America* **5**, 753–779.

Rietschel, E. Th., Kirikae, T., Schade, F. U. & 9 other authors (1996a). Bacterial endotoxin: molecular relationships of structure to activity and function. *FASEB Journal* **218**, 217–225.

Rietschel, E. Th., Brade, H., Holst, O. & 19 other authors (1996b). Bacterial endotoxin: chemical constitution, biological recognition, host response, and immunological detoxification. *Current Topics in Microbiology and Immunology* **216**, 39–81.

Robinson, M. J. & Cobb, M. H. (1997). Mitogen-activated protein kinase pathways. *Current Opinion in Cell Biology* **9**, 180–186.

Romulo, R. L. C., Palardy, J. E. & Opal, S. M. (1993). Efficacy of anti-endotoxin monoclonal antibody E5 alone or in combination with ciprofloxacin in neutropenic rats with *Pseudomonas* sepsis. *Journal of Infectious Diseases* **167**, 126–130.

Rozalski, A., Brade, L., Kosma, P., Appelmelk, B. J., Krogmann, C. & Brade, H. (1989). Epitope specificities of murine monoclonal and rabbit polyclonal antibodies against enterobacterial lipopolysaccharides of the Re chemotype. *Infection and Immunity* **57**, 2645–2652.

Sarabia-Garcia, F., Lopez-Herrera, F. J. & Pino-Gonzalez, M. S. (1994). A new synthesis for 2-deoxy-Kdo, a potent inhibitor of CMP-KDO synthetase. *Tetrahedron Letters* **35**, 6709–6712.

Schlag, G., Redl, H., Davies, J. & Scannon, P. (1999). Protective effect of bactericidal/permeability-increasing protein (rBPI21) in baboon sepsis is related to its antibacterial, not antiendotoxin, properties. *Annals of Surgery* **229**, 262–271.

Schromm, A. B., Brandenburg, K., Loppnow, H., Zähringer, U., Rietschel, E. Th., Carroll, S. F., Koch, M. H. J., Kusumoto, S. & Seydel, U. (1998). The charge of endotoxin molecules influences their conformation and interleukin-6 inducing capacity. *Journal of Immunology* **161**, 5464–5471.

Schumann, R. R., Leong, S. R., Flaggs, G. W., Gray, P. W., Wright, S. D., Mathison, J. C., Tobias, P. S. & Ulevitch, R. J. (1990). Structure and function of lipopolysaccharide binding protein. *Science* **249**, 1431–1433.

Schütt, C., Schilling, T., Grunwald, U., Schönfeld, W. & Krüger, C. (1992). Endotoxin-neutralizing capacity of soluble CD14. *Research in Immunology* **143**, 71–78.

Schwartz, R. W., Arden, W. A., Pofahl, W., Derbin, M., Oremus, R., Greenberg, R. N. & Gross, D. R. (1993). Effect of anti-lipid A monoclonal antibody (E5) on microcirculatory function during lipopolysaccharide shock. *Journal of Surgical Research* **54**, 474–479.

Seydel, U., Brandenburg, K. & Rietschel, E. T. (1994). A case for an endotoxic conformation. *Progress in Clinical and Biological Research* **388**, 17–30.

Shapira, L., Takashiba, S., Champagne, C., Amar, S. & Van Dyke, T. E. (1994). Involvement of protein kinase C and protein tyrosine kinase in lipopolysaccharide-induced TNF-alpha and IL-1 beta production by human monocytes. *Journal of Immunology* **153**, 1818–1824.

Siber, G. R., Kania, S. A. & Warren, H. S. (1985). Cross-reactivity of rabbit antibodies to lipopolysaccharides of *Escherichia coli* J5 and other gram-negative bacteria. *Journal of Infectious Diseases* **152**, 954–964.

Stegmayr, B., Björck, S., Holm, S., Nisell, J., Rydvall, A. & Settergren, B. (1992). Septic shock induced by group A streptococcal infection: clinical and therapeutic aspects. *Scandinavian Journal of Infectious Diseases* **24**, 589–597.

Strieter, R. M., Remick, D. G., Ward, P. A., Spengler, R. N., Lynch, J. P., Larrick, J. & Kunkel, S. L. (1988). Cellular and molecular regulation of tumor necrosis factor-alpha production by pentoxifylline. *Biochemical and Biophysical Research Communications* **155**, 1230–1236.

Sunshine, M. G., Gibson, B. W., Engstrom, J. J., Nichols, W. A., Jones, B. D. & Apicella, M. A. (1997). Mutation of the *htrB* gene in a virulent *Salmonella typhimurium* strain by intergeneric transduction: strain construction and phenotypic characterization. *Journal of Bacteriology* **179**, 5521–5533.

Swierzko, A., Brade, L., Brabetz, W., Zych, K., Paulsen, H. & Brade, H. (1994). Monoclonal antibodies against the heptose region of enterobacterial lipopolysaccharides. *Journal of Endotoxin Research* **1**, 38–44.

Tate, W. J., Douglas, H., Braude, A. I. & Wells, W. W. (1966). Protection against lethality of *E. coli* endotoxin with "O" antiserum. *Annals of the New York Academy of Sciences* **133**, 746–762.

Tebo, J. M., Chaoqun, W., Ohmori, Y. & Hamilton, T. A. (1994). Murine inhibitory protein-kappa B alpha negatively regulates kappa B-dependent transcription in lipopolysaccharide-stimulated RAW 264.7 macrophages. *Journal of Immunology* 153, 4713–4720.

Ulmer, A. J., Feist, W., Heine, H. & 8 other authors (1992). Modulation of endotoxin-induced monokine release in human monocytes by lipid A partial structures that inhibit binding of ^{125}I-lipopolysaccharide. *Infection and Immunity* 60, 5145–5152.

Unger, F. U. (1981). The chemistry and biological significance of 3-deoxy-D-*manno*-octulosonic acid (Kdo). *Advances in Carbohydrate Chemistry and Biochemistry* 38, 323–388.

Van Deventer, S. J. H., Ten Cate, J. W. & Tytgat, G. N. J. (1988). Intestinal endotoxaemia. *Gastroenterology* 94, 825–831.

Vincent, J. L. (1996). Definition and pathogenesis of septic shock. *Current Topics in Microbiology and Immunology* 216, 1–13.

Vogel, S. N. (1990). The role of cytokines in endotoxin-mediated host response. In *Immunopharmacology—the Role of Cells and Cytokines in Immunity and Inflammation*, pp. 238–258. Edited by J. J. Oppenheim & E. M. Shevack. New York: Oxford University Press.

Vogel, S. N., Johnson, D., Perera, P.-Y., Medvedev, A., Lariviere, L., Qureshi, S. T. & Malo, D. (1999). Functional characterization of the effect of the C3H/HeJ defect in mice that lack an *LPS*n gene: in vivo evidence for a dominant negative effect. *Journal of Immunology* 162, 5666–5670.

Weinstein, S. L., Sanghera, J. S., Lemke, K., DeFranco, A. L. & Pelech, S. L. (1992). Bacterial lipopolysaccharide induces tyrosine phosphorylation and activation of mitogen-activated protein kinases in macrophages. *Journal of Biological Chemistry* 267, 14955–14962.

Wenzel, R. P., Pinsky, M. R., Ulevitch, R. J. & Young, L. (1996). Current understandings of sepsis. *Clinical Infectious Diseases* 22, 407–413.

Westphal, O. & Lüderitz, O. (1954). Chemische Erforschung von Lipopolysacchariden Gram-negativer Bakterien. *Angewandte Chemie* 66, 407–417.

Whitfield, C. & Valvano, M. A. (1993). Biosynthesis and expression of cell-surface polysaccharides in Gram-negative bacteria. *Advances in Microbial Physiology* 35, 135–246.

Wiese, A., Brandenburg, K., Carroll, S. F., Rietschel, E. T. & Seydel U. (1997a). Mechanisms of action of bactericidal/permeability-increasing protein BPI on reconstituted outer membranes of gram-negative bacteria. *Biochemistry* 36, 10311–10319.

Wiese, A., Brandenburg, K., Lindner, B., Schromm, A. B., Carroll, S. F., Rietschel, E. T. & Seydel, U. (1997b). Mechanisms of action of the bactericidal/permeability-increasing protein BPI on endotoxin and phospholipid monolayers and aggregates. *Biochemistry* 36, 10301–10310.

Wiese, A., Münstermann, M., Gutsmann, T., Lindner, B., Kawahara, K., Zahringer, U. & Seydel, U. (1998). Molecular mechanisms of polymyxin B–membrane interactions: direct correlation between surface charge density and self-promoted transport. *Journal of Membrane Biology* 162, 127–138.

Winkelhake, J. L., Gauny, S. S., Senyk, G., Piazza, D. & Stevens, P. (1992). Human monoclonal antibodies to glycolipid A that exhibit complement species-specific effector functions. *Journal of Infectious Diseases* 165, 26–33.

Wood, D. M., Parent, J. B., Gazzano-Sontoro, H. & 7 other authors (1992). Reactivity of monoclonal antibody E5 with endotoxin. I. Binding to lipid A and rough lipopolysaccharides. *Circulatory Shock* 38, 55–62.

Wright, S. D., Ramos, R. A., Tobias, P. S., Ulevitch, R. J. & Mathison, J. C. (1990). CD14, a receptor for complexes of lipopolysaccharide (LPS) and LPS binding protein. *Science* 249, 1431–1433.

Wurfel, M. M., Kunitake, S. T., Lichenstein, H., Kane, J. P. & Wright, S. D. (1994). Lipopolysaccharide (LPS)-binding protein is carried on lipoproteins and acts as a cofactor in the neutralization of LPS. *Journal of Experimental Medicine* 180, 1025–1035.

Xie, Y., von Gavel, S., Cassady, A. I., Stacey, K. J., Dunn, T. L. & Hume, D. A. (1993). The resistance of macrophage-like tumour cell lines to growth inhibition by lipopolysaccharide and pertussis toxin. *British Journal of Haematology* 84, 392–401.

Yang, R. B., Mark, M. R., Gray, A. & 7 other authors (1998). Toll-like receptor-2 mediates lipopolysaccharide-induced cellular signalling. *Nature* 395, 284–288.

Yu, B. & Wright, S. D. (1996). Catalytic properties of lipopolysaccharide (LPS) binding protein. Transfer of LPS to soluble CD14. *Journal of Biological Chemistry* 271, 4100–4105.

Zabel, P., Wolter, D. T., Schönharting, M. & Schade, F. U. (1989). Oxpentifylline in endotoxaemia. *Lancet* 2, 1474–1477.

Zähringer, U., Lindner, B. & Rietschel, E. Th. (1994). Molecular structure of lipid A. The endotoxic center of bacterial lipopolysaccharides. *Advances in Carbohydrate Chemistry and Biochemistry* 50, 211–276.

Ziegler, E. J., Douglas, H., Sherman, J. E., Davis, C. E. & Braude, A. I. (1973). Treatment of *E. coli* and *Klebsiella* bacteremia in agranulocytic animals with antiserum to a Udp-Gal epimerase-deficient mutant. *Journal of Immunology* 111, 433–438.

Ziegler, E. J., McCutchan, J. A., Douglas, H. & Braude, A. I. (1975). Prevention of lethal pseudomonas bacteremia with epimerase-deficient *E. coli* antiserum. *Transactions of the Association of American Physicians* 88, 101–108.

Ziegler, E. J., McCutchan, J. A., Fierer, J., Glauser, M. P., Sadoff, J. C., Douglas, H. & Braude, A. I. (1982). Treatment of gram-negative bacteremia and shock with human antiserum to a mutant *Escherichia coli*. *New England Journal of Medicine* 307, 1225–1230.

Ziegler, E. J., Fisher, C. J., Jr., Sprung, C. L., 12 other authors & HA-1A sepsis study group (1991). Treatment of gram-negative bacteremia and septic shock with HA-1A human monoclonal antibody against endotoxin. *New England Journal of Medicine* 324, 429–436.

Biological terrorism

Paul Taylor

Director, CBD Porton Down, Salisbury SP4 0JQ, UK

Introduction

Weapons of mass destruction have been divided into three classes (nuclear, chemical and biological) depending upon the kinds of agents involved. Biological weapons are micro-organisms such as bacteria, fungi or viruses, or their products such as toxins, which are used against an opponent to inflict damage. The use of such weapons by states is banned under the 1972 Convention on the Prohibition of the Development, Production and Stockpiling of Biological and Toxin Weapons. However, terrorist groups do not respect such agreements, and may see biological weapons as a way of achieving their ends.

The last decade of the 20th century gave us the first modern example of terrorism with chemical or biological warfare agents when the Japanese Aum cult killed 12 people in Tokyo using the nerve agent sarin. The Aum were also probably working on botulinum toxin and anthrax as potential weapons for the future.

Since this incident, the subject of bioterrorism has become much more widely discussed, with many works of fiction being written on the subject and a greatly increased level of public concern and government funding, especially in the USA (Mangold & Goldberg, 1999). It is the nature of society these days that we seem to be witnessing the emergence of many small extremist groups, many of whom are showing a local interest in the acquisition of weapons of mass destruction. There is also the possibility of the use of these materials by terrorists supported by a foreign state, though this would invite a very serious response to the state sponsor from the attacked country.

Utility of biological agents to terrorists

Some features of micro-organisms may make biological agents attractive to terrorists. Many bacterial pathogens are relatively easy to grow in large volumes, using basic readily available media. Biological agents are also relatively easy to deliver to the target by means of simple sprays, which may be carried by individuals or on vehicles or aeroplanes, or explosive devices; examples of simple delivery methods are catapults, garden pesticide sprayers, agricultural jet sprayers, umbrellas and motorized pesticide sprayers. The agents themselves may be infectious and cause a large number of casualties by transmission from one person to another. Moreover, the time delay between activation of a device and the first experience of symptoms gives the perpetrators time to disappear. The mere idea that terrorists might have used biological agents may cause fear in the public, which may inspire panic sufficient to have achieved the aims of the hoaxers.

Getting hold of pathogenic micro-organisms is becoming more difficult, but determined individuals or groups will still try to gain access to national and international collections, and those with state sponsorship will have materials supplied to them. One thing that the experience with Aum Shinrikyo has shown us is that we should expect the unusual. Organisms that may be used by terrorists may not be those which we accept as biological warfare agents at all.

The Internet is replete with references to chemical and biological terrorism and is full of advice on how to grow and disseminate micro-organisms. Fortunately, not all the advice is correct, or even safe for the perpetrators to use. The views of many involved in preparing to counter bioterrorism are that people should not be attempting to construct these devices in the first place. Putting the e-world to one side for a moment, much of the information needed to cultivate these organisms is also available in standard microbiological textbooks.

Examples of bioterrorism

There are examples of incidents where micro-organisms or toxins arising from micro-organisms have been used with the intent of causing harm. In the USA in 1998, a right-wing extremist called Larry Wayne Harris threatened to release anthrax in Las Vegas. This incident received enormous coverage in the US media and led to many copycat anthrax hoaxes, mainly directed against government organizations.

In 1984, there were 750 cases of gastro-enteritis resulting from the release of *Salmonella typhimurium* in salad bars in the USA by followers of the Bhagwan Shree Rajneesh. In 1996, a disaffected member of staff in a US organization poisoned 12 of his colleagues with *Shigella dysenteriae*. These examples demonstrate that terrorist groups do have the technical know-how and the motivation to undertake such attacks, so governments must take steps to protect their populations.

Finally, an unusual example of a biological attack occurred in the UK, when the Bulgarian dissident Georgi Markov was assassinated by the Bulgarian KGB using a pellet containing the toxin ricin. This pellet was introduced into his leg by stabbing him with an umbrella tip. Markov became unwell and eventually died. The pellet shown in Fig. 1 was isolated from his leg and shown to contain ricin. It is the only

Fig. 1. Pellet from Markov's leg compared with a pinhead. © British Crown copyright 2000/DERA.

proven example of assassination of an individual by a biological weapon we have been able to identify.

Dealing with the threat

The most important thing to recognize is that the threat of bioterrorism should be kept in proportion. The likelihood of an event is low, although the interest being shown by terrorist groups and others is growing. There are a whole range of disasters and emergencies that can arise and bioterrorism must be seen in the wider context. There are a number of activities that can be pursued to reduce the likelihood of such events occurring and to minimize the effects if they do occur.

Firstly, it is important to control proliferation of materials which may be used by terrorists. The movement of precursor chemicals for the production of chemical weapons is effectively controlled and monitored by the Chemical Weapons Convention (CWC) with more than 170 nations now involved. This year, negotiations on the implementation of a comparable system, the Biological and Toxin Weapons Convention, are due to move to a close. This will impose similar requirements to those under the CWC with relevance to biological materials.

Secondly, it is important to alert others who may be used as alternative sources of supply to the changes. This is where the vigilance of those who keep the national and international culture collections is so important. Also, there is an individual responsibility on all those who work, or supervise work, with dangerous pathogens. Thirdly, governments rely upon intelligence to give warnings of imminent attacks or of increased activity amongst groups or individuals of concern.

In addition, there are many things that can be done to mitigate the potential problem. These range from specialist response teams through to national awareness and training programmes and, in particular, targeting responsible professional bodies such as the Society for General Microbiology and the Society for Applied Microbiology and raising awareness amongst their members.

Conclusion

The threat from bioterrorism is real, but is at a low level. Most disease outbreaks, such as the recent spate of influenza, are natural national occurrences and will continue to be so. However, we should be prepared for any eventuality. CBD Porton Down (Carter, 1992) offers advice and support to government and other organizations aimed at preventing or reducing the impact of any bioterrorism incident should it occur.

References

Carter, G. B. (1992). *Porton Down: 75 years of Chemical and Biological Research.* London: HMSO.

Mangold, T. & Goldberg, J. (1999). *Plague Wars: a True Story of Biological Warfare.* London: MacMillan.

Characterization of bacterial isolates with molecular techniques: multilocus sequence typing

Martin C.J. Maiden

The Wellcome Trust Centre for the Epidemiology of Infectious Disease, Department of Zoology, University of Oxford, South Parks Road, Oxford OX1 3FY, UK

Introduction

In the fight against disease-causing bacteria, epidemiology plays a central role, by generating a knowledge of diseases that enables the rational development and implementation of public health interventions. Such interventions, even in the absence of other technologies, can dramatically reduce or eliminate the burden of disease experienced by human populations. For example, the principles behind John Snow's famous intervention at the Broad Street pump in London during the 1854 cholera epidemic (Tanihara *et al.*, 1998) still enable the populations of large cities to avoid unacceptable levels of enteric infection. Epidemiological studies remain of central importance in identifying, characterizing and combating infectious diseases, playing a prominent role in routine public health work. Accurate isolate characterization is a mainstay of present-day bacterial epidemiology and this chapter will discuss recent developments in the molecular techniques available for this work.

Epidemiological investigations can be divided into three categories: short term, the identification of a given outbreak of disease; medium term, the determination of the extent of an outbreak and its relationships with other disease outbreaks; and long term, the monitoring of national and global trends in the spread of a particular disease agent over extended periods. The objectives of, and approaches employed by, investigations in the different categories can be highly disparate, with short-term epidemiological studies oriented towards local diagnosis and the rapid discrimination of clinical isolates, and studies of the evolution and global spread of pathogens applying approaches derived from evolutionary and population biology. The sometimes wide gap in the techniques and perspectives of those investigating these disparate areas can hinder progress in understanding infectious disease. For example, the data collected from a study with clinical objectives may have little applicability to evolutionary studies and vice versa. This chapter will argue that a multi-disciplinary approach, which unifies clinical and evolutionary perspectives, has a major role to play in improving our effectiveness in fighting bacterial infection. This will be illustrated by describing how improvements in models of bacterial population structure, combined with advances in nucleotide sequence technology and the growth of the Internet, provide opportunities for the development of isolate characterization techniques that unify bacterial epidemiology.

Population biology of bacterial pathogens

For a number of years the clonal model was the predominant paradigm for bacterial population structure. Clonality assumes that genetic exchange is sufficiently rare among members of a given bacterial population that it does not affect its structure (Selander & Levin, 1980; Levin, 1981). In the absence of genetic exchange among bacteria, any variation that arises by mutational processes is confined to the descendants of the bacterial cell in which the mutation first occurred. For epidemiological purposes this provides two powerful simplifying assumptions: first, all members of the bacterial population can be linked to each other by a bifurcating tree, the branches of which are generated by mutation (Dykhuizen *et al.*, 1993); and second, this phylogeny will be recorded in the genetic variation present at all loci of the chromosome. Consequently, an accurate phylogeny establishing the relationships among isolates can, in principle, be reconstructed with data from any locus (Boyd *et al.*, 1996).

More recently, an accumulation of molecular data has indicated that horizontal genetic exchange occurs in many, perhaps most, bacterial species at rates sufficient to disrupt clonal population structures (Maynard Smith, 1991; Maynard Smith, Dowson & Spratt, 1991; Maynard Smith *et al.*, 1993). The transfer of genetic material among bacterial cells that do not necessarily share a recent common ancestor is most clearly observed in bacteria that are naturally transformable, such as those belonging to the genera *Neisseria* (Maiden, 1993), *Streptococcus* (Coffey *et al.*, 1991) and *Haemophilus* (Kroll & Moxon, 1990), but also occurs in other species (DuBose, Dykhuizen & Hartl, 1988), presumably by the processes of transduction and conjugation (Davison, 1999). For epidemiology, relatively frequent recombination has two important consequences: the relationships among members of the population will not be modelled accurately by a bifurcating tree (Guttman & Dykhuizen, 1994); and genetic variation can move from the lineage in which it arose into other lineages (Spratt & Maiden, 1999). Absence of a tree-like phylogeny, together with the mobility of genetic material, makes isolate characterization for the purposes of epidemiological investigation very much more difficult. The phylogenies recorded at different loci are likely to be conflicting, so that analysis of genetic variation at different loci will produce non-congruent relationships among isolates (Holmes, Urwin & Maiden, 1999).

The need for new ways of characterizing bacterial isolates

All bacterial strain characterization ultimately relies on distinguishing genetic variation among isolates accurately. Early successes in the typing of the enteric bacteria, particularly the Kauffmann–White scheme for *Salmonella enterica* (Popoff,
Bockemuhl & Brenner, 1998), established serological typing as a principal means of
isolate differentiation. Many of the schemes subsequently developed, which relied
on polyclonal antisera, were extended with the advent of monoclonal antibodies
(mAbs) (Frasch, 1994); however, immunological typing schemes can have drawbacks. These problems include: a reliance on a small number of antigens, and hence
a small and perhaps unrepresentative sample of the genome; the unpredictable
reactivities of certain typing antibodies—including mAbs—with different antigenic variants (Suker, Feavers & Maiden, 1996); and incongruence of antigenic
type and clonal lineage as a result of diversifying selection by host immune
responses (Li *et al.*, 1994; Whatmore *et al.*, 1994).

From a practical viewpoint, immunological typing schemes rely on panels of
reagents that are expensive to produce and which are frequently not comprehensive. Further, as most serotyping reagents did not originate in humans, they may
not accurately reflect relevant immunological differences among isolates (Delvig *et
al.*, 1995), undermining much of the attractiveness of immunological typing
schemes as a basis for vaccine development. The reagent panels may also have a
built-in redundancy: as the bacterial surface antigens evolve in response to human
immune selection, new reagents are continually required (Maiden & Feavers,
1994). These schemes can be difficult to deploy, requiring laboratory staff with
specialized training to interpret the results accurately, and in some cases these considerations have limited the number of laboratories which can satisfactorily char-
acterize isolates of a given bacterial species. These caveats do not imply that all
immunological typing schemes are flawed, and indeed schemes that are flawed
from a population genetic viewpoint may still provide useful epidemiological
information in particular circumstances, for example in the investigation of local
disease outbreaks. However, these problems limit the general applicability of the
approach, particularly for the investigation of the global epidemiology of
pathogens with non-clonal population structures.

Population biology based typing schemes

The most reliable means of determining the genetic relationships among bacterial
isolates is the characterization of multiple genetic markers located in unlinked
parts of the chromosome, avoiding those genes subject to strong selective pressures
by indexing neutral genetic variation. In practice, this means examining the variation present at a number of housekeeping loci, particularly those genes encoding

intracellular enzymes of intermediary metabolism, an approach established and exemplified by multilocus enzyme electrophoresis (MLEE) (Selander *et al.*, 1986). This technique identifies variants of metabolic enzymes by starch gel electrophoresis and employs cluster analyses of the resultant allelic profiles (electrophoretic types, ETs) to determine relationships among isolates. These relationships are assumed to reflect the genetic relatedness of the isolates. Pioneering MLEE studies of bacterial pathogens laid the foundations of bacterial population biology and revealed the population structures of a number of pathogens (Ochman *et al.*, 1983; Ochman & Selander, 1984; Selander *et al.*, 1986).

Although MLEE is a powerful approach which has had a pivotal role in establishing long-term bacterial epidemiology, it is not without some weaknesses. First, it is not a genetic typing method *sensu stricto*, relying as it does on phenotypic information to infer alleles. In addition, the resolution of starch gel electrophoresis is low, as only those mutations that change migration of the enzyme are detected. Consequently, polymorphisms that are synonymous or that result in biochemically conservative amino acid changes, both of which are ideal candidates for neutral variation, are not detected by this method (Feil, Carpenter & Spratt, 1996). A further practical problem is that MLEE is a rather cumbersome technique that has only been adopted routinely by a handful of laboratories. Nucleotide sequence determination provides a solution to these problems as it is a generic technology that can measure directly and definitively the genetic variation present in any gene of any organism, but until recently a nucleotide sequence based approach was impractical and prohibitively expensive.

Advances in nucleotide sequence technology

The development of automated nucleotide sequence determination technologies that has accompanied the progress of large-scale genome sequencing projects has dramatically increased the speed, and reduced the costs, of nucleotide sequencing. Novel methods are continually under development, but the widespread availability and reliability of dideoxy sequencing (Sanger, Nicklen & Coulson, 1977) are likely to ensure that this approach continues to occupy a central position in nucleotide sequence determination for some time to come. The ability to detect extension products of nucleotide sequence reactions by non-radioactive means was an important advance in this technology (Smith *et al.*, 1986; Prober *et al.*, 1987), particularly the development of dye-terminator chemistries that employ different colours for each base and which can be detected at very low levels (Lee *et al.*, 1997). These reduce sample handling and processing costs, by enabling a complete sequence to be determined on both strands quickly, with virtually no chemical or radioactive hazard. Ready-made sequence reactions are now commercially available, so that the mixing of three components (reaction mix, primer and template) followed by thermocycling is all that is required to generate the sequence extension products. Cycle sequencing with thermostable DNA polymerases (Murray, 1989) ensures that essentially any piece of DNA that can be amplified by PCR can be sequenced. Large-scale sequencing is now performed in microtitre plate format enabling 96 or 384 sequence extension reactions to be performed simultaneously.

Other important advances have been made in the technology available to separate and detect reaction products. There are several systems now available at a range of costs with various levels of throughput, including slab gel and capillary based systems, both of which employ laser and electronic camera technology controlled by computers. For the highest throughput with very low cost, capillary based electrophoresis has the advantage as it eliminates the need to manipulate or analyse slab gels. The most recent instruments have arrays of capillaries capable of 96 simultaneous runs. Additional automation by the integration of robots with the separation instrumentation, which is most easily achieved with capillary based systems, means that production line equipment is now in existence, potentially working 24 h a day, 365 days a year with minimal intervention by operators. Clearly, this holds out the prospect of substantial savings in the staff costs associated with the generation of nucleotide sequence data. In addition to the high-throughput—and expensive—instrumentation, lower cost automated sequencers are also becoming available, and it is not unreasonable to suppose that in the near future automated nucleotide sequencers will be as ubiquitous as thermocyclers are today.

Software improvements have streamlined the manipulation of nucleotide sequence data and a number of powerful suites of programs are available from commercial and academic sources (Staden, 1996) at a variety of prices. The latest sequencing instruments can download data directly to computers equipped with the necessary software for compilation and analysis, further reducing the staff costs required for these processes. Together these advances have delivered very substantial increases in the nucleotide sequence determination capacity available worldwide in the last 2 years, and further substantial increases in this capacity are likely to follow.

The Internet

The Internet has become an essential tool for biomedical research and is also a commonly used resource for epidemiology. Large biological data sets, such as GenBank or complete genome sequences, are ideally accessed and interpreted with systems available through the World Wide Web (www). This avoids the necessity of individual users implementing locally large databases, which have to be updated, and specialized analysis software, which has to be maintained. The elevated interest in the analysis of nucleotide sequence data that has accompanied the proliferation of genome sequencing projects has resulted in an increasing number of sequence manipulation programs, written in the Java computer language, that can be added to web pages. Many of these are made freely available for academic use, and they represent a valuable resource that can be incorporated in new web sites without the necessity of writing new code.

The ability to update and distribute data by email is ideal for the exchange of epidemiological information via systems such as the Program for Monitoring Emerging Infectious Diseases (PROMED; http://www.fas.org/promed/). This is particularly important for the monitoring of diseases which spread globally, and epidemiological data can be disseminated simultaneously with public health advice. An increasing number of resources are available to epidemiologists worldwide on numerous web pages located at centres such as the World Health Organization in Geneva (the Weekly Epidemiological Record provides information on disease outbreaks worldwide, http://www.who.int/wer/) and the Centers for Disease Control in Atlanta (which provides both its Morbidity and Mortality Weekly Report and the journal *Emerging Infectious Diseases* through its web site, http://www.cdc.gov). The decrease in the costs of personal computers and increasing access to the Internet provide the prospect that these resources will continue to be widely available to the epidemiological community.

Harnessing these advances: multilocus sequence typing

Paradoxically, many applications of molecular and recombinant DNA technology to the characterization of pathogen isolates have served to fragment, rather than unify, bacterial epidemiology. It is not difficult to clone or amplify a bacterial gene and index its variation by many of the indirect techniques available, such as restriction fragment length polymorphism. The data obtained from multiple isolates, often in the form of 'fingerprint' patterns generated by the electrophoretic separation of DNA fragments, are commonly interpreted as phylogenetic trees, regardless of whether a tree is an appropriate model for the data. Isolate characterization techniques of this kind have been described jocularly as YATMs (*Yet Another Typing Method*) (Achtman, 1996), but this concept makes serious points: YATMs are of limited epidemiological value and, typically, they are never used outside of the laboratories in which they were developed. In many cases they are difficult to standardize and almost all of them are only suited to short-term epidemiology, that is to say in defining whether the isolates obtained from a suspected disease outbreak are indistinguishable or not, which is relatively easily achieved by a variety of means.

The aim of multilocus sequence typing (MLST) (Maiden *et al.,* 1998) was to provide a unified approach to isolate characterization that could be deployed worldwide to provide data appropriate for clinical diagnosis, epidemiological monitoring and population studies. It was designed to be comprehensive and generic, in that it could be applied in many laboratories, and to accurately reflect the global population structure of bacterial pathogens. A major aim was that the technique would not be dependent on either specialized reagents, which become outdated as bacterial populations evolve, or species-specific data interpretation, which would require training for each species examined. Finally, the data had to be easily portable and comparable among laboratories worldwide.

The MLST approach builds on the success of MLEE by indexing the variation present at multiple housekeeping loci. Unlike MLEE, definitive identification of genetic variation at these loci is achieved by direct nucleotide sequence determination of PCR products. All distinct sequences are given a unique allele number, assigned in order of discovery, and the alleles present at all loci are combined into an allelic profile, or sequence type (ST), for each isolate examined. Relationships among isolates are then apparent by comparing the STs: the members of a clonal lineage will have identical STs, or STs that differ by only two or three loci, and unrelated isolates will have unrelated STs.

A central feature of MLST systems, and one which excites much comment, is that for the purposes of determining STs there is no weighting, such that an allele with a difference of two base pairs from an existing allele is not treated differently from one with 10 or 50 base pair differences. This is important because much, in some species most, bacterial variation is the result of recombination rather than point mutation (Spratt & Maiden, 1999). The consequence of this is that, although an allele with 50 base pair differences may appear to be more diverse from a putative ancestor than an allele with two, it is entirely possible that the two base pair differences are due to two relatively rare events (point mutations) and the 50 base pair differences to a single more common event (horizontal genetic exchange).

From the outset it was intended that the data would be available on the www, ensuring that alleles or allelic profiles determined anywhere in the world could be compared to those already determined and deposited in a central database. As such, MLST www sites (see, for example, http://mlst.zoo.ox.ac.uk) are virtual isolate collections, permanently available to anyone with access to the Internet. These databases have to be specialized, as they organize data in a completely different way to sequence depositories such as GenBank, and they need to be curated. If an MLST database is to be effective for epidemiology, the data have to be accurate as sequence errors will generate non-existent alleles and STs. Fortunately, the advances in nucleotide sequence technology discussed above mean that errors are both less likely to occur and easier to identify than formerly. To ensure accuracy, each sequence from each isolate is determined on both DNA strands. An additional safeguard is the employment of PCR direct nucleotide sequence determination from killed cell suspensions or chromosomal DNA preparations for allele identification. This approach is unlikely to generate sequence errors as there are many templates present in each amplification and sequencing reaction and no cloning is involved. In terms of data analysis, the sequence chromatograms (or 'trace files') produced by automated sequencers are highly reproducible and can, if necessary, be sent by email to a curator to be checked against a database of all known trace files. If this does not resolve the problem, a sample can be sent to the curator for resequencing, but in practice this happens rarely as it is unusual for a base to be both polymorphic in a target sequence and difficult to resolve. Other internal checks are whether the nucleotide differences that define a new allele are novel, that is to say have never been seen before in other alleles, and whether they are non-synonymous.

A final advantage of MLST is that samples required for determination of a sequence type are highly portable. All that is required is a DNA preparation or killed cell suspension, in which nucleases have been inactivated by boiling, both of which are easily transported in the regular mail without the problems associated with transporting potentially infective material or deterioration of the sample. It is also possible to send dried amplified MLST loci or complete sequence reactions by this means.

The design of MLST systems

The size of the nucleotide sequences that define MLST alleles is based on practical constraints, principally the length of nucleotide sequence that can be readily determined with one primer in each direction. This is around 450 bp for many automated nucleotide sequencing instruments and experience with a range of bacterial species is that fragments of housekeeping genes of this size have sufficient genetic variation for MLST. The ideal target gene encodes an intracellular metabolic enzyme, for example one involved in intermediary metabolism, preferably flanked by similar genes. The availability of genome sequences facilitates the choice of genes, but the first MLST systems were developed in the absence of such data.

Experience has shown that for most purposes six or seven housekeeping loci are sufficient for routine isolate characterization (Enright & Spratt, 1998; Maiden *et al.*, 1998). Fewer loci make chance associations of loci too likely and additional loci add little resolution, although it is possible to add more loci to an MLST data set if necessary. During development it is wise to have more candidate loci, say 10–15, than are envisaged for the final MLST system, to allow for unexpected problems. For example, a housekeeping gene could, for unknown reasons, be under a strong selective pressure and therefore unsuitable. This would be evident by an unexpectedly high diversity, perhaps accompanied by a high ratio of non-synonymous to synonymous substitutions (d_N/d_S) (Sharp, 1991). Alternatively, especially if a genome sequence was not available during the choice of loci, the candidate MLST locus may be adjacent to a locus experiencing a strong selection pressure. Increased recombination rates around the selected locus could then increase the variation present in the adjacent candidate locus. This happened during the development of the first MLST system for *Neisseria meningitidis*. A number of candidate loci were adjacent to *opa* loci which are regularly involved in inter- and intragenic recombination. Consequently, these housekeeping loci added little resolution to the MLST scheme and were not used in the final system (Maiden *et al.*, 1998).

Once the loci have been chosen, primers have to be designed and it is highly desirable that a nested strategy is used for this. With this strategy, a larger DNA fragment than is required for the final sequence is amplified by PCR (with amplification primers), which permits the use of distinct primers for sequencing (sequencing primers), which are internal to the amplified fragment. This has a number of advantages. First, sequencing primers internal to an amplified sequence generally produce a better quality nucleotide sequence; second, the possibility of sequencing spurious amplification products is eliminated; and third, lower stringencies, in both the annealing temperature and primer design, can be used for primary amplification. The latter point is important in both overcoming any polymorphism in the regions of the gene used for the amplification primers and to increase sensitivity, for example in applying MLST directly to clinical or environmental samples. Primer design is aided by comparison of the gene sequence with homologues present in other genera to identify those regions most highly conserved.

It is advisable to choose as diverse an isolate collection as possible for the initial stages of development. This should comprise about 100 isolates to ensure that the primers developed will be applicable to as many isolates as possible. In practice, it may be necessary to alter the primers used several times during development and even to use degenerate or multiple primers for amplification: this is particularly the case for diverse species or those with especially high or low GC content, where primer design can be problematic. Once the initial database of 100 isolates has been obtained, it is possible to examine candidate loci for evidence of unexpected selection and to choose those that are most readily amplified and sequenced from most isolates. The involvement of the scientific community by posting primers, protocols and preliminary data on a web site as soon as possible greatly speeds the development process, as increasing the diversity of the isolates examined increases the robustness of the system. Once primer sequences and protocols have been developed, the data can be deposited into an MLST www database system and the scheme becomes operational. After this it is necessary that a curator is appointed and that the database is both maintained and receives regular data submissions to enable up-to-date global epidemiology to be carried out and to ensure that the data submitted and stored in the database are accurate.

Applications of MLST

MLST can be used for epidemiological monitoring of any bacterial pathogen with sufficient variation to index by nucleotide sequencing. The system is equally effective for bacteria with population structures ranging from strictly clonal to non-clonal, but there are a number of limitations to the approach. Some pathogens, for example *Mycobacterium tuberculosis* (Sreevatsan *et al.*, 1997) and *Yersinia pestis* (Achtman *et al.*, 1999b), have been shown by large-scale sequencing studies to be highly uniform in their housekeeping genes. This is probably because they represent single lineages that have evolved a pathogenic lifestyle recently, in evolutionary terms, and sufficient variation has not yet accumulated in their housekeeping genes. An MLST system for these organisms would simply achieve the same level of discrimination as conventional speciation techniques. In such cases it is necessary to index more rapidly evolving genes or characters for epidemiological purposes, such as genes encoding antigens and antibiotic-resistance determinants (Musser, 1995), or the location of insertion sequences (van Helden, 1998).

Other bacteria, for example *Helicobacter pylori* (Suerbaum *et al.*, 1998), are sufficiently diverse that epidemiology is difficult no matter what technique is used (Achtman *et al.*, 1999a). Although all bacteria have a clonal population over the very short term, for example during transmission from one host to another (Suerbaum *et al.*, 1998), it seems that *H. pylori* strains recombine so frequently that clones do not persist within transmission systems, although variation among distinct geographical regions can be recognized (Achtman *et al.*, 1999a), which illustrates the population genetic insights that can be made when sequence data are available. In those bacterial species where MLST is effective in identifying clonal lineages, this may not provide sufficient isolate discrimination for short-term epidemiology, as distinct variants of the same clonal lineage may be circulating in a given human population at a given time. In this case nucleotide sequence determination of one or more highly diverse genes, such as those encoding antigens or antibiotic-resistance determinants, can be used to supplement the MLST data (Vogel *et al.*, 1998; Bygraves *et al.*, 1999; Feavers *et al.*, 1999).

The data from MLST studies can be used to investigate the evolution and population structure of pathogens, especially by referring to allele sequences to recover the information lost in the simplification of the variation present in the housekeeping loci to allele numbers. While it is not necessary to know the population structure of a bacterial population prior to the development of an MLST system, this structure will become apparent as the MLST data become available (Achtman *et al.*, 1999a, b). For clonal population structures, phylogenetic trees constructed for the allele sequences will be both robust and congruent with each other and with a tree built from allelic profiles (Dykhuizen *et al.*, 1993; Boyd *et al.*, 1996). For non-clonal organisms there will be extensive evidence for recombination within the allele sequences and a lack of congruence among phylogenies obtained for individual loci and allelic profiles (Holmes *et al.*, 1999). For some evolutionary analyses, the MLST loci are too short and contain insufficient variation for robust analyses, but this can be overcome by combining data from more than one locus or by using strain collections that have been defined by MLST as a basis for further studies. Once clonal lineages have been identified by MLST, it is possible to analyse the data to establish the most common mechanisms of diversification among those loci (Guttman & Dykhuizen, 1994; Feil *et al.*, 1999).

Conclusions

The MLST approach provides an effective nucleotide sequence based bacterial characterization method that is widely applicable, and which produces data that can be used to address all levels of epidemiological investigation. While this is an important end in itself, it has perhaps more significance as an essential first step in developing population studies of bacterial pathogens that will enable biological diversity to be exploited in studies of bacterial pathogenesis.

On a population level, there are a number of areas still to be addressed for bacterial pathogens. These include the relationships of isolates from cases of invasive disease to the bacterial population as a whole. Such relationships are only straightforward if pathogenesis in humans is a necessary part of the life cycle of the organism in question. For many pathogens, for example those that are principally commensals of humans, pathogens or commensals of other animals, or environmental organisms, current isolate collections are not ideal for population studies as they contain an over-representation of isolates from cases of severe disease and few from non-disease situations. For a number of bacterial species we know that some lineages are very much more likely to be isolated from cases of disease than others (Whittam, 1995; Enright & Spratt, 1998; Maiden *et al.*, 1998), but for most bacteria we have a limited understanding of the population of organisms of low human virulence from which pathogenic lineages have arisen. Of the many such examples, it is salutary to reflect that we know very little concerning the population diversity and structure of an organism as familiar as *Escherichia coli*, as virtually all of our insights come from isolates associated with human disease (Nataro & Kaper, 1998). The commensal state and the mechanisms whereby bacteria can maintain intimate associations with hosts without causing overt disease are also poorly understood, limiting our ability to understand disease.

Many parameters are undetermined for most bacterial populations, including their size, growth rate and generation time. This lack of information inhibits modelling the dynamics of bacterial populations and makes it difficult to estimate evolutionary rates. It is to be hoped that the widespread adoption of sequence based typing techniques, such as MLST, will encourage studies of large representative isolate collections that will generate these necessary data. Hypothesis-driven analysis of very large nucleotide sequence data sets represents a new challenge in infectious disease biology as we enter an era where it is data interpretation, rather than data generation, that limits our understanding of populations of pathogens. As with whole genomes, it is probable that bacteriology will be at the forefront of this effort, and it is likely that it will be a bacterial species which is the first to be investigated by the comparison of many genomes from members of the same species. The challenge is to exploit these new data resources and to deploy them effectively in the fight against infection.

Acknowledgements

My first debt of gratitude is to those members of my research group, past and present, who have contributed to my work in this area over the last 12 years: Rachel Urwin, Janet Suker, Jane Bygraves, Joanne Russell, Keith Jolley, Man-Suen Chan, Emily Thompson, Kate Dingle, Frances Colles, Martin Callaghan and Julia Bennet. I am also indebted to the following collaborators who contributed to the development and implementation of some of the concepts described here: Mark Achtman, Dominique Caugant, Ian Feavers, Eddie Holmes, Brian Spratt and Ed Feil. Finally, I gratefully acknowledge the financial support of: The Wellcome Trust; The Alexander von Humboldt Stiftung; The Meningitis Research Foundation; The National Meningitis Trust; and the Ministry of Agriculture, Fisheries and Food.

References

Achtman, M. (1996). A surfeit of YATMs? *Journal of Clinical Microbiology* 34, 1870.

Achtman, M., Azuma, T., Berg, D. E. & 7 other authors (1999a). Recombination and clonal groupings within *Helicobacter pylori* from different geographical regions. *Molecular Microbiology* 32, 459–470.

Achtman, M., Zurth, K., Morelli, G., Torrea, G., Guiyoule, A. & Carniel, E. (1999b). *Yersinia pestis*, the cause of plague, is a recently emerged clone of *Yersinia pseudotuberculosis*. *Proceedings of the National Academy of Sciences, USA* 96, 14043–14048.

Boyd, E. F., Wang, F. S., Whittam, T. S. & Selander, R. K. (1996). Molecular genetic relationships of the salmonellae. *Applied and Environmental Microbiology* 62, 804–808.

Bygraves, J. A., Urwin, R., Fox, A. J., Gray, S. J., Russell, J. E., Feavers, I. M. & Maiden, M. C. J. (1999). Population genetic and evolutionary approaches to the analysis of *Neisseria meningitidis* isolates belonging to the ET-5 complex. *Journal of Bacteriology* 181, 5551–5556.

Coffey, T. J., Dowson, C. G., Daniels, M., Zhou, J., Martin, C., Spratt, B. G. & Musser, J. M. (1991). Horizontal transfer of multiple penicillin-binding protein genes, and capsular biosynthetic genes, in natural populations of *Streptococcus pneumoniae*. *Molecular Microbiology* 5, 2255–2260.

Davison, J. (1999). Genetic exchange between bacteria in the environment. *Plasmid* 42, 73–91.

Delvig, A., Koumaré, B., Glaser, R. W., Wang, J.-F., Jahn, S. & Achtman, M. (1995). Comparison of three human–murine heteromyeloma cell lines for the formation of human hybridomas after electrofusion with human peripheral blood lymphocytes from meningococcal cases and carriers. *Human Antibodies and Hybridomas* 6, 42–46.

DuBose, R. F., Dykhuizen, D. E. & Hartl, D. L. (1988). Genetic exchange among natural isolates of bacteria: recombination within the *phoA* gene of *Escherichia coli*. *Proceedings of the National Academy of Sciences, USA* 85, 7036–7040.

Dykhuizen, D. E., Polin, D. S., Dunn, J. J., Wilske, B., Preac Mursic, V., Dattwyler, R. J. & Luft, B. J. (1993). *Borrelia burgdorferi* is clonal: implications for taxonomy and vaccine development. *Proceedings of the National Academy of Sciences, USA* 90, 10163–10167.

Enright, M. & Spratt, B. G. (1998). A multilocus sequence typing scheme for *Streptococcus pneumoniae*: identification of clones associated with serious invasive disease. *Microbiology* 144, 3049–3060.

Feavers, I. M., Gray, S. J., Urwin, R., Russell, J. E., Bygraves, J. A., Kaczmarski, E. B. & Maiden, M. C. J. (1999). Multilocus sequence typing and antigen gene sequencing in the investigation of a meningococcal disease outbreak. *Journal of Clinical Microbiology* 37, 3883–3887.

Feil, E., Carpenter, G. & Spratt, B. G. (1996). Electrophoretic variation in adenylate kinase of *Neisseria meningitidis* is due to inter- and intraspecies recombination. *Proceedings of the National Academy of Sciences, USA* 92, 10535–10539.

Feil, E. J., Maiden, M. C. J., Achtman, M. & Spratt, B. G. (1999). The relative contribution of recombination and mutation to the divergence of clones of *Neisseria meningitidis*. *Molecular Biology and Evolution* 16, 1496–1502.

Frasch, C. E. (1994). Serogroup and serotype classification of bacterial pathogens. *Methods in Enzymology* 235, 159–174.

Guttman, D. S. & Dykhuizen, D. E. (1994). Clonal divergence in *Escherichia coli* as a result of recombination, not mutation. *Science* 266, 1380–1383.

van Helden, P. D. (1998). Bacterial genetics and strain variation. *Novatis Foundation Symposium* 217, 178–190.

Holmes, E. C., Urwin, R. & Maiden, M. C. J. (1999). The influence of recombination on the population structure and evolution of the human pathogen *Neisseria meningitidis*. *Molecular Biology and Evolution* 16, 741–749.

Kroll, J. S. & Moxon, E. R. (1990). Capsulation in distantly related strains of *Haemophilus influenzae* type b: genetic drift and gene transfer at the capsulation locus. *Journal of Bacteriology* 172, 1374–1379.

Lee, L. G., Spurgeon, S. L., Heiner, C. R. & 7 other authors (1997). New energy transfer dyes for DNA sequencing. *Nucleic Acids Research* 25, 2816–2822.

Levin, B. R. (1981). Periodic selection, infectious gene exchange and the genetic structure of *E. coli* populations. *Genetics* 99, 1–23.

Li, J., Nelson, K., McWhorter, A. C., Whittam, T. S. & Selander, R. K. (1994). Recombinational basis of serovar diversity in *Salmonella enterica*. *Proceedings of the National Academy of Sciences, USA* 91, 2552–2556.

Maiden, M. C. J. (1993). Population genetics of a transformable bacterium: the influence of horizontal genetical exchange on the biology of *Neisseria meningitidis*. *FEMS Microbiology Letters* 112, 243–250.

Maiden, M. C. J. & Feavers, I. M. (1994). Meningococcal typing. *Journal of Medical Microbiology* 40, 157–158.

Maiden, M. C. J., Bygraves, J. A., Feil, E. & 10 other authors (1998). Multilocus sequence typing: a portable approach to the identification of clones within populations of pathogenic microorganisms. *Proceedings of the National Academy of Sciences, USA* 95, 3140–3145.

Maynard Smith, J. (1991). The population genetics of bacteria. *Proceedings of the Royal Society of London. Series B: Biological Sciences* 245, 37–41.

Maynard Smith, J., Dowson, C. G. & Spratt, B. G. (1991). Localized sex in bacteria. *Nature* 349, 29–31.

Maynard Smith, J., Smith, N. H., O'Rourke, M. & Spratt, B. G. (1993). How clonal are bacteria? *Proceedings of the National Academy of Sciences, USA* 90, 4384–4388.

Murray, V. (1989). Improved double-stranded DNA sequencing using the linear polymerase chain reaction. *Nucleic Acids Research* 17, 8889.

Musser, J. M. (1995). Antimicrobial agent resistance in mycobacteria: molecular genetic insights. *Clinical Microbiology Reviews* 8, 496–514.

Nataro, J. P. & Kaper, J. B. (1998). Diarrheagenic *Escherichia coli*. *Clinical Microbiology Reviews* 11, 142–201.

Ochman, H. & Selander, R. K. (1984). Evidence for clonal population structure in *Escherichia coli*. *Proceedings of the National Academy of Sciences, USA* 81, 198–201.

Ochman, H., Whittam, T. S., Caugant, D. A. & Selander, R. K. (1983). Enzyme polymorphism and genetic population structure in *Escherichia coli* and *Shigella*. *Journal of General Microbiology* 129, 2715–2726.

Popoff, M. Y., Bockemuhl, J. & Brenner, F. W. (1998). Supplement 1997 (no. 41) to the Kauffmann–White scheme. *Research in Microbiology* 149, 601–604.

Prober, J. M., Trainor, G. L., Dam, R. J., Hobbs, F. W., Robertson, C. W., Zagursky, R. J., Cocuzza, A. J., Jensen, M. A. & Baumeister, K. (1987). A system for rapid DNA sequencing with fluorescent chain-terminating dideoxynucleotides. *Science* 238, 336–341.

Sanger, F., Nicklen, S. & Coulson, A. R. (1977). DNA sequencing with chain-terminating inhibitors. *Proceedings of the National Academy of Sciences, USA* 74, 5463–5467.

Selander, R. K. & Levin, B. R. (1980). Genetic diversity and structure in *Escherichia coli* populations. *Science* 210, 545–547.

Selander, R. K., Caugant, D. A., Ochman, H., Musser, J. M., Gilmour, M. N. & Whittam, T. S. (1986). Methods of multilocus enzyme electrophoresis for bacterial population genetics and systematics. *Applied and Environmental Microbiology* 51, 873–884.

Sharp, P. M. (1991). Determinants of DNA sequence divergence between *Escherichia coli* and *Salmonella typhimurium*: codon usage, map position, and concerted evolution. *Journal of Molecular Evolution* 33, 23–33.

Smith, L. M., Sanders, J. Z., Kaiser, R. J., Hughes, P., Dodd, C., Connell, C. R.,

Heiner, C., Kent, S. B. & Hood, L. E. (1986). Fluorescence detection in automated DNA sequence analysis. *Nature* 321, 674–679.

Spratt, B. G. & Maiden, M. C. (1999). Bacterial population genetics, evolution and epidemiology. *Proceedings of the Royal Society of London. Series B: Biological Sciences* 354, 701–710.

Sreevatsan, S., Pan, X., Stockbauer, K. E., Connell, N. D., Kreiswirth, B. N., Whittam, T. S. & Musser, J. M. (1997). Restricted structural gene polymorphism in the *Mycobacterium tuberculosis* complex indicates evolutionarily recent global dissemination. *Proceedings of the National Academy of Sciences, USA* 94, 9869–9874.

Staden, R. (1996). The Staden sequence analysis package. *Molecular Biotechnology* 5, 233–241.

Suerbaum, S., Maynard Smith, J., Bapumia, K., Morelli, G., Smith, N. H., Kunstmann, E., Dyrek, I. & Achtman, M. (1998). Free recombination within *Helicobacter pylori*. *Proceedings of the National Academy of Sciences, USA* 95, 12619–12624.

Suker, J., Feavers, I. M. & Maiden, M. C. J. (1996). Monoclonal antibody recognition of members of the P1<PT>10 variable region family: implications for serological typing and vaccine design. *Microbiology* 142, 63–69.

Tanihara, S., Morioka, S., Kodama, K., Hashimoto, T., Yanagawa, H. & Holland, W. W. (1998). Snow on cholera—the special lecture in the Second British Epidemiology and Public Health Course at Kansai Systems Laboratory on 24 August 1996. *Journal of Epidemiology* 8, 185–194.

Vogel, U., Morelli, G., Zurth, K., Claus, H., Kriener, E., Achtman, M. & Frosch, M. (1998). Necessity of molecular techniques to distinguish between *Neisseria meningitidis* strains isolated from patients with meningococcal disease and from their healthy contacts. *Journal of Clinical Microbiology* 36, 2465–2470.

Whatmore, A. M., Kapur, V., Sullivan, D. J., Musser, J. M. & Kehoe, M. A. (1994). Non-congruent relationships between variation in *emm* gene sequences and the population genetic structure of group A streptococci. *Molecular Microbiology* 14, 619–631.

Whittam, T. S. (1995). Genetic population structure and pathogenicity in enteric bacteria. In *Population Genetics of Bacteria*, pp. 217–245. Society for General Microbiology Symposium 52. Edited by S. Baumberg, J. P. W. Young, E. M. H. Wellington & J. R. Saunders. Cambridge: Cambridge University Press.

Lessons from the first antibiotic era

Alexander Tomasz

The Rockefeller University, 1230 York Avenue, New York, NY 10021, USA

The armaments race of the antibiotic era

The tumultuous century that has just passed has also given birth to one of the most important discoveries in microbiology, a discovery which radically altered the relationship between Man and Microbe. A brief report by Alexander Fleming in a 1929 issue of the *British Journal of Experimental Pathology*, 'On the antibacterial action of cultures of a penicillium . . .' (Fleming, 1929), opened up the era of antibiotics, magic bullets that promised to provide control of infectious diseases that throughout human history—both in peace but particularly in wartime—had remained major causes of mortality. It is ironic that the devoted and ingenious efforts of scientists to isolate, purify and mass produce the medicine that was to save millions of human lives took place in the midst of the most murderous war in human history, World War II. The availability of preparations of penicillin by 1942–1943 undoubtedly saved the lives of tens of thousands of allied soldiers who would have died from infections of deep wounds elicited by the WW II armaments. The first successful mouse protection experiment was completed and described in a brief Note to *Lancet* during August of 1940, at the time of the great German offensive and Dunkirk. The first truly miraculous cures of human disease were also completed and reported shortly afterwards, including the successful cure of a streptococcal meningitis (Fleming, 1943) and the case of a disseminated *Staphylococcus aureus* and *Streptococcus pyogenes* infection. 'A 43-year-old policeman in Oxford, England, was admitted to the hospital in early October of 1940, with disseminated *Staphylococcus aureus* and *Streptococcus pyogenes* infection. His disease began as a sore at the corner of his mouth. He failed all local drainage therapy and also systematically administered sulfapyridine. When the penicillin was begun on February 12, 1941, the infection had already spread to involve most of his face, both orbits, his lungs, and his right arm (with osteomyelitis). Between February 12–17, 4.4 grams of penicillin were administered. This caused dramatic improvement. Infection of the face and arm disappeared, and the policeman's fever subsided. His white blood cell count fell from 20,000 at the start of the therapy to 8,400 at the end. Treatment had to be discontinued because the supply of penicillin was exhausted. The patient relapsed and died of overwhelming staphylococcal infection on March 15, 1941.' (Abraham *et al.*, 1941).

In little more than a year after this promising but ultimately tragic episode, penicillin had become recognized as the new miracle drug. It was available in abundant supply, which undoubtedly would have saved the life of the policeman. Discovery, mass production and introduction into chemotherapeutic practice of penicillin signalled the true beginning of the antibiotic era, which included, in rapid sequel, the discovery and introduction of a large number of generically distinct antibacterial agents matching or even surpassing the efficacy of penicillin in combating disease-causing bacteria. The impact of new antimicrobial agents on the mortality of bacterial infections was immense and, indeed, appeared to be near miraculous in an era in which fatality by bacterial infections was common. This is documented in Table 1 for two bacterial pathogens — *Staphylococcus aureus* and *Streptococcus pneumoniae* — that have remained major causative agents of human disease, both in the pre-antibiotic as well as in the antibiotic era; and the multiresistant and pandemic strains of which have ultimately emerged as major threats in our current era (Table 1). Antibacterial agents such as cephalosporins, tetracycline, chloramphenicol, aminoglycosides, erythromycin, rifampin, trimethoprim, fluoroquinolones, carbapenems and vancomycin were powerful additions to an immense and sophisticated antibacterial armamentarium, which led many to believe that ultimate control over bacterial disease was within reach. However, this particular goal was never reached. Instead, the era of antibiotics was gradually and disappointingly transformed into the era of an armaments race: introduction of each new antibacterial agent was followed by the emergence of a matching bacterial antibiotic-resistance mechanism. Physicians in hospitals all over the world began to report on the appearance of disease-causing bacteria against which formerly

Table 1. Mortality (%) of diseases caused by *Staphylococcus aureus* and *Streptococcus pneumoniae*.

	Treated	Untreated
Staphylococcus aureus		
Pneumonia	10–15	50–70
Bloodstream infection	10	70–80
Wound infection	–	20
Endocarditis	10–15	100
Osteomyelitis	5	30
Streptococcus pneumoniae		
Pneumonia	10 (+)	50
Bloodstream infection	5–10	70
Meningitis	20	100
Otitis media	–	–

usable antibiotics seemed to have lost their efficacy. Particularly troublesome were strains of S. aureus, the same bacterial species that infected the British policeman in 1941. Within 6–8 years of extensive use of penicillin, staphylococcal isolates with decreasing susceptibility to penicillin began to be reported. By the end of the 1940s and early 1950s, penicillin had become virtually useless against S. aureus because of the widespread occurrence of a resistance mechanism among clinical isolates worldwide. This resistance mechanism involved the acquisition by staphylococci of a plasmid from an unknown extra-species source that carried the genetic determinant of penicillinase, a protein that was capable of destroying the penicillin molecule before it could have reached its intracellular targets in the bacterium. By the mid-1950s, this 'plasmid epidemic' had reached global proportions and penicillin had to be abandoned for the therapy of S. aureus infections. The acquisition of the plasmid-born penicillinase by over 95% of all S. aureus has effectively defeated penicillin as a chemotherapeutic agent against these bacteria. Thus ended a stage in the armaments race from which S. aureus emerged as the winner.

In order to overcome the problem posed by the penicillinase-based resistance mechanism, chemists at Beecham Laboratories redesigned the penicillin molecule. The first such penicillinase-resistant compound was introduced in 1957, called celbenin — later renamed methicillin. Initially, this new class of antibiotic turned out to be highly efficient against penicillin-resistant staphylococci, raising the hope that control with this modified antibiotic over staphylococcal infections could be achieved again. However, within 1 year of its introduction, reports began to appear in the clinical microbiology literature describing a new type of staphylococcal antibiotic resistance which made the bacteria resistant against this new class of antibiotics as well. Initially, the S. aureus isolates equipped with this second type of resistance, called methicillin resistance, were rare. However, methicillin-resistant S. aureus (MRSA) soon began to spread, reaching epidemic proportions in Europe by the mid-1970s. Just as this first wave of MRSA seemed to recede from Europe, a second wave of MRSA strains emerged, literally invading southern Europe, the US, Australia and the Far East as well. While the penicillin-resistance mechanism only affected the use of a single type of antibiotic, the methicillin-resistance mechanism was much more devastating: it soon became apparent that this mechanism made useless not only penicillin, but virtually all members of the large family of β-lactam antibiotics, which represented the single most useful and most abundant group of antimicrobial agents developed by the pharmaceutical industry. The only similarity between the penicillinase type of resistance and the mechanism of methicillin resistance was that, in this second case too, the staphylococci resorted to importing a piece of DNA from an unknown extra-species source. This piece of imported DNA — the so-called *mec* element — carried the genetic blueprint (the *mecA* gene) encoding a low-affinity target protein — penicillin-binding protein (PBP) 2A —

which is the critical element of this second type of resistance mechanism in the MRSA strains.

By the beginning of the 1990s, a second period of the armaments race began to close, with the massive spread of MRSA strains, which in hospitals in several European countries, the US and Japan had reached a prevalence of 40–60% of all *S. aureus* isolates. The overwhelming majority of clinical isolates of *S. aureus* have not only retained the penicillinase type of resistance, but have now also equipped themselves with the *mecA* gene. After penicillin, methicillin had to be abandoned too, followed by all the new, more sophisticated, penicillins and cephalosporins, and macrolides, and aminoglycosides. The new and promising DNA inhibitor ciprofloxacin, introduced in the mid-1980s, became useless against many staphylococcal strains within an alarmingly short time after its introduction into therapeutic practice. Surprisingly, perhaps because this new fluoroquinolone was used primarily against MRSA, 80–90% of ciprofloxacin-resistant strains have turned out to be MRSA. This observation was of major concern since new antimicrobials will undoubtedly always be used where they are most needed, i.e. against bacteria already resistant to other antibiotics. This practice may ensure that resistance to a new drug will also preferentially emerge in already multiresistant pathogens.

MRSA strains have continued to pick up resistance genes to many generically different antibacterial agents, and by the mid-1990s, the abbreviation 'MRSA' could stand for '<u>M</u>ultidrug <u>R</u>esistant <u>S</u>. <u>a</u>ureus', against which chemotherapeutic choices were reduced to the glycopeptide antibiotic vancomycin developed specifically as an anti-staphylococcal agent. From the early 1990s on, vancomycin became the antibiotic of choice worldwide in controlling infections caused by resistant strains of *S. aureus*.

Vancomycin-resistant enterococci

In the meantime, as a surprise in the antibiotic resistance armaments race, a powerful vancomycin-resistance mechanism first appeared, not in staphylococci, but in another bacterium, *Enterococcus faecium*. Shortly following their first identification in Europe in 1986, vancomycin-resistant *E. faecium* isolates were detected in hospitals in New York and began to spread in an epidemic fashion in hospitals on the Eastern seaboard of the US. The experience of a hospital in New York City may illustrate this.

Vancomycin-resistant enterococci had never been detected among patient isolates at the New York Hospital prior to the spring of 1990. The first vancomycin-resistant *E. faecium* (VREF) isolate was detected in March of 1990, which was followed by the recovery of 14 (9%) resistant strains among 164 *E. faecium* isolates later during the same year, 29 (14%) of 213 *E. faecium* isolates in 1991, and 158

(53%) of a total of 299 cases in 1992. These surveys included both inpatient and outpatient isolates. The enterococcal isolates were also resistant to most other chemotherapeutically useful antibacterial agents, except for tetracycline and chloramphenicol.

Enterococci (both resistant and susceptible strains together) had become the second most common cause of nosocomial urinary tract infections, and the third most frequent cause of surgical site infections and bloodstream infections in US hospitals belonging to the National Nosocomial Infection System (NNIS) by 1990–1992. Vancomycin-resistant enterococci (both *Enterococcus faecalis* and *E. faecium*) accounted for 7·9% of all nosocomial enterococcal isolates in US hospitals in 1993 (Emori & Gaynes, 1993). The glycopeptide antibiotic vancomycin, used for treatment of severe infections due to Gram-positive bacteria, and the rapid spread of *E. faecium* carrying a resistance mechanism to this antibiotic, has become a major concern.

The enterococcal vancomycin resistance is only similar to the two staphylococcal β-lactam resistance mechanisms in that here too, the resistance genes are 'imported' from an as yet unidentified source. Through the acquired resistance genes, enterococci modify an entire portion of the pathway of cell wall biosynthesis in such a way that the normal building blocks that can be recognized, captured and inactivated by the vancomycin molecules are changed to building blocks that are 'unfamiliar' to the antibiotic molecule. Using these 'camouflaged' cell wall building blocks, vancomycin-resistant enterococci can proceed to synthesize their essential cell wall components and go on multiplying unhindered in the presence of the vancomycin molecule.

Penicillin-resistant *S. pneumoniae*

All clinical isolates of *S. pneumoniae* were exquisitely sensitive to penicillin at the time when penicillin was first introduced into chemotherapeutic practice. In contrast to staphylococci, which quickly learned how to cope with this antibiotic, resistance to penicillin only appeared among clinical isolates of pneumococci nearly a quarter century later. In 1965, an *S. pneumoniae* isolate showing low-level resistance to penicillin was identified in a child in a remote village in Papua New Guinea. Penicillin resistance in this dangerous pathogen was first regarded as a curiosity, until a little more than 10 years later, when, in 1977, the antibiotic-resistant *S. pneumoniae* made a dramatic new appearance. *S. pneumoniae* strains equipped not only with high-level penicillin resistance but with resistance to a number of generically different antibiotics were reported to cause a major epidemic that was sweeping through several hospitals in South Africa. Following that date, penicillin-resistant pneumococci and, more recently, multiresistant strains have begun to be reported with increasing frequency all over the world. By the mid-

1990s, antibiotic-resistant *S. pneumoniae* had spread globally, accounting, in some countries, for 20–40% of all disease-causing isolates.

Similarly to the two β-lactam-resistance mechanisms in *S. aureus* and the vancomycin resistance in enterococci, in *S. pneumoniae* too the mechanisms of penicillin resistance involve acquisition of genetic material from 'abroad'. The pneumococcal resistance mechanism also shows some similarity to the mechanism of methicillin resistance in *S. aureus*. In both cases, it is the antibiotic target molecules in the cells that the resistant bacteria modify in such a way that they become less sensitive to inactivation by the antibiotic molecule. While MRSA strains import a complete resistance gene, *mecA*, encoding a novel, low-affinity target protein, penicillin-resistant pneumococci use foreign DNA sequences to 'patch up' their native penicillin target genes in such a way that the products of these 'mosaic' target genes—the PBPs—have reduced affinity for penicillin. The precise source of the foreign DNA used for the remodelling of the penicillin target genes is not known.

The shrinking armamentarium of effective antimicrobial agents is illustrated in Table 2 for a select group of strains of staphylococci, *S. pneumoniae* and *E. faecium* which have managed to arm themselves with a nearly complete set of resistance mechanisms against currently used antimicrobial agents. *E. faecium* is not a highly virulent organism, but its extensive drug-resistance profile can cause complications and prolongation of therapy—in surgery or in patients with primary disease in the immune system. On the other hand, coagulase-negative staphylococci are frequent causes of foreign-body-related infections in many hospitals, and the MRSA strain shown in the table is one of the major epidemic strains—a representative of the Brazilian clone of MRSA first identified as a widely disseminated clone in hospitals in Brazil and subsequently shown to have spread to other Latin American countries and also to hospitals in Portugal and the Czech Republic.

MRSA isolates, drug-resistant pneumococci and even enterococci recovered at different geographic/infection sites may differ widely in their degrees of resistance to antimicrobial agents. Nevertheless, the fact that strains such as those shown in Table 2 did appear in clinical specimens documents the capacity of bacterial pathogens to match the ingenuity of the pharmaceutical chemist by their own virtually limitless genetic resources in producing mechanisms of antibiotic resistance. The table is focused on Gram-positive pathogens which appear to be dominant among disease-causing bacteria in the current era. However, similar antibiotic-resistance charts may also be constructed with a wide variety of Gram-negative pathogens and mycobacteria as well.

The resistance profile most carefully watched is that of staphylococcal vancomycin susceptibility. Given the virtually exclusive use of this drug against multiresistant *S. aureus*, appearance of a clinically relevant resistance mechanism (e.g. by transfer of the vancomycin-resistance complex from enterococci) could precip-

Table 2. Antibiotic-resistance patterns of some contemporary clones of Gram-positive pathogens.

	Staph. aureus ATCC 6538 (1930)	Strep. pneumoniae 6B Dallas, Texas (1992)	MRSA Brazilian epidemic clone (1994)	Methicillin-resistant Staph. epidermidis New York (1996)	Enterococcus faecium vanA+ F-clone New York (1993)
Amikacin	S	–	R	R	R
Amp/sulbactam	–	R	R	R	R
Ampicillin	S	R	R	R	R
Cephalothin	S	–	R	R	R
Cefotaxime	S	R	–	–	–
Chloramphenicol	S	R	R	R	S
Ciprofloxacin	S	R	R	R	R
Clindamycin	S	R	R	R	R
Erythromycin	S	R	R	R	R
Gentamicin	S	–	R	R	R
Imipenem	S	S	R	R	R
Oxacillin	S	R	R	R	R
Rifampin	S	–	R	R	R
Vancomycin	S	S	S	S	R
Teicoplanin	S	S	S	–	R
Tetracycline	S	R	R	R	S
Trimeth/sulfa	S	R	R	–	R
Mupirocin (topical)	S	–	R	R	–

itate a genuine public health crisis. It is for this reason that a report from Japan in early 1997 generated considerable alarm.

Japanese colleagues reported the failure of vancomycin therapy in some patients who had infections with MRSA and who received the formerly highly efficient vancomycin in therapy (Hiramatsu *et al.*, 1997). The isolates of staphylococci recovered from the patients showed decreased susceptibility to vancomycin (minimal inhibitory concentration value increased from the expected 0·5 to 4 or $8\,\mu g\,ml^{-1}$). Shortly afterwards, similar reports appeared from several hospitals in the US. One of these reports was judged ominous enough to be described for the readership of the Sunday issue of the *New York Times Magazine* in August of 1998, even before the results of a full microbiological and molecular characterization in the laboratory had been published (Sieradzki *et al.*, 1999). The nature of this significant but relatively modest increase in resistance is not known. It may represent the first move in an 'evolutionary experiment' by staphylococci to come up with a mechanism of resistance against vancomycin.

Are there lessons to be learned?

The landscape of infectious diseases at the beginning of the year 2000

One should view the appearance of antibiotic resistance with the background of changes in the landscape of infectious diseases in our era. Infectious diseases staged a worldwide comeback in the 1990s. According to World Health Organization statistics, 1/3 of the 57 million deaths in 1997 were caused by infectious diseases. In the US, re-evaluation of health statistics has led to the move of infectious diseases from fifth to third position as the most frequent cause of mortality in the US (Pinner *et al.*, 1996). Several factors have contributed to this phenomenon. One is the great increase in the size of human populations. Some time during the 1990s, the human population of Earth reached 6 billion people. Out of these, 2 billion were added in the last 25 years (Cohen, 1995). Only approximate figures exist as to the mass displacements and movements of this human population, but it is estimated that at least 100 million people now live in a geographic area different from their birthplace; the projection for the number of people involved with international travel in the early 2000s is about 600 million. On a global scale, dissemination of infectious agents, including multidrug-resistant bacterial strains, has become particularly rapid during the last decade, partly due to disturbances of planetary ecosystems, globalization of food sources, and to the immensely increased mobility of human populations related to political upheaval, emigration and tourism. The appearance of megalopoles with very large concentrations of inhabitants, often living below the poverty line and with less than adequate sanitary facilities, considerably increases the potential threat of infectious diseases, particularly when they emerge in forms resistant to antibiotics.

There is a great increase in the size of human populations at high risk of infections due to immunocompromised status. Modern medical care allows prolonged survival of many patients in hospitals and/or nursing homes who have chronic, debilitating diseases, and requires extensive use of antibiotics to compensate for a weakened immune system. Some of the most advanced achievements of modern medicine involve surgical procedures, which increase the risk of surgical infections. This is also the case with other invasive medical interventions in hospitals, such as the extensive use of foreign bodies (artificial heart valves, hip replacements, indwelling catheters, etc.).

Antibiotics in an evolutionary context

The introduction of antimicrobial agents into therapy introduced a radically new element into the aeons-old face-off between pathogenic microbes of the prokaryotic world and the human host. With the availability of antimicrobial agents, an episode of infectious disease was no longer a 'private affair' between the host defences of an individual and an invading microbe, since the patient could reach for the powerful and selective antimicrobial agents which would settle the conflict in the overwhelming majority of cases in favour of the human host. The discovery and introduction of antimicrobial agents have opened up a kind of second front against the prokaryotic world, and through this the conflict has shifted from the arena of natural selection between human and microbe to a completely novel field: a warfare between microbes and highly toxic chemicals (Fig. 1). In a sense, one may consider the birth of the anti-infective industry as a stage in human evolution in which this arm of the pharmaceutical industry has emerged as a kind of institutionalized host defence—a societal structure ready to replace host defences with a different kind of violent intervention against the invading microbes. The complex and sophisticated weaponry of virulence factors and the counter-weaponry of host defences have evolved on a million year timescale. As compared to that, the antibiotic era, the birthdate of which most would agree was in the early 1940s with a history of barely 60 years, represents a millisecond on this evolutionary timescale. It is all the more remarkable how rapidly and effectively the prokaryotic world has fought back on this second front by the invention and acquisition of resistance genes.

It is important to realize that the overwhelming majority of antimicrobial agents are products of the microbial world (Table 3), and so are many of the antimicrobial-resistance mechanisms that have evolved among the producers of antimicrobial agents as protective mechanisms against a suicidal effect of their own products. The ecological role of antimicrobial agents for their microbial producers may be in the control of a microscopic-size environment against intrusion of other species.

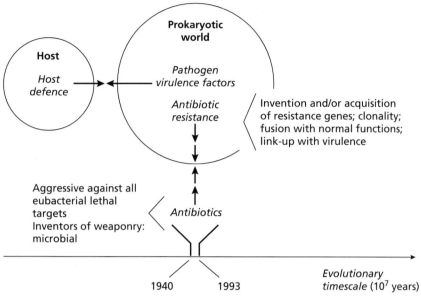

Fig. 1. Emergence of antibiotic resistance in an evolutionary context.

Table 3. Microbial producers of antibiotics.

Antibiotic	Producing organism(s)
Penicillin G	*Penicillium notatum, Penicillium chrysogenum*
Penicillin V	*Penicillium chrysogenum*
Cephalosporin C	*Cephalosporium acremonium*
Cephamycin	*Streptomyces lactamdurans*
Thienamycin	*Streptomyces cattleya*
Clavulanic acid	*Streptomyces clavuligerus*
Sulfazecin (monobactam)	*Pseudomonas acidophila*
SQ 26,180 (monobactam)	*Chromobacterium violaceum*
Vancomycin	*Streptomyces orientalis*
Ristocetin (glycopeptide)	*Nocardia lurida*
Streptomycin	*Streptomyces griseus*
Erythromycin	*Streptomyces erythreus*
Oleandomycin	*Streptomyces antibioticus*
Chloramphenicol	*Streptomyces venezuelae*
Lincomycin	*Streptomyces lincolnensis*
Tetracycline	*Streptomyces viridifaciens, Streptomyces aureofaciens*
Polymyxin	*Bacillus polymyxa*

The reintroduction of these highly toxic chemicals into the environment on a gigantic scale has amplified a local warfare among microbes in a few grams of soil into a global planetary war between Man and Microbe. If one looks for the seeds of the antibiotic armaments race, it was in this removal of antimicrobial agents from their ecological context, both in terms of the amplification in their quantity and also in sharpening the range of efficacy of these agents in the laboratories of the pharmaceutical industry, so that the original agents, which usually are of less wide inhibitory spectrum, are perfected to antimicrobial killers covering as wide a range of bacterial species as possible.

Economy of the antibiotics

The spectacular success of the early antibiotic era has led to the emergence and rapid growth of an anti-infective industry. In 1995, the total pharmaceutical market had an estimated size of 223 billion US dollars out of which 10%, about 22 billion dollars, was in antibiotics, with the major consumers being North America, Europe and Japan, each representing about 1/3 of the total market. With the introduction of enormous economic gains, increasing the sales of antimicrobial agents has become a legitimate (albeit in the long run a dangerous) undertaking of marketing departments of pharmaceutical industries, which began to compete with one another for the sales of their particular antimicrobial products and for the market.

Overuse, unnecessary use, misuse

Extensive literature exists on the amounts of antimicrobial agents introduced into the environment through human and veterinary medicine, agrobusiness, and as growth promoters in the macro-farming industry. A Task Force of the Spanish Ministry of Health has recently provided an overview of antibiotic consumption and its impact on resistance to various antimicrobial agents in that country. In 1976, over 360 tons of antimicrobial agents were used in Spain and this amount increased in 1983 to 340 tons in human use and 250 additional tons in animal use as growth promoters (Baquero & The Task Force of the General Direction for Health Planning of the Spanish Ministry of Health, 1996). In order to appreciate the impact of these quantities, one has to remember that as few as 10 molecules of penicillin per molecule of its protein target (a penicillin-binding protein) in *S. pneumoniae* is sufficient to irreversibly inhibit the vital function of that protein and put the bacterium on a pathway of lethality. The Spanish Task Force also provided a gross estimate of the percentage of bacterial pathogens that showed resistance to various antimicrobial agents. The fraction of resistant bacteria ranged from in the vicinity of 20 up to 80% of isolates. Extensive data are also available documenting the prescription habits of medical practitioners in various countries. It is well established that antimicrobial agents are often prescribed against afflictions which, in the overwhelming majority of cases, are viral in nature or which represent self-limiting diseases. Such prescriptions, particularly in the case of upper respiratory infections in children, represent a very substantial portion of total antibiotic sales.

Besides contributing to the total selective pressure helping the spread of antibiotic-resistant strains, extensive use of antimicrobial agents as growth promoters and as prophylactic agents against plant parasites has recently come under scrutiny upon the realization that antibiotics used in the animal growth industry can select for resistance mechanisms and resistant bacteria, which can make their way through mechanisms of interspecific gene transfer and/or through the zoonotic route into humans, jeopardizing the therapeutic efficacy of antibiotics used in human medicine. As a result of this, the European Union has recently outlawed the use of a number of antimicrobial agents as growth promoters.

Exclusive use

In the mid-1800s, one of the most frequent infections with mortality rates staggering by our contemporary standards was puerperal fever. Tables 4–6 reproduce some figures from the diary of the great Austro-Hungarian physician, Ignacz Semmelweis (Benedek, 1980), who practised in the mid-1800s at the first gynaecological clinic in Vienna, Austria, a hospital that was considered the most advanced of the era. Of the 2000 women who came to give birth at the hospital between March and October of 1846, over 300 died at childbirth, generating a staggering close to 15% mortality in which both the mother and the newborn died. The recognition that the cause of these fatalities was a strictly nosocomial phenomenon (infection

Table 4. Mortality of puerperal fever at the gynaecological clinic I of the public hospital in Vienna, Austria (from the notebook of I. Semmelweis) (Benedek, 1980).

Date (1846)	No. of births	No. of deaths	%
March	311	48	15·4
April	253	48	18·9
May	305	41	13·4
June	266	27	10·1
July	252	33	13·1
August	216	39	18·0
September	271	39	14·3
October	254	38	14·9
Total	2128	313	14·6

Table 5. Introduction of chlorinated-water handwash.

Date (1848)	No. of births	No. of deaths	%
January	283	10	3·5
February	291	2	0·7
March	276	0	0·0
April	305	2	0·6
May	313	3	0·99
June	264	3	1·1
July	269	1	0·3
August	261	0	0·0
September	312	3	0·9
October	299	7	2·3
November	310	9	2·9
December	373	5	1·3
Total	3556	45	1·27

Table 6. Mortality of puerperal fever.

Year	Location	Mortality (%)
1850s	München	16
	Stockholm	22
	Prussia	12
	St Petersburg, Russia	9
	Vienna, Austria	12–18
1906	Vienna, Austria	0·04
1935 (sulfa antibiotics)	Vienna, Austria	0·005

of the pregnant women by careless medical staff, who also participated in autopsies in the same clinic) led to the introduction of chlorinated-water handwashing, which resulted in the precipitous decline in the mortality of puerperal fever. The perfection of such simple hygienic measures eventually led to the decline of the fatality rate (most likely caused by streptococci and/or staphylococci) from 15 to 0.04%. There were no antimicrobial agents used. Introduction of sulfa drugs in the mid-1930s made only a relatively small additional decline in mortality. Interestingly, however, between 1950 and 1970, the rate of infectious complications at childbirth increased into the range of 5–20% while mortality remained low (Table 7). These data together illustrate one of the main problems of the antimicrobial era, namely, the replacement of hygienic methods by treatment with antimicrobial agents, a fact almost certainly responsible for the observation shown in Table 7: the abandonment of strict hygienic measures has again allowed the appearance of infections which nevertheless were prevented from becoming the cause of mortality by the use of antibiotics.

Table 7. Change in the mortality of puerperal fever.

Location	Year	Mortality (%)	Infectious complications at childbirth (%)
Europe-wide	1850	4–8	–
Vienna	1906	0·04	–
Budapest (gynaecological clinic)	1950–1970	0·06	5–20

The virtually complete shift to the use of antimicrobial agents in the control of infectious diseases has led not only to the decline in the observance of hygiene, but to the abandonment of other preventive methods such as vaccinations, and to the disassembly of surveillance systems for monitoring the number and spread of infectious agents. Data from the US available through the National Notifiable Reporting System indicate that, in 1992, the total amount of federal funds spent for monitoring antibacterial and antiviral drug resistance was barely above $48 000 (quoted in Tomasz, 1994), in spite of the fact that by this time extensive reportage on the rising tide of resistant pathogens was available in the professional literature and even the public media.

Wide-spectrum versus narrow-spectrum antibacterial agents

The antibiotic-resistance problem has reignited the debate about the selective power of antimicrobial agents. Particular attention was paid to wide-spectrum antimicrobial agents extensively used in empirical therapy. The possibility of developing narrow-spectrum antimicrobial agents with activity focused on a particular group or on a single pathogen would soon become feasible with the development of rapid molecular-based diagnostics. Wide-spectrum antimicrobial agents are clearly capable of causing wider damage to the commensal flora. It is less clear if wide-spectrum agents also cause more frequent and more widespread resistance. We do not really understand the rate with which different bacterial pathogens develop resistance in response to the same antimicrobial agent. Penicillin, which at the time of its introduction was clearly a wide-spectrum antibiotic, caused the rapid appearance of resistance in staphylococci, but the first resistant strain in the equally penicillin susceptible *S. pneumoniae* only appeared after a quarter of a century, and penicillin-resistant isolates are notoriously and perplexingly still lacking from among group A streptococci, an exquisitely penicillin-susceptible microbe against which penicillin therapy has been used in enormous quantities for the past several decades as a prophylactic measure to prevent rheumatoid complications. Mycobacteria did develop resistance against isoniazid, a truly narrow-spectrum antimicrobial agent, and clinically significant resistance against vancomycin is still lacking among clinical isolates of MRSA, against which this antimicrobial agent has been selectively used in enormous quantities and over considerable periods of time.

The ecology of antibiotic use

The single most often overlooked and perhaps most important 'dark side' of antibiotic use is the indiscriminate nature of these powerful agents (Ciba Foundation, 1997). Use of antibiotics against a specific pathogen causing a specific illness not only brings about the killing of the pathogen, but also causes massive damage to the commensal flora of bacteria that colonize the human body. It is this damage to mostly non-invasive species that offers a window of opportunity to antibiotic-resistant strains to move in and establish a foothold in the ecological niche vacated by the antibiotic treatment. The longer and more extensive the antibiotic treatment, the longer this window of opportunity remains open. This often overlooked aspect of the ecological impact of antibiotic use may be one of the main culprits responsible for the spread of antibiotic-resistant strains and for the success of resistant strains in establishing large reservoirs in a variety of epidemiological settings, such as hospitals or day care centres, where such resistant strains can become a significant part of the flora carried by healthy individuals.

How serious is it?

In the 1941 episode (Abraham *et al.*, 1941), the patient died of overwhelming sepsis because there was not enough penicillin left for continued treatment. Is there a parallel now—running out of therapeutic agents not because of lack of supply, but because of the altered multiresistance pathogen? Mortality of a disease directly attributable to a bacterial infection is extremely difficult to define. Nevertheless, serious, life-threatening problems with the therapy of bacterial disease have been reported in the clinical literature in cases in which resistance to multiple antibacterial agents has played a major role in complications and even in therapeutic failure (Armstrong *et al.*, 1995). The change in the mood and expectations surrounding the triumphs and disappointments of the antibiotic era may be best illustrated by juxtaposing the optimistic statement of the US Surgeon General in his report to the US Congress in 1970 ('we can close the books on infectious diseases') with the gloomy remark by Joshua Lederberg in 1993 ('...barring genosuicide, the human dominion is challenged only by the pathogenic microbes for whom we remain the prey, they the predator'). In 1993, a small group of international experts—microbiologists, physicians and public health experts—gathered at The Rockefeller University for a workshop to survey data on the accelerating spread of resistant pathogens (Tomasz, 1994). The central question posed was this: is the information available sufficient to call the increased rates of resistance among bacterial pathogens a threat to public health? The participants in the workshop felt that the situation had reached a stage which made it necessary for public health agencies to take steps to prevent a potential public health crisis. Specific examples of genetic events that could bring such a crisis situation closer would be the acquisition and spread of high-level vancomycin resistance among MRSA and/or drug-resistant pneumococci, or the acquisition of a β-lactamase plasmid by group A streptococci. The workshop also proposed specific recommendations: efforts to increase awareness of the problem; increased funding for basic research into the mechanisms and epidemiology of antibiotic resistance; improvement of surveillance systems; and encouragement of the pharmaceutical industry to produce new antimicrobial agents by providing a fast track for their approval.

Several of these recommendations have since been implemented. Particularly effective were the public media and governmental agencies in raising public awareness of the problem, and redefining the image of antimicrobials from that of safe commodities to drugs that should only be available through prescription since their overuse or misuse may have long-term deleterious consequences. Also impressive was the surge in putting together national and international surveillance systems, with the participation of both the public and private sectors, including the pharmaceutical industry.

Status of the armaments race at the beginning of the new century

Table 8 shows the balance of the armamentarium: the current initiatives and capa-

Table 8. Balance of armamentarium.

Pharmaceutical chemistry	Pathogens
Drug design targeted on:	More than 100 distinct resistance mechanisms in place: in
Resistance mechanism	pathogens, in commensals
In vivo physiology of pathogen	
Multicomponent novel drugs	
'Smart' drugs, targeted on:	
Specific pathogens	
Virulence genes	Paying 'biological price' of resistance: rapid selection for
	'fitness' genes
Sequence-based (ultrafast) diagnostics	
Reconstitution therapy	Rapid dissemination routes:
	For strains: mass travel
	For genes: plasmids, transposons, integrons
New targets through genome sequences	
Fast-track drug development via:	
Molecular genetics	
Combinatorial chemistry	
Robotic screening	
Reinvesting in prevention, hygienic culture	Multiresistance
Molecular epidemiology	International epidemic clones
Global intelligence networks (surveillance)	Disturbed ecological boundaries

bilities of modern pharmaceutical chemistry as it is facing the antibiotic-resistant bacterial pathogens.

It seems that the current scenario is ripe with the possibility of a major change in the strategy of control of infectious diseases. The frustrations and cost of the accelerating armaments race between antibiotics and antibiotic-resistant bacteria allow one to draw a few sobering conclusions concerning the options available for the control of infectious diseases in the near future.

One option is to continue the armaments race. The availability of new science and technology, such as genomic information on bacterial pathogens, a better understanding of microbial regulatory mechanisms, the appearance of combinatorial chemistry, robotic screening, etc., has injected new confidence into a reawakened pharmaceutical industry and allied biotech companies to resume the armaments race with novel drugs and innovative strategies. Multidrug-resistant bacteria are targeted by multicomponent (two-component) drugs. One component of augmentin is aimed at silencing the β-lactamase activity, which then enables the second drug component to penetrate and inhibit bacterial targets. Similar strategies are being developed to resensitize tetracycline-resistant bacteria by inhibitors of an antibiotic efflux system or to suppress β-lactam resistance by inhibitors of 'auxiliary genes' that are essential for the expression of high-level resistance. Nevertheless, looking back at the history of the antibiotic era, there is little doubt that ultimately the virtually infinite genetic potential of bacteria will produce resistance mechanisms matching in complexity the mechanism of action of the novel drugs. It may be instructive to consider more closely the armaments race metaphor which, in the arena of human against human conflict, has produced an infinite spiral of weapons and counter-weapons, consumed enormous proportions of national budgets, and has led to the dispersal of means of mass destruction to large as well as small nations. It is hard to avoid the analogy of antimicrobial agents of increasing sophistication, widening antibacterial spectrum and killing power; the rapid dispersal of resistance mechanisms over the species barriers of human and animal pathogens; and the establishment of reservoirs of antibiotic-resistance genes in commensal bacteria. It is estimated that the cost of bringing a new antibacterial agent to the market approaches half a billion dollars. Is it not timely to consider the economy of keeping a drug 'alive', through more restricted and prudent use?

These observations and thoughts bring up a second option: to replace the *strategy of bacterial killing* (which is the core philosophy of the armaments race) with a more versatile strategy in which antimicrobial therapy is only one of several components of the *management of microbial diseases*, the other equally important components being prevention (vaccination), renewed emphasis on infection control and general hygiene; monitoring regional use of antibiotics and prescription habits; establishment of national and international surveillance systems; and educational efforts to illuminate the dangers of excessive and unnecessary use of antimicrobial agents.

Slowing down the antibiotic armaments race is particularly urgent since the increasingly indiscriminate use of antimicrobials (as growth promoters; in macro-farming of fish and poultry; as prophylactics in agrobusiness; etc.) is bound to expand the ecological damage of these agents beyond the immediate human ecosphere, spreading the antibiotic-resistance armaments race to more microbial species.

It seems that proponents of the management philosophy are increasing in number, echoing the voice of the late René Dubos (Moberg, 1996), who urged acceptance of the microbial world for what it is, namely, a planetary partner of mankind, composed overwhelmingly by 'friendly' species, many of which are now recognized as critical for life on our planet and some of which are beginning to be used with success in such important tasks as recycling and waste management. It seems that this second strategy is more attuned to the general ecologically conscious thinking in other spheres of human activity as well. The central dilemma in adopting this second strategy seems to be to change the thinking and attitude of the public, the antibiotic prescribers, and the pharmaceutical industry itself. Another important roadblock is the prevailing human attitude, which tends to react to crises rather than trying to prevent them. In spite of the tendency to exaggerate, the public media has done a service in keeping the issue of resurgent infectious diseases and antibiotic resistance alive.

References

Abraham, E. P., Gardner, A. D., Chain, E., Heatley, N. G., Fletcher, C. M., Jennings, M. A. & Florey, H. W. (1941). Further observations on penicillin. *Lancet* ii, 177–189.

Armstrong, D., Neu, H., Peterson, L. R. & Tomasz, A. (1995). The prospects of treatment failure in the chemotherapy of infectious diseases in the 1990s. *Microbial Drug Resistance* 1, 1–4.

Baquero, F. & The Task Force of the General Direction for Health Planning of the Spanish Ministry of Health (1996). Antibiotic resistance in Spain: what can be done? *Clinical Infectious Diseases* 23, 819–823.

Benedek, I. (1980). *Semmelweis*. Hungary: Gondolat.

Ciba Foundation (1997). *Antibiotic Resistance: Origins, Evolution, Selection and Spread*. Edited by D. J. Chadwick & J. Goode. New York: Wiley.

Cohen, J. E. (1995). *How Many People can the Earth Support?* New York: Norton.

Emori, T. G. & Gaynes, R. P. (1993). An overview of nosocomial infections, including the role of the microbiology laboratory. *Clinical Microbiology Reviews* 6, 428–442.

Fleming, A. (1929). On the antibacterial action of cultures of a penicillium with special reference to their use in the isolation of *B. influenzae*. *British Journal of Experimental Pathology* X, 226–236.

Fleming, A. (1943). Streptococcal meningitis treated with penicillin. *Lancet* ii, 434.

Hiramatsu, K., Aritaka, N., Hanaki, H., Kawasaki, S., Hosoda, Y., Hori, S., Fukuchi, Y. & Kobayashi, I. (1997). Dissemination in Japanese hospitals of strains of *Staphylococcus aureus* heterogeneously resistant to vancomycin. *Lancet* 350, 1670–1673.

Moberg, C. L. (1996). René Dubos: a harbinger of microbial resistance to antibiotics. *Microbial Drug Resistance* 2, 287–297.

Pinner, R. W., Teutsch, S. M., Simonsen, L., Klug, L. A., Graber, J. M., Clarke, M. J. & Berkelman, R. L. (1996). Trends in infectious diseases mortality in the United States. *Journal of the American Medical Association* 275, 189–193.

Sieradzki, K., Roberts, R. B., Haber, S. W. & Tomasz, A. (1999). The development of vancomycin resistance in a patient with methicillin-resistant *Staphylococcus aureus* infection. *New England Journal of Medicine* **340**, 517–523.

Tomasz, A. (1994). Multiple-antibiotic-resistant pathogenic bacteria. A report on the Rockefeller University workshop. *New England Journal of Medicine* **330**, 1247–1251.

New strategies for identifying and developing novel vaccines: genome-based discoveries

Ling Lissolo and Marie-José Quentin-Millet

Aventis Pasteur, Campus Mérieux, 1541 avenue Marcel Mérieux, 69280 Marcy L'Etoile, France

Introduction

At the beginning of the 21st century, the potential breakthroughs in the domain of vaccines and vaccination have never been greater. This situation stems from four areas of research that scientists replenish constantly with their new findings: (1) improvements in the quality of vaccines to increase their effectiveness and to diminish side effects and adverse reactions; (2) reduction in the number of injections by combining the existing vaccines in the same vial without affecting the immunogenicity of each component or by the development of new delivery systems such as transdermal, needle-less injection, intranasal and oral or, even better, by blending the potential antigens into edible plants; (3) development of new vaccines benefiting from the progress in biotechnology; and (4) developing the new concept of using vaccines not only to prevent diseases but also to cure established chronic diseases and even cancer.

Vaccination represents the most efficacious and probably one of the most cost-effective measures for the control of infectious disease. Its impact is evident, and since the first vaccination against smallpox by Jenner about 200 years ago, vaccination has had major impacts on at least 10 diseases. Polio will be eradicated at the beginning of this century, and will be the second infectious agent eradicated after smallpox in 1977. Eradication of these diseases would not be possible without a global immunization programme and of course efficacious vaccines. Despite the progress made in vaccines and vaccination, the basic technologies applied for developing new vaccines were, for the most part, still conventional, that is, until the entire genome of a pathogenic bacterium was first unveiled in 1995. The advances in genome-related technologies exploded and allowed at least some areas of vaccine development to enter a new era that biologists had never encountered before.

The present chapter is not intended to be a complete review of the entire field of new vaccine development, but rather focuses on the initial step of this development, which is antigen discovery. This fundamental step has been revolutionized in the last few years, especially in the field of bacterial vaccines, where the availability

217

of complete bacterial genome sequences has led to dramatic changes in antigen discovery strategies.

Vaccinology then and now

Historical milestones

Vaccinology is today almost a science on its own. The term vaccinology was introduced successfully by Jonas Salk in 1977 (Salk & Salk, 1977), almost 100 years after Pasteur had proposed the principle of vaccination in his lecture to the Academy of Science in 1881, in which he had chosen the word 'vaccination' in honour of Jenner and his 'vaccinia'. Vaccinology has grown over the years and has drawn from advances in other disciplines, namely bacteriology, virology, immunology, biochemistry, molecular biology and genetics. It is clear that the history of vaccines is inextricably linked with the advances made in these disciplines (for a historical review on vaccines and vaccinology, see Plotkin & Plotkin, 1999; see also the chapter by Hendriksen in this volume). While the first vaccine developed by Jenner was a live attenuated viral vaccine, other vaccines have been developed based on different technologies. These include the attenuation of viruses through reverse genetics, inactivation of whole micro-organisms, purified subunit or polysaccharide vaccines, conjugate protein–polysaccharide vaccines, purified subunit protein vaccines made by recombinant DNA technology, recombinant live virus or bacteria expressing antigens of interest, even recombinant plants expressing the antigen of interest, and more recently DNA vaccines. The research has been intensive; many are under clinical trials but all the experimental vaccines have not reached the market.

As in most historical developments, major milestones that have been significant landmarks in the field have punctuated the history of vaccines; while the list is certainly not exhaustive, some of these milestones are summarized in Table 1 (Ajjan, 1995).

Pre-genomics: hypothesis-driven research

In the last 25 years or so before the explosion of genome sequences, bacterial vaccine development could be considered as hypothesis-driven. It seems fair at this stage to differentiate between viral vaccines and bacterial vaccines because most of the virus genomes, which are much smaller in size, were partially or even totally sequenced before mass sequencing was available. To develop a new vaccine, it was necessary for the scientist to collect a series of observations in order to deduce a potential vaccine target, which often was based on isolation of virulence factors. The virulence factor, once isolated, would be the basis of the vaccine development

Table 1. Milestones in the history of human vaccine discovery (adapted from Ajjan, 1995).

1798	Publication of Jenner's book
1885	Pasteur's rabies vaccine and post-exposure treatment of Joseph Meister
1896	Typhoid vaccine (Wright)
1921	Tuberculosis vaccine (Calmette and Guérin; BCG)
1923	Diphtheria toxoid (Ramon and Glenny)
1923	Pertussis vaccine (Madsen)
1932	Yellow fever vaccine (Sellard and Laigret)
1937	Yellow fever 17 D vaccine (Thelier)
1937	First inactivated influenza vaccine (Salk)
1938	Clinical evidence for protection with purified pneumococcal capsular polysaccharides (Felton, Ekwursel)
1949	Mumps vaccine (Smorodintsev) (live attenuated vaccine)
1954	Inactivated poliomyelitis vaccine (Salk)
1957	Live attenuated oral poliomyelitis vaccine (Sabin)
1960	Measles vaccines (Edmonston, then Schwarz)
1966	Mumps vaccine (Weibel, Buynach, Hillemann, then Takahashi)
1968	Meningococcus C vaccine (Gotschlich)
1971	Meningococcus A vaccine (Gotschlich)
1976	First administration of the vaccine against hepatitis B (Maupas, then Hillemann)
1978	Vaccine against pneumococcal infection
1980	Vaccine against *Haemophilus influenzae* type b
1982	Hepatitis B produced in yeast
1984	Typhoid Vi vaccine

rationale (or hypothesis), meaning that if one was to develop an immune response that would interfere with this virulence factor or mechanism, one would be likely to protect against the microbial infection. A good example of this hypothesis-driven vaccine research is the development of acellular pertussis vaccines.

In 1984, M. Pittman published a famous article, 'The concept of pertussis as a toxin-mediated disease' (Pittman, 1984), which opened a whole new era in pertussis vaccine development. Although pertussis toxin had been isolated prior to this date (Morse & Morse, 1976) and preliminary results suggested that immunity to the protein interfered with bacterial infection (Munoz, Arai & Cole, 1981), this article exemplifies the hypothesis-driven rationale. Today, after many years of research and development, several acellular pertussis vaccines are on the market or close to it and they all contain detoxified pertussis toxin. Another example, which has not led so far to the development of vaccines that are presently on the market, but which has generated a lot of research, is the field of protein receptors involved in iron uptake (Cornelissen & Sparling, 1994). In the late 1980s, it was demonstrated that pathogenic *Neisseria* grown in iron-deprived culture media expressed novel proteins which bound specifically human iron-loaded proteins such as transferrin and lactoferrin (Schryvers & Morris, 1988; Blanton *et al.*, 1990). Many efforts have focused on using these receptors as vaccine antigens, not only for *Neisseria*, but also for many other bacteria such as *Haemophilus influenzae* or *Moraxella catarrhalis* (Myers *et al.*, 1998). The hypotheses are: firstly, if in encapsulated bacteria these molecules could bind large proteins such as human transferrin without the polysaccharide capsule interfering, they would also be accessible to antibodies; secondly, if antibodies were able to interact with these surface-exposed molecules, they could promote bactericidal killing and opsonophagocytosis, and also interfere with iron uptake and therefore with bacterial growth. It is at this stage too early to predict whether such a vaccine will be licensed soon, but clearly results obtained using animal models have been encouraging (Webb & Cripps, 1999).

The vaccine candidates identified based on *hypothesis-driven research* are numerous; the goal is not to be exhaustive but rather to compare this approach to the new approach, which is based on *discovery-driven research*. This is a method by which all possible vaccine candidates are screened without a pre-defined idea about the mechanism behind it. This new approach is today possible because of the enormous technological leaps made in the last few years in large-scale sequencing, giving access to the total genome sequence of bacteria.

Genomic era: discovery-driven research

The word 'genomics' is recent and immensely popular in scientific literature, biotechnology news and beyond. As with any fashionable expression, the more often it is used, the more it gets abused and its real definition gets blurred. Constructed in a similar way to 'electron*ics*' or 'robot*ics*', genomics in its broad definition refers to the science of genomes. In this chapter, a classical and somewhat narrower definition will be used. Genomics is the part of molecular biology that deals with the determination of the entire genome sequence of a particular organism and the exploitation of the information contained in this sequence.

Genomic sequencing

All biological phenomena involve, at one moment or another, protein activities. As the protein activity depends upon the tridimensional structure of the protein, which itself is the result of the primary structure, the protein sequence is of paramount importance in understanding protein behaviour. This fact was recognized relatively early by biochemists, but with the technology constraints at this time, determination of a protein sequence was out of reach except on very rare occasions.

This epic effort was achieved in the 1950s by Sanger, who sequenced the 51-amino-acid human insulin. At about the same time and on the DNA front, the situation seemed to be more complicated and apparently hopeless. The determination of the complete sequence of the bacteriophage phiX174 (5386 bp) was considered as a unique and somewhat definitive success (Sanger *et al.*, 1977). It was not, however, perceived as the beginning of a new era, the genomic era. The emergence of genomics as a specific field was an important event in which many individuals made significant contributions. This, however, is not the scope of this chapter, but it is important to mention at least two breakthroughs and a general pattern common to all the advancements in science.

The first breakthrough came from the publication of Sanger and others in 1977, which described a method of utilizing dideoxynucleotides as chain terminator for DNA sequencing (Sanger, Nicklen & Coulson, 1977). This method was so successful that it even surprised these authors, who revolutionized everyday life in biochemistry and molecular biology laboratories. To obtain a complete protein sequence, it became easier and faster to isolate and sequence the gene encoding the protein of interest rather than to determine directly the amino acid sequence. This technology allowed nearly every molecular biology laboratory to sequence genes corresponding to proteins of interest.

The second breakthrough appeared in the issue of *Science* on October 15th 1995, when a team from The Institute for Genomic Research (TIGR) under the leadership of C. Venter published the first complete genome sequence of a bacterium, *H. influenzae* (Fleischmann *et al.*, 1995). This extraordinary accomplishment was possible because of the huge technological efforts in mass sequencing undertaken as part of the programme to sequence the human genome (3.5×10^9 bp). The relatively small bacterial genome of *H. influenzae* (1.6×10^6 bp) was used to test the whole shotgun strategy on a real battlefield. However, the results greatly exceeded expectation and the whole sequencing community adopted sooner or later the TIGR strategy. This success was the cumulative result of several technological and conceptual breakthroughs. One of these was the availability of a largely automated DNA sequencer with some fluorescence-based gel readers connected to a computer. In the case of *Haemophilus* sequencing, eight such machines worked around the clock for 6 months. It is one of the rare examples where industrial methods were successfully applied to basic research. Other breakthroughs were the concept of 'shotgun' sequencing and the availability of computers and, above all, computer software which assembled short pieces of sequence reads (typically 450 bp) into large DNA fragments up to 500 000 bp termed 'contigs' (for a comprehensive review, refer to Fraser & Fleischmann, 1997). The assembly of some contigs enables the construction of a complete genome, typically around 1·5 Mb for many bacteria, up to 4·7 Mb for *Escherichia coli*. The use of adequate computer software for the assembly of the contigs was a key success factor in this strategy.

Since 1995, the pace of genome sequencing has run with increasing speed and the list of micro-organisms studied is growing as well (for a complete list, see the review article of Moir *et al.*, 1999). However, these lists are almost immediately outdated by the time they are published and the best solution to have an updated view is to visit the web sites listed in the review of Moir *et al.* (1999) and to consult many public web sites available, including TIGR (http://www.tigr.org/tdb/mdb/mdb.html). Approximately half of the microbial genomes sequenced were done at TIGR, and others by organizations and consortia around the world. One powerful engine behind the sequencing effort of bacterial genomes is the pharmaceutical industry, which partially funded research institutions to sequence the whole genome of many bacterial pathogens, including *Helicobacter pylori*, *Chlamydia pneumoniae* and *Mycobacterium tuberculosis*. Whether by careful planning or random choice, a delicate balance has so far been maintained between microbes of medical or industrial (*Bacillus subtilis*) interest and micro-organisms that interest basic microbiology, such as the extremophile archaeon *Pyrococcus horikoshii* or the methanogen *Methanococcus jannaschii*. Knowledge of the complete genome of these exotic organisms may shed new light on the origin of life and the mechanisms of molecular evolution, whereas the sequence of pathogenic bacteria provides a gold mine and new hope for the discovery of novel bacterial proteins as molecular targets for new antimicrobial drugs (Jones & Fitzpatrick, 1999; Smith, 1996) or as vaccine antigens.

Bioinformatics: the use of genomic information
It may sound strange to people outside the genomics field, but the human brain is absolutely not adapted to extract information harboured in DNA sequences. This work is naturally reserved for DNA-dependent RNA polymerase and computers are well adapted to decipher the same information. History will mark that the progress in genomics and in high-speed computers emerged simultaneously, and not independently, at the end of the last century. Genomics could not have emerged without the computational power of modern microcomputers and the usefulness of sequence data has been greatly improved concomitantly by the development of the Internet, which allowed immediate information sharing.

The scientific profits are, so far, astonishing for basic sciences as well as for industrial or medical applications, as exemplified by the following. (1) The establishment of phylogenetic relationships among species by comparing the genome of different organisms. Closely related species share most of their genes, which display a high degree of identity, and even the global arrangement of the genes in the genome is conserved. On the other hand, distant species do not share many genes, except those common in all living organisms such as those encoding the potassium transporter. (2) The targeting of the cellular localization of a protein (Horton & Nakai, 1996; and their web site, http://psort.nibb.ac.jp). The localization of proteins in different cellular compartments involves complex sorting mechanisms. In Gram-negative bacteria, proteins can be found in not less than six different cellular locations and the leader sequence of each protein determines its cellular destination. Knowledge gained from complete genome sequences enables prediction with a reasonable confidence about a protein's cellular localization. This is of great interest, for instance, to classify and later to study the potential outer-membrane proteins for adherence or host–pathogen interactions. (3) The study of metabolism in a more systematic fashion. With the picture of the whole genome, it becomes possible to predict the metabolism of bacteria. The case of *Mycobacterium leprae* is fascinating. This pathogen, responsible for leprosy, cannot grow outside a limited number of hosts. For generations after generations, medical microbiologists tried to create a medium to grow this bacterium but always failed. Genomic information revealed the reason why (Smith *et al.*, 1997). Many metabolic pathways are not functional in this bacterium due to the lack of one or several enzymes. This bacterium is therefore rather a parasite for humans. Of greater surprise to the investigators, the *M. leprae* genome contains an unusually large number of genes involved in lipid metabolism although not as many as in *M. tuberculosis*. The enzymes encoded by these genes are probably used to synthesize the complex lipids of the cell envelope or to degrade the host cell membranes.

Functional genomics

The acquisition of the complete genome sequence of an organism represents the end of the first stage of the genomic era. This is now relatively easy in the case of prokaryotes and was achieved for 90% of the human genome in the year 2000. The next step will be to understand what these genes do and how their expression impacts on the functionality of a single cell, and of the whole organism at the level of higher organisms. To reach these objectives, new powerful techniques have been developed. One of them is two-dimensional polyacrylamide gel electrophoresis coupled with mass spectrometry analysis (2D PAGE-MS). This tool allows the study of a total set of proteins expressed by a given cell under given conditions. This set has been termed proteome by M. Wilkins (Wilkins *et al.*, 1996) and the study of the proteome becomes 'proteom*ics*'.

Another new technique arose from the merger of solid state chemistry and silicon engraving technology used in the microelectronic industry. Basically, a ligand is covalently attached to a surface and can bind to a labelled molecule. This booming technology is termed 'microarray' or chip technology and detailed explanation of this is reviewed by Schena *et al.* (1998). From an outsider's perspective, it can be noted that these new fields generate a lot of excitement (Marchall & Hodgson, 1998), the main investigators operating from inside are new biotechnology start-up companies rather than academic institutions and the input from the microelectronic industry plays a vital role. For example, one of the leading companies in the field, Affymetrix, is located in Santa Clara, California, in Silicon Valley.

As a likely future achievement of this technique, it is possible to apply this to the study of global gene expression. With the knowledge accumulated from the genome sequence, it is possible to synthesize 25–30-mer oligonucleotides specific to each potential gene of a genome, and to covalently attach these oligonucleotides on microarrays. The cells or tissues of interest are then grown under given conditions, mRNA is isolated and fluorescent-labelled cDNAs are synthesized. This single-stranded cDNA is allowed to hybridize the oligomer on the grid, and by looking at the fluorescence intensity of each microarray, it is possible to determine the nature and amount of each mRNA in the cell and so access directly the expression level of each gene. If a different fluorescent dye is used to make the cDNAs for each physiological state, direct comparison of gene expression on a single chip is possible. When a whole collection of antibodies corresponding to all the proteins of a genome becomes available, it will be possible to detect directly all the proteome on a protein chip. It will be the ultimate development of ELISA technology, and will probably out-compete the 2D PAGE-MS technique.

Other techniques which allow the study of pathogenicity-associated genes and give insights as to whether they are expressed during the course of the infection and whether they influence pathology in a given animal model would probably be used if the technology has the potential to be automated. Simply to cite a few familiar terms and current technologies: '*in vivo* expression technology' (IVET), reverse transcriptase PCR (RT-PCR) and signature-tagged mutagenesis (STM) (detailed techniques are reviewed by Handfield & Levesque, 1999).

Genome sequence exploitation: *H. pylori* as a model system

Helicobacter infections

H. pylori is now recognized as the causative agent of chronic gastritis and most gastric and duodenal ulcers and it has been shown to be involved in the development of gastric cancer. *Helicobacter* infection affects about 50% of the world population, ranging from 20 to 90% depending upon the geographic zones involved and their respective socioeconomic conditions. The concerns of *Helicobacter* infection for public health are such that the National Institutes of Health published a consensus statement in 1994 recommending that all patients with duodenal or gastric ulcers who are infected with *H. pylori* should be treated with antimicrobials (National Institutes of Health, 1994). *H. pylori* is also classified by the World Health Organization International Agency for Research on Cancer as a class I carcinogen like hepatitis B and C viruses.

Although antibiotic treatments have been shown to be efficacious, with as high as 85–95% cure rate (Mégraud, 1994), these treatments are frequently linked to some adverse reactions and problems of compliance. More importantly, a major drawback for antibiotic utilization is the emergence of resistant strains. Vaccination thus appears to be a preferred approach, and like all infectious diseases with known pathogens, vaccination is by far the best and most economical way to prevent and control diseases in the long term.

Pre-genomic approach

The search for vaccines to combat *Helicobacter* has been considered by some to be uncertain since the host natural immunity generated after infection does not eliminate the bacterium. Pallen & Clayton (1990) suggested that urease would be a vaccine candidate based on the observation that immunization of mammals with jack-bean urease induced a response which suppressed ammonia production in the intestine by ureolytic bacteria. Later, Czinn & Nedrud (1991) demonstrated that oral immunization of mice and ferrets with a whole-cell sonicate of *H. pylori* elicited specific immunoglobulins G and A to *H. pylori* antigens in both sera and gastrointestinal secretions. Subsequently, the same group (Czinn, Cai & Nedrud, 1993) and Chen and colleagues (Chen, Lee & Hazell, 1992) independently demonstrated that sonicated *Helicobacter felis* whole-cell lysate, when administered orally together with a mucosal adjuvant, cholera toxin (CT), protected mice from an acute *H. felis* infection. In mice who were previously infected with *H. felis*, active immunization with whole-cell sonicates and CT resulted in eradication of infection in 90% of the animals. These encouraging findings proved that immunization against *H. pylori* may be a feasible approach to protect hosts and constituted proof of the concept that both prophylactic and therapeutic vaccinations against *H. pylori* infection were within our reach. Great efforts have then been focused on the development of protective subunit vaccines against *H. pylori* infection.

The conventional approach for the identification of protective antigens from bacterial pathogens requires a thorough understanding of bacterial physiology, bacterium–host interactions and host immunology. To be included in a final vaccine composition, an antigen must have several characteristics: for antibody-mediated protection, it should be located at the surface of the bacterium to be accessible to the host effectors; it should be expressed by all strains with minimal structural variations among isolates; and, finally, it should elicit protective immune responses in an appropriate animal model of infection. Often the antigen turns out to be a virulence factor. For *H. pylori* antigens, Labigne, Cussac & Courcoux (1991) first isolated the gene encoding urease, a virulence factor responsible for the production of ammonia and facilitating colonization by the bacterium (Eaton & Krakowka, 1992). Michetti *et al.* (1990) and Ferrero *et al.* (1994) have independently demonstrated that a vaccine preparation containing urease conferred protection in mice against *H. felis* infection when administered orally in the presence of mucosal adjuvant CT or *E. coli* heat-labile toxin (LT). A large amount of data were accumulated from different laboratories confirming that urease can be a good vaccine candidate for *H. pylori* infection after either oral or systemic immunization (Lee *et al.*, 1995; Marchetti *et al.*, 1995; Batchelder *et al.*, 1996; Lee, Soike & Tibbits, 1996; Guy *et al.*, 1998). After nearly 10 years of research, only a few other antigens (Table 2) have been described to have protective activities in animals; among these

Table 2. Protective antigens identified prior to genome exploitation.

Antigen	Animal model	Adjuvant/route*	Challenge strain	Reference
Urease and subunits	Mouse	CT/orogastric	*H. felis*	Michetti *et al.* (1990)
		CT, LT/orogastric	*H. felis*	Lee *et al.* (1995)
		LT/orogastric	*H. pylori*	Marchetti *et al.* (1995)
		Various adjuvants/ subcutaneous	*H. pylori*	Guy *et al.* (1998)
	Cat	LT/orogastric	*H. felis*	Batchelder *et al.* (1996)
	Monkey	LT/orogastric	*H. pylori*	Lee, Soike & Tibbits (1996)
Hsp and subunits	Mouse	CT/orogastric	*H. felis*	Ferrero *et al.* (1995)
VacA	Mouse	CT/orogastric	*H. pylori* (type I)	Marchetti *et al.* (1995)
Catalase	Mouse	CT/orogastric	*H. pylori*	Radcliff *et al.* (1997)
32 kDa protein	Mouse	LT/orogastric	*H. pylori*	Lissolo *et al.* (1998)

*CT, cholera toxin; LT, *E. coli* heat-labile toxin.

are heat-shock proteins (Hsp) (Ferrero *et al.*, 1995), vacuolating cytotoxin VacA (Marchetti *et al.*, 1995), catalase (Radcliff *et al.*, 1997) and a protein of 32 kDa (Lissolo *et al.*, 1998). Most of these proteins have been identified through initial work on fractionation of bacterial total proteins and a search for membrane-associated antigens. Other approaches looking for exported proteins or bacterial adhesin molecules have led to the discovery of adhesin-like lipoproteins (AlpA, AlpB) by Odenbreit *et al.* (1999) and blood group antigen-binding adhesins (BabA and BabB) by Boren and colleagues (Boren *et al.*, 1993; Ilver *et al.*, 1998). These latter antigens have not been produced in large quantities and so far less information is available concerning their protective role. These studies again exemplify hypothesis-driven research, and although conducted by logical thinking and rational design, the result of this conventional approach remains empirical. Moreover, protective antigen discovery represents only the first step of a long process of vaccine development that spans, as for most drugs, a period of at least 10 years.

Genomic approach

In August 1997, J.-F. Tomb and his colleagues at TIGR published the complete genome sequence of *H. pylori* strain 26695 (Tomb *et al.*, 1997), which was only 15 years after the first discovery of the bacterium by Marshall and Warren (Marshall & Warren, 1984). While information embedded in the entire genome sequence is immense, the sequence does not appear to directly resolve problems linked to clinical manifestations or to provide direct information on bacterial pathogenesis. However, the information stored in this 1·67 Mb genome, once decoded, constitutes a huge database that benefits the whole scientific community in terms of exploitation and data mining. The 1590 open reading frames (ORFs) encode potential proteins that may serve as new molecular targets for antimicrobials and vaccines. The availability of these data significantly changed the way for vaccine research. Scientists quickly switched from an empirical conventional approach based on hypotheses to a much more systematic approach by testing these potential protein antigens one after another in the same pre-designed format.

Choice of methods

Genetic immunization or protein immunization? This becomes a basic question to be addressed very early on. DNA or genetic immunization has moved rapidly from a novel concept to an important strategy for the development of vaccines for numerous indications. As opposed to the classical practice through protein immunization, DNA vaccination has been shown to provide a stable and long-lived source of the protein vaccine in certain cases. Theoretically, it is a simple, robust and effective means of eliciting both antibody (Wild *et al.*, 1998) and mostly cell-mediated immune responses (Wang *et al.*, 1998). DNA vaccines provide a number of potential advantages, such as they are relatively inexpensive and easy to produce, they are much more stable than proteins and no special equipment for cold storage is required.

However, as for *H. pylori* antigens, naked DNA encoding UreA or UreB, the most promising vaccine candidate, did not consistently show protection in all the immunized animals, in contrast to UreA or UreB administered as protein (unpublished results). In the specific case of *H. pylori* genome exploitation, the authors chose to produce proteins rather than DNA as antigens to screen the genome (Sodoyer *et al.*, 1998).

Recombinant-protein-based antigen production. This was in fact not more time-consuming than using DNA directly as it first appears to be. The choice of expression system is important and one can argue the advantage of using different expression systems for proteins of different cellular localization. However, proteins cloned in different vectors would not have the same fusion partners and they would not be purified the same way. For example, it will be even difficult to make a direct comparison of the same protein in a hexa-His fusion and in a GST fusion without considering different proteins with different fusions. The basic idea was to limit as much as possible the number of variables during the screening procedures. The expression system used for obtaining protein antigens has another advantage, which is that the size of the expected antigen can be determined by simply running a polyacrylamide gel and measuring the size of the expressed protein.

Animal models and evaluation. It is important to recognize that animal models used for genomic screening do not necessarily have to be the ones that most resemble the human or the natural disease. The animal should be manageable in size and easy to handle and should allow an evaluation of the efficacy of the tested antigens. In the case where surrogate markers of protection exist, they can be used for final read-out, otherwise protection rate will be the most pertinent read-out. In the case of *H. pylori*, a murine challenge model was used. The criterion for selection was that a novel antigen must protect at least as efficaciously as urease using the same regimen.

Experimental work and results

One of the major objectives of shifting to genomic screening for vaccine antigen discovery is to gain time and reduce time to market. To achieve this goal, groups of scientists with complementary expertise (computation, genetic engineering, bacteriology, biochemistry and immunology) must band together into interactive teams. Genomic exploitation for vaccine target identification depends as much on coordination and high team spirit as it does upon highly mastered technical skills.

The *H. pylori* experimental work, with the available genomic information and tools at that time, was divided amongst four groups: (1) computational analysis; (2) molecular biology; (3) biochemistry; and (4) animal testing. In such an organization, the deliverables of the first group will be the starting material for the next group and the final results will be the result of global efforts. A great advantage of this organization is to have teams of scientists from different disciplines working together and promoting exchange.

1 Computational analysis. This was the first step of screening. Its goal was to select ORFs that display the defined criteria and to bring down the number of genes to a workable size. Criteria are based either on prior knowledge of the pathogen, its physiology and metabolic habits or on specific characteristics of proteins, such as membrane proteins, potential toxins or receptors. Many computer programs are currently available for this *in silico* selection, mostly based on two major categories of tools: (i) sequence similarity search and alignment; (ii) pattern and profile detection.

For the *H. pylori* genome, ORFs were identified by standard computer packages and translated into protein sequences. The algorithms used included prediction of type I and type II signal sequences or transmembrane domains. The PSORT program was used to pick up membrane-associated proteins and PROSITE was used to search for specific protein motifs. Furthermore, the transmembrane proteins were verified by the presence of phenylalanine at the C-terminus. DNASTAR protein analysis was used to group the selected ORFs into families. The BLAST homology score for each selected protein was compared with major antigens and with known virulence factors (adhesins, toxins, iron-regulated proteins, etc.) from other micro-organisms. A database of proprietary protective antigens was very useful and essential from an industrial company perspective. After the work of computation, about 400 ORFs out of 1590 potential ORFs were selected and their characteristics are listed in Table 3 (H. Kleanthous and others, unpublished results).

2 Molecular biology. The main function of this group was to clone all the above selected ORFs in the same vector for high expression. Since the fusion protein product was to be used for immunization, only one cloning vector was used to avoid the introduction of different fusion partners into the final product. The principal activities of this group were: (i) identification of genes encoding the potential candidates; (ii) definition and synthesis of oligonucleotide primers; (iii) PCR amplification from the strain which was to be used for challenge, in this case strain X47-2AL (Kleanthous *et al.*, 1995); (iv) cloning into pET28 vector with His-Tag (Novagen); and (v) heterologous expression in *E. coli* BL21λDE3.

A major difficulty encountered was the strain variability at the genomic level. Frequently, the genome is sequenced from one strain while the biologists have more data on another strain. This raises great concerns during *in vivo* testing, especially when challenge is required. For *H. pylori*, the PCR amplification was made by using DNA from the challenge strain X47-2AL, although the probes were designed from sequenced strain 26695. It was found that about 20% of the selected genes could not be amplified. These genes were amplified later from a strain similar to that of TIGR's sequenced strain, but no information is available for the variability of the gene products from both strains. Interpretation of protection data was therefore more difficult. Recently, Alm *et al.* (1999) compared genomic sequences from two unrelated *H. pylori* strains and showed that the two strains were quite similar, with only 6–7% of the genes specific for each strain. Curiously enough, about 20% variability of the selected genes was found among another two unrelated strains (Sodoyer *et al.*, 1998). This may be a result of the variability that is clustered in the surface-exposed elements. One other thing to note: while the overall positive expression rate was high, over 70%, the clones that showed low or no expression were often membrane-associated proteins.

3 Biochemistry. The main function of this group was to provide, for animal

Table 3. Characteristics of the deduced proteins after *in silico* selection from the *H. pylori* genome.

	Percentage of total selected	Percentage of total genome
PSORT analysis		
Periplasmic or outer-membrane protein	17	7·8
Inner membrane	48	22·0
Cytoplasmic	34	16·0
Type II signal sequence		
Lipoproteins	5	1·3
Phenylalanine at C-terminus	21	9·7

testing, purified and characterized fusion proteins in sufficient quantities. To be efficient, two standardized purification protocols were defined for all ORFs to minimize variations amongst novel unknown antigens. One was used to purify proteins from inclusion bodies and the other was used for cytosoluble proteins. Recombinant *E. coli* was cultured in shake flasks and harvested after induction with IPTG. Cells were disrupted and fractionated by sonication and centrifugation. Protein purification was carried out on IMAC/Ni columns with or without solubilization of fusion protein, depending on the situation. Product renaturation and conditioning were performed prior to immunization. Without optimization of the culture conditions, it was found that the majority of the proteins were expressed in an aggregated form in the insoluble fraction after cell disruption. In this case, proteins were denatured, purified under denaturing conditions and renatured. In some cases, proteins were difficult to renature in buffers without chaotropic agents that might affect conformational epitopes.

4 Animal testing. The ability of each novel protein to prevent colonization of the mouse stomach by *H. pylori* was evaluated and compared with the protective efficacy afforded by recombinant urease, the gold standard. The purified recombinant proteins were evaluated individually in a group of mice and recombinant urease was included in each experiment as a positive control. The negative control of protection was a group of non-immunized but infected mice. Groups of 10 8-week-old outbred OF1 mice were immunized by the rectal route four times, 1 week apart, with 25 µg purified antigen mixed with 0·5 µg LT in a total volume of 25 µl. The immunized mice were challenged by the oral route 2 weeks after the last boost with 10^6 c.f.u. *H. pylori* strain X47-2AL, and protection was assessed 2 weeks later by evaluating the urease activity in the stomach and by quantitative bacterial culture when necessary. Statistical analysis was performed by using the test of Kruskall–Wallis. In each experiment, the best hits, meaning those conferring a protection similar to that of urease, were selected and re-tested in another independent experiment for a head-to-head comparison.

Discussion

Genomic screening for vaccine candidates is clearly feasible. In less than a year, a genome-based screening for protective antigens against *H. pylori* infection was completed; this would not have been possible with conventional methods. Furthermore, this method allowed the selection of the highly protective antigens and therefore the elimination of many partially protective antigens, and reduced the number of vaccine candidates. Compared to conventional methods, by which five protective antigens were identified by different laboratories after many years of work, in only a year, genome-based screening allowed the identification of about three times that number (Table 4).

Table 4. Experimental results from *H. pylori* genome exploitation.

Step	Expected result	Successful rate (%)	
		Per step	Accumulated
Genome	–	–	100
Computational	Potential candidates selected	25	25
Molecular biology	Positive expression of proteins of correct molecular mass	72	18
Biochemistry	Proteins purified of sufficient quantity and quality for immunization	77	14
Animal testing	Highly protective antigens	7	1

From a functional point of view, it is important to validate the expression of these protective antigens during infection in humans. Techniques like RT-PCR may be used to compare transcripts of bacteria grown on agar, in infected animals and in humans at different stages of disease. Knockout mutants can also be constructed to verify the presumed function of these protective antigens based on sequence analysis or to elucidate whether the gene encodes an essential function. From a product development point of view, it is still difficult to carry out development for more than four or five antigens; therefore stricter criteria were set up to further select the most protective antigens for clinical development. These include variability of these genes among clinical isolates, the potential production level of the mature proteins and the complexity of the downstream process. This changes the way vaccine development used to be carried out. Before genomics, scientists focused on one of their favourite antigens and spent all their efforts to get around the obstacles at each level of development. Today, the choice from genomic screening is multiple, and at each stage of the development new selection criteria have to be included to eliminate 'the weaknesses'.

It should be mentioned that, although genomics provides new hopes and apparently shortens the period for vaccine candidate identification, this approach concerns only protein-derived antigens. In the case of bacterial vaccines, surface components such as capsular polysaccharides and lipopolysaccharides, which are not identified by genomics, have demonstrated their efficacy as vaccine antigens (Austrian, 1989; Cadoz, 1998; Moxon, Hood & Richards, 1998) and are relatively simple for manufacturing. Genomic information can be helpful, however, in understanding and leading to useful modifications of capsular polysaccharides or lipopolysaccharides (Hood *et al.*, 1996).

At the speed at which computer science and software is advancing, it is fair to say that everything is out of date shortly after it is accomplished. For example, the computational analysis made so far was by looking at linear sequences but structural information would greatly help in the selection. The progress in genomics is growing exponentially and what was done 2 years ago appears to be archaic now. There is probably no ideal way of doing it, but one should try to implement newer technology or methodology in each new genome exploitation so as to be able to draw conclusions from each approach. Nevertheless, the objective of genomic screening was not to have an exhaustive list of protective antigens but to have an efficacious vaccine for human usage as early as possible. It is worth mentioning that *H. pylori* is the first micro-organism for which complete genomes have been sequenced from two independent strains (Alm *et al.*, 1999), showing that the medical and pharmaceutical implications are huge. In addition, the scientific community can benefit from the comparison of these two genomes to assess genetic diversity and strain-specific plasticity. With all the efforts made in vaccine discovery on a genomic scale for *H. pylori*, a vaccine is certainly feasible in the foreseeable future.

Conclusions

In the past few years, significant advances in biological sciences and technologies have been made which provide new hope for vaccine development. The biological sciences have evolved from genetic or recombinant manipulations to a much larger scale, genomic scale. This evolution allows biological research to expand from a reductionist point of view where an individual phenomenon is focused on, to a discovery-oriented stage where possibilities and choices are numerous. However, scientists are so overwhelmed with the huge quantity of information coming from genomic sequences that their immediate utilization does not appear to be evident: sophisticated tools are required to retrieve, compare and analyse the massive quantity of genomic data. Technology enabling an interpretation of genomic data will certainly lead to the next major breakthrough in understanding the relationship between genomic structure and biological function.

In a very similar manner, information technology evolved almost parallel to the progress of biological sciences and it greatly influenced the way that scientists conducted their experiments. Until now, biological scientists were trained to observe a phenomenon, put forward a hypothesis, test it by trial-and-error, and try again due to the lack of an overall picture at the beginning. The joy of serendipity is the greatest reward. With the speed of genomic sequencing, it is now time to act differently and to rethink experimental design in order to benefit from the vast information in genome databases.

Discovery of vaccines against infectious diseases followed the same paradigm of research until the publication of the first complete bacterial genome sequence, that of *Haemophilus influenzae*. From there, more systematic approaches have been undertaken as described for the case of *H. pylori*. One example was the identification of vaccine candidates against serogroup B meningococcus (Pizza *et al.*, 2000).

Two other examples, tuberculosis and malaria, are worth mentioning. BCG has been in use for many years, yet over 7 million new cases of tuberculosis are reported per year with 3 million deaths. There is a need for a new vaccine. Information from the genome sequence is being processed in several laboratories to design a new vaccine; one of these approaches consists of using comparative hybridization experiments on a DNA microarray to better understand the differences between *M. tuberculosis, Mycobacterium bovis* and various strains of BCG (Behr *et al.*, 1999). Understanding the genetic differences between these closely related strains might lead to the design of improved vaccines.

Malaria kills an estimated 2–3 million people each year and drug-resistant parasites are spreading. Initiatives have been undertaken around the world to accelerate the development of a malaria vaccine. Genomics is playing a large part. In November 1998, Drs Gardner, Hoffman and their collaborators reported the complete sequence of chromosome 2 of *Plasmodium falciparum* (Gardner *et al.*, 1998). An international consortium of agencies is supporting efforts to sequence the other 13 chromosomes of the parasite. The challenge here, as for vaccine discovery, will be the development of new methods to efficiently filter the number of candidates to test to a small enough size. This is yet another step in complexity by comparison to bacterial genomes.

In the present chapter, the importance of genomics in the early step of vaccine design, i.e. antigen identification or identification of gene function, is emphasized. As vaccinology is a multidisciplinary field, it is obvious that this is a very restrictive view. Concomitantly, nanobiotechnology is an emerging area of opportunity that seeks to fuse nano- or micro-fabrication and biosystems for the benefit of both. In parallel, a considerable amount of research is being conducted in the field of immunology to understand how to generate optimal humoral and cellular responses and how to increase memory, to identify correlates of immunity to reduce trial times, while chemists are designing new molecules that trigger specific responses. Although genomics seems to shift biological research to a more systematic approach or to an apparently pre-designed pattern as cited in this chapter, scientists should take advantage of this flood of information and let their imagination work. If so, the future of vaccine discovery can only be brighter.

Acknowledgements

The *H. pylori* genomic exploitation was funded by Aventis Pasteur (formerly Pasteur Mérieux Connaught), Marcy L'Etoile, France, and OraVax Inc., Cambridge, Massachusetts. The scientific input, the constructive discussions and the excellent computational and laboratory works from V. Mazarin, R. Oomen, R. Sodoyer, C. Manin, B. Guy, B. Rokbi, F. Rizvi, H. Kleanthous, A. Al-Garawi, C. Miller, G. Myers, W. Thomas, T. Monath and the scientific staff at TIGR and HGS are greatly acknowledged. The authors are grateful to B. Guy, R. Sodoyer and P. Meulien for their critical readings and suggestions.

References

Ajjan, N. (1995). History. In *Vaccination*, pp. 6–7. Edited by Pasteur Mérieux MSD. Pasteur Mérieux MSD.

Alm, R. A., Ling, L.-S. L., Moir, D. T. & 20 other authors (1999). Genomic-sequence comparison of two unrelated isolates of the human gastric pathogen *Helicobacter pylori*. *Nature* 397, 176–180.

Austrian, R. (1989). Pneumococcal polysaccharide vaccines. *Reviews of Infectious Diseases* 3, S598–602.

Batchelder, M., Fox, J. G., Monath, T. & 8 other authors (1996). Oral vaccination with recombinant urease reduces gastric *Helicobacter pylori* colonization in the cat. *Gastroenterology* 110, A58.

Behr, M. A., Wilson, M. A., Gill, W. P., Salamon, H., Schollnick, G. K., Rane, S. & Small, P. M. (1999). Comparative genomics of BCG vaccines by whole-genome DNA microarray. *Science* 284, 1520–1523.

Blanton, K. J., Biswas, G. D., Tsai, J., Adams, J., Dyer, D. W., Davis, S. M., Koch, G. G., Sen, P. K. & Sparling, P. F. (1990). Genetic evidence that *Neisseria gonorrhoeae* produces specific receptors for transferrin and lactoferrin. *Journal of Bacteriology* 172, 5225–5235.

Boren, T., Falk, P., Roth, K. A., Larson, G. & Normark, S. (1993). Attachment of *Helicobacter pylori* to human gastric epithelium mediated by blood group antigens. *Science* 262, 1892–1895.

Cadoz, M. (1998). Potential and limitations of polysaccharide vaccines in infancy. *Vaccine* 16, 1391–1395.

Chen, M., Lee, A. & Hazell, S. (1992). Immunization against gastric *Helicobacter* infection in a mouse *Helicobacter felis* model. *Lancet* 339, 1120–1121.

Cornelissen, C. N. & Sparling, P. F. (1994). Iron piracy: acquisition of transferrin-bound iron by bacterial pathogens. *Molecular Microbiology* 14, 843–850.

Czinn, S. J. & Nedrud, J. G. (1991). Oral immunization against *Helicobacter pylori*. *Infection and Immunity* 59, 2359–2363.

Czinn, S. J., Cai, A. & Nedrud, J. C. (1993). Protection of germ-free mice from infection by *Helicobacter felis* after active oral or passive IgA immunization. *Vaccine* 11, 637–642.

Eaton, K. A. & Krakowka, S. (1992). Chronic active gastritis due to *Helicobacter pylori* in immunized gnotobiotic piglets. *Gastroenterology* 103, 1580–1586.

Ferrero, R. L., Thiberge, J. M., Huerre, M. & Labigne, A. (1994). Recombinant antigens prepared from the urease subunits of *Helicobacter* spp: evidence of protection in a mouse model of gastric infection. *Infection and Immunity* 62, 4981–4989.

Ferrero, R. L., Thiberge, J. M., Huerre, M. & Labigne, A. (1995). The GroES homolog of *Helicobacter pylori* confers protective immunity against mucosal infection in mice. *Proceedings of the National Academy of Sciences, USA* 92, 6499–6503.

Fleischmann, R. D., Adams, M. D., White, O. & 37 other authors (1995). Whole-genome random sequencing and assembly of *Haemophilus influenzae* Rd. *Science* 269, 496–512.

Fraser, M. & Fleischmann, R. D. (1997). Strategies for whole microbial genome sequencing and analysis. *Electrophoresis* 18, 1207–1216.

Gardner, M. J., Tettelin, H., Carucci, D. J. & 24 other authors (1998). Chromosome 2 sequence of the human malaria parasite *Plasmodium falciparum*. *Science* 282, 1126–1132.

Guy, B., Hessler, C., Fourage, S., Haensler, J., Vialon-Lafay, E., Rokbi, B. & Quentin-Millet, M. J. (1998). Systemic immunization with urease protects mice against *Helicobacter pylori* infection. *Vaccine* 16, 850–856.

Handfield, M. & Levesque, R. C. (1999). Strategies for isolation of *in vivo* expressed genes from bacteria. *FEMS Microbiology Reviews* 23, 69–91.

Hood, D. W., Deadman, M. E., Allen, T. & 7 other authors (1996). Use of the complete genome sequence information of *Haemophilus influenzae* strain Rd to investigate lipopolysaccharide biosynthesis. *Molecular Microbiology* 22, 951–965.

Horton, P. & Nakai, K. (1996). A probabilistic classification system for predicting the cellular localization sites of proteins. *Intelligent Systems for Molecular Biology* 4, 109–115.

Ilver, D., Arnqvist, A., Ogren, J. & 7 other authors (1998). *Helicobacter pylori* adhesin binding fucosylated histo-blood group antigens revealed by retagging. Science 279, 373–377.

Jones, D. A. & Fitzpatrick, F. A. (1999). Genomics and the discovery of new drug targets. *Current Opinion in Chemical Biology* 3, 71–76.

Kleanthous, H., Tibbits, T., Bahios, T. J., Georgokopoulos, K., Myers, G., Ermak, T. H., Fox, J. & Monath, T. (1995). *In vivo* selection of a highly adapted *H. pylori* isolate and the development of an *H. pylori* mouse model for studying vaccine efficacy. *Gut* 37, A94.

Labigne, A., Cussac, V. & Courcoux, P. (1991). Shuttle cloning and nucleotide sequence of *Helicobacter pylori* genes responsible for urease activity. *Journal of Bacteriology* 173, 1920–1931.

Lee, C. K., Weltzin, R., Thomas, W. D., Kleanthous, H., Ermak, T., Soman, G., Hill, J. E., Ackerman, S. & Monath, T. (1995). Oral immunization with recombinant *Helicobacter pylori* urease induces secretory IgA antibodies and protects mice from challenge with *Helicobacter felis*. *Journal of Infectious Diseases* 172, 161–172.

Lee, C. K., Soike, K. & Tibbits, T. (1996). Urease immunization protects against reinfection by *Helicobacter pylori* infection in rhesus monkeys. *Gut* 39, A43.

Lissolo, L., Fourrichon, L., Kleanthous, H., Guinet, F., Myers, G. & Quentin-Millet, M. J. (1998). Purification and preliminary characterization of a novel membrane protein from *Helicobacter pylori* and its use as a vaccine candidate. *Abstracts of the 98th General Meeting of the American Society for Microbiology, Atlanta*. Abstract B382. Washington D.C. American Society for Microbiology.

Marchall, A. & Hodgson, J. (1998). DNA chips: an array of possibilities. *Nature Biotechnology* 16, 27–31.

Marchetti, M., Arico, B., Burroni, D., Figure, N., Rappuoli, R. & Ghiara, P. (1995). Development of a mouse model of *Helicobacter pylori* infection that mimics human disease. *Science* 267, 1655–1658.

Marshall, B. J. & Warren, J. R. (1984). Unidentified curved bacilli in the stomach of patients with gastritis and peptic ulceration. *Lancet* 1, 1311–1315.

Mégraud, F. (1994). *Helicobacter pylori* resistance to antibiotics. In *Helicobacter pylori: Basic Mechanisms to Clinical Cure*, pp. 570–583. Edited by R. H. Hunt & G. N. J. Tytgat. Dordrecht: Kluwer Academic Publishers.

Michetti, P., Corthésy-Theulaz, I., Davin, C. & 7 other authors (1990). Immunization of BALB/c mice against *H. felis* infection with *H. pylori* urease. *Gastroenterology* 107, 1002–1011.

Moir, D. T., Shaw, K. J., Hare, R. S. & Vovis, G. F. (1999). Genomics and antimicrobial drug discovery. *Antimicrobial Agents and Chemotherapy* 43, 439–446.

Morse, S. I. & Morse, J. H. (1976). Isolation and properties of the leucocytosis- and lymphocytosis-promoting factor of *Bordetella pertussis*. *Journal of Experimental Medicine* 143, 1483–1502.

Moxon, E. R., Hood, D. & Richards, J. (1998). Bacterial lipopolysaccharides: candidate vaccines to prevent *Neisseria meningitidis* and *Haemophilus influenzae* infections. *Advances in Experimental Medicine and Biology* 435, 237–243.

Munoz, J. J., Arai, H. & Cole, R. L. (1981). Mouse-protecting and histamine-sensitizing activities of pertussigen and fimbrial hemagglutinin from *Bordetella pertussis*. *Infection and Immunity* 32, 243–250.

Myers, L. E., Yang, Y. P., Du, R. P., Wang, Q., Harkness, R. E., Schryvers, A. B., Klein, M. H. & Loosmore, S. M. (1998). The transferrin binding protein B of *Moraxella catarrhalis* elicits bactericidal antibodies and is a potential vaccine antigen. *Infection and Immunity* 66, 4183–4192.

National Institutes of Health (1994). Consensus

development conference statement: *Helicobacter pylori* in peptic ulcer disease. *JAMA (Journal of the American Medical Association)* **272**, 65–69.

Odenbreit, S., Till, M., Hofreuter, D. & Hass, R. (1999). Genetic and functional characterization of the AlpAB gene locus essential for the adhesion of *Helicobacter pylori* to human gastric tissue. *Molecular Microbiology* **31**, 1537–1548.

Pallen, M. J. & Clayton, C. L. (1990). Vaccination against *Helicobacter pylori* urease. *Lancet* **336**, 186–187.

Pittman, M. (1984). The concept of pertussis as a toxin-mediated disease. *Pediatric Infectious Disease Journal* **3**, 467–486.

Pizza, M., Scarlato, V., Masignani, V. & 33 other authors (2000). Identification of vaccine candidates against serogroup B meningococcus by whole-genome sequencing. *Science* **287**, 1816–1820.

Plotkin, S. L. & Plotkin, S. A. (1999). A short history of vaccination. In *Vaccines*, pp. 1–12. Edited by S. A. Plotkin & W. A. Orenstein. The Curtis Center, Philadelphia: W. B. Saunders.

Radcliff, F. J., Hazell, S. L., Kolesnikow, T., Doidge, C. & Lee, A. (1997). Catalase, a novel antigen for *Helicobacter pylori* vaccination. *Infection and Immunity* **65**, 4668–4674.

Salk, J. & Salk, D. (1977). Control of influenza and poliomyelitis with killed virus vaccines. *Science* **195**, 834–847.

Sanger, F., Air, G. M., Barrell, B. G., Brown, N. L., Coulson, A. R., Fiddes, C. A., Hutchison, C. A., Slocombe, P. M. & Smith, M. (1977). Nucleotide sequence of bacteriophage phi X174 DNA. *Nature* **265**, 687–695.

Sanger, F., Nicklen, S. & Coulson, A. R. (1977). DNA sequencing with chain-terminating inhibitors. *Proceedings of the National Academy of Sciences, USA* **74**, 5463–5467.

Schena, M., Heller, R. A., Theriault, T. P., Konrad, K., Lachenmeier, E. & Davis, R. W. (1998). Microarrays: biotechnology's discovery platform for functional genomics. *Trends in Biotechnology* **16**, 301–306.

Schryvers, A. B. & Morris, L. J. (1988). Identification and characterization of the human lactoferrin-binding protein from *Neisseria meningitidis*. *Infection and Immunity* **56**, 1144–1149.

Smith, D. S. (1996). Microbial pathogen genomes—new strategies for identifying therapeutics and vaccine targets. *Trends in Biotechnology* **14**, 290–293.

Smith, D. R., Richterich, P., Rubenfield, M. & 34 other authors (1997). Multiplex sequencing of 1.5 Mb of the *Mycobacterium leprae* genome. *Genome Research* **7**, 802–819.

Sodoyer, R., Mazarin, V., Guy, B. & 17 other authors (1998). Post-genomic strategies for vaccine target identification against *Helicobacter pylori*. In *Abstracts of the 98th General Meeting of the American Society for Microbiology*. Abstract E6. Washington D.C.: American Society for Microbiology.

Tomb, J. F., White, O., Kerlavage, A. R. and 39 other authors (1997). The complete genome sequence of the gastric pathogen *Helicobacter pylori*. *Nature* **388**, 539–547.

Wang, R., Doolan, D. L., Le, T. P. & 12 other authors (1998). Induction of antigen-specific cytotoxic T lymphocytes in humans by a malaria DNA vaccine. *Science* **282**, 476–480.

Webb, D. C. & Cripps, A. W. (1999). Immunization with recombinant transferrin binding protein B enhances clearance of nontypeable *Haemophilus influenzae* from the rat lung. *Infection and Immunity* **67**, 2138–2144.

Wild, J., Gruner, B., Metzqer, K., Kuhrober, A., Pudollek, H. P., Hauser, H., Schirmbeck, R. & Reimann, J. (1998). Polyvalent vaccination against hepatitis B surface and core antigen using a dicistronic expression plasmid. *Vaccine* **16**, 353–360.

Wilkins, M. R., Sanchez, J. C., Williams, K. L. & Hochstrasser, D. F. (1996). Current challenges and future applications for protein maps and post-translational vector maps in proteome projects. *Electrophoresis* **17**, 830–838.

Are molecular methods the optimum route to antimicrobial drugs?

Frank C. Odds

Department of Molecular and Cell Biology, Institute of Medical Sciences, University of Aberdeen, Foresterhill, Aberdeen AB25 2ZD, UK

Introduction

A topic whose title ends with a question mark is surely one whose reader deserves some immediate idea of the likely answer before reading further. In this author's personal opinion, the short answer to the question 'Are molecular methods the optimum route to new antimicrobial drugs?' is 'not yet'. That conclusion assumes that the expression 'molecular methods' means the use of entirely subcellular approaches (*in vitro* and *in silico*) to discover lead molecules for development as clinically useful antimicrobial agents. It must be stressed, however, that the comments in this chapter are inevitably a *personal* view. The comments are also concerned exclusively with *antimicrobial* agents and should not be taken as relevant to any other area of medicine.

The purpose of this chapter is to review the contribution of rational molecular target screening and drug design to the field of antimicrobial drug discovery, to look at the potentially exciting developments in novel, molecular approaches to finding such drugs, and to appraise the extent to which these technologies have so far aided the drug discovery process. The scientific literature relevant to this topic is so vast that review articles and examples of publications in support of comments will inevitably be cited selectively, rather than in the form of a comprehensive bibliography. Since the author's particular expertise lies in the domain of antifungal agents, the reader may occasionally sense a bias towards the antifungal literature.

There is no doubt of the considerable need for novel antimicrobial agents, which means agents with mechanisms of action different from those of already established antimicrobial drugs. The facility with which micro-organisms develop resistance to inhibitory agents, either by destroying them, pumping them out or altering target structure and expression to minimize their inhibitory impact, is impressive. Infectious diseases today are as big a problem as ever—some might even say they are becoming more so—and what the medical world needs is as diverse a range of antimicrobial agents as it can possibly obtain: diverse in the sense of agents that work by different mechanisms to eradicate pathogenic organisms.

What we currently have available in the clinic is a large number of agents with relatively minor differences in functional properties, and which belong to a restricted number of classes in terms of their mechanisms of action. What we urgently seek, therefore, are scientific short cuts that will facilitate and accelerate the drug discovery process. There is no doubt that the massive build-up of novel disciplines that have gained names such as 'genomics', 'proteomics' and 'computer-assisted molecular design' provides a formidable theoretical background for discovery of new antimicrobial drugs. The pragmatic question is: to what extent does the published hyperbole around such cutting-edge technologies pay off in accelerating antimicrobial drug discovery?

Definitions

The expression 'molecular methods' as used in the title of this chapter is capable of many shades of meaning and requires definition. Historically, almost all antimicrobial agents have been discovered serendipitously through screening of natural products or synthetic chemicals for inhibitors of microbial growth. Chance discoveries of novel inhibitory compounds ('hits' or 'leads') are followed by the rational synthesis of analogous molecules with variations in chemical structure that facilitate determination of structure–activity relationships for the class of inhibitors concerned. Experimentation with established selective inhibitors of microbial growth usually results in the discovery of a single macromolecule that is the primary subcellular target for the inhibitors and explains their mode of action. Knowledge of the drug target and its structure has contributed subsequently both to general understanding of microbial physiology and behaviour, and to the advancement of efforts to make novel chemicals within an established class of inhibitors. However, the parallel use of whole-cell screening tests and experimentation to gain deeper understanding of a molecular target is not what is meant by 'molecular methods' for the present chapter.

The strategy underlying current 'molecular' approaches to novel antimicrobial discovery is to turn the traditional process on its head. Starting with knowledge of the structure and biochemistry of a validated subcellular macromolecule as a potential target for an inhibitory agent, small chemical molecules can be specifically designed as inhibitors of that target. Candidate inhibitors can be synthesized, tested and optimized for use as antimicrobial agents *before* any tests are done in intact microbial cells. 'Molecular methods' therefore refers to a combination of exclusively molecular approaches to target selection, inhibitor design and, ultimately, optimization of inhibitors for clinical development. There is nothing inherently new about this concept: indeed, advances in knowledge of protein structures have been used for decades to refine understanding of the interactions between macromolecules and small-molecule substrates and inhibitors in all branches of biological science. Medicinal chemists can gain inspiration for syntheses of novel inhibitory molecules from the structures of substrates and intermediates in enzyme reactions. However, there are many micro-organisms pathogenic for humans, each of which is a highly complex mixture of possible macromolecular targets. It is therefore only comparatively recently that biotechnology has reached the point at which it is possible to suppose that a single microbial macromolecule may be designated a novel target and its structure used as the conceptual basis for the design of a new antimicrobial agent.

In an ideal world where all scientific theories rapidly become applied reality, we are talking about a process in which comparison of DNA base sequences from microbial and mammalian genes defines targets that are both specific and essential to the microbe. Computers are then used to display the tertiary structure of a chosen molecular target and techniques of computational chemistry allow the scientist to design *in silico* the structure of an ideal inhibitory drug for that target. The chosen molecule can then be synthesized, checked for activity against the target molecule and developed for medical use, perhaps with further input from molecular and computer technologies along the way to help predict properties of the new agent, such as solubility, toxicity and pharmacokinetic behaviour.

Even a starry-eyed advocate for the concept that genomics, proteomics and molecular design can solve all known problems in biology and medicine will recognize that this utopian description of molecular drug discovery is still a pipe dream. Certainly, if we had just one novel, clinically useful antimicrobial agent for every ten literature claims of a new antimicrobial target we would indeed have an impressive armoury of agents to combat infection. However, discovering antimicrobial agents is not at all the simple biochemical challenge it is sometimes suggested to be. At the present time of writing, the expectations for 'molecular' discovery of microbial inhibitors exceed the tangible evidence for success of the approach. However, science occasionally makes unexpected leaps forward, which means that it is a matter of pure conjecture to speculate how many decades in the future such an approach to antimicrobial drug discovery will truly become commonplace.

Targets

The expression 'drug target' is very frequently encountered nowadays in publications concerning pharmaceutical and biomedical research. It is perhaps worth reminding ourselves from time to time that, for antimicrobial agents, the real target is the infected patient. We cannot screen new molecules for activity in infected patients, but we certainly can move one step backwards and use infected animals as a vehicle for screening and evaluation of novel agents. However, animal testing is tedious, expensive, riddled with irreproducibilities and requires large amounts of test compounds (even outwith ethical considerations). That is why the traditional discovery process usually moves a second step further away from the target and uses the screening of candidate molecules against micro-organisms in test tubes as a surrogate for detection of potential antimicrobial activity in a sick human. The subcellular screening of potential inhibitors against one isolated macromolecule from a single strain of one microbial species is therefore placed three conceptual leaps away from the human patient. This situation with antimicrobial discovery differs from the discovery of many other therapeutic agents, where the (human) molecular target is usually only two conceptual stages away from the patient. The molecular approach therefore needs to offer significant means of short-circuiting direct antimicrobial testing *in vitro* and *in vivo* if it is to confer real scientific and economic benefits to the discovery process.

What is wrong with serendipitous drug discovery?

Publications describing 'molecular' approaches to antimicrobial drug discovery sometimes tend towards evangelistic zeal in claiming how a new DNA base sequence is going to benefit medicine. With such tangible excitement at the level of scientific publications, it is perhaps unreasonable to blame the popular media for breathlessly describing almost any and every new finding as 'a breakthrough that will lead to new treatments'. However, the enthusiasm for novel approaches itself makes a clear case that older, 'tried and true' discovery methods are nowadays no longer producing sufficient medical benefits.

In the case of antimicrobial agents, such a conclusion is eminently reasonable. Between 1945 and 1965 the screening of synthetic chemicals and natural products for serendipitous discovery of new microbial inhibitors proceeded at a pace that resulted in an explosive expansion of novel drugs, particularly of antibacterial antibiotics. Since 1970, however, the pace of such discoveries in bacteriology has slowed to a trickle, and a law of diminishing returns may now be operating, particularly in the screening of fermentation products from microbial broths. Most antimicrobial research during the past 20–30 years has concentrated on new structural variations on already proven antimicrobial themes: one hears of 'generations' of cephalosporins, quinolones and other classes of agents whose prototypes were in fact discovered by chance a very long time ago. In virology, the learning curve with screening against intact viruses began more recently, and products such as acyclovir, gancyclovir and nevirapine exemplify the several clinically registered successes with antiviral agents discovered this way.

With the current urgent need for new agents in the clinic, it is obvious that scientists will seek to increase the power of their discovery research by adopting new approaches to the problem. The cost of present-day research based in the molecular biology field is, however, very high, and there have been few signs so far of efforts to optimize and refine the ways in which fundamental research at the molecular level is directed to antimicrobial discovery. The time may be approaching when the economics of the whole discovery process will start to dictate the most cost-effective routes of exploration.

Theoretical advantages of molecular target-based discovery

There are two particular benefits of screening for inhibitors of isolated molecular drug targets *in vitro*. One is the small scale of such tests (only minuscule quantities of compounds are required), which makes them eminently suitable for synthetic approaches such as combinatorial chemistry and the automation of high-through-put screening procedures. The other is the knowledge that a well-chosen molecular drug target can represent a highly specific point of attack for an antimicrobial molecule. A lot of energy is now expended on 'validating' potential antimicrobial targets, i.e. proving that the chosen target is a vital molecule found specifically in the microbial invader but not in the mammalian host. Its absence in the host means the target is one whose inhibition is unlikely to be paralleled by equivalent effects in the infected host. The presumed advantages of a selective target should not, of course, be confused with the selectivity of an inhibitory chemical. Just because the target is unique to a microbe, it cannot be automatically assumed that an inhibitory molecule will not exhibit toxic pharmacological effects against other, unanticipated mammalian targets as well.

When a molecular drug target has been sufficiently characterized to allow representation of its tertiary structure on a computer screen, then rational design of inhibitors becomes possible. The term 'rational design' obviously refers to a process in which a novel molecular structure is logically proposed for synthesis rather than being stumbled upon, by accident, in nature. It is sometimes important to differentiate, however, between a capable chemist jotting novel structural ideas on paper and a scientist using computer-based technologies to arrive at a (statistically) optimized candidate molecule for synthesis. This second process, computer-assisted molecular design, is still to come of age, but if it progresses satisfactorily it may become a very powerful tool for designing structures of new agents at some stage in the future.

The main attraction of designing drugs to interact with specific molecular targets is the sheer scientific logic of the entire process. With rational design of, at the very least, initial lead compounds, the tedious process of experimentally screening thousands of molecules to see which ones affect a target can be eliminated. Only a few tests with the most carefully designed candidate molecules need to be done. It may therefore seem paradoxical to note that, alongside huge investment in molecular target research, there is also massive investment nowadays both in high-throughput screening systems and in combinatorial chemistry (synthesis of mixtures of molecules that can be screened once as a 'cocktail' then separated when necessary into submixtures to determine which component confers activity). Both of the latter are designed to increase the numbers of molecules that are blindly tested for activity and represent the precise opposite of truly rational drug design. However, the paradox is apparent rather than real. At present there are very few microbial target proteins for which the crystal structure is known. Without a reliable tertiary structure there is no possibility of designing inhibitors with computers, which makes large-scale experimental screening inevitable as the only means of taking rapid advantage of target discovery and validation. Even with rational molecular design, it is possible that no single entity can be accurately predicted on theoretical grounds as the optimum inhibitory structure, so that a means of rapid screening of large numbers of chemical analogues remains necessary.

Success stories of antimicrobial molecular design

Considering the theoretical power of the 'molecular' approach to antimicrobial drug discovery, it is perhaps disappointing that, to date, relatively few clear positive examples of results of the process have made their way into clinical development. Indeed, most publications concerning rational antimicrobial drug design are still only at the stage of proposing and validating targets. That inhibitors of microbial targets have been discovered by rational design and by screening compounds against purified target proteins is not in doubt. However, to be of real value, the approach has to lead to inhibitors that can be used in the *patient*. To quote other recent reviewers (Setti, Quattrocchio & Micetich, 1997) '…none of the antimicrobial agents presently in clinical use were developed using [rational design]…'.

Anti-HIV compounds

The notable exception is in the area of antiviral agents. The simplicity of virus genomes, with minimal open reading frames that encode very few proteins, makes the identification of target macromolecules much easier than with complex prokaryotic and eukaryotic organisms. The proteinase and reverse transcriptase enzymes of the human immunodeficiency virus (HIV) are compelling examples of molecular targets whose function is vital to virus replication. Considerable effort has been expended in designing chemical inhibitors based on theoretical interactions of small molecules with the tertiary structures of these proteins (Bold *et al.*, 1998; Das *et al.*, 1996; Ding, Hughes & Arnold, 1997; Wlodawer & Erickson, 1993; Wlodawer & Vondrasek, 1998). Compounds in clinical use or development such as indinavir (Dorsey, Vacca & Huff, 1997; Lacy & Abriola, 1996), nelfinavir (Kaldor *et al.*, 1998) and saquinavir (Noble & Faulds, 1996), among others, all began life on a rational design drawing board, albeit based on previously discovered molecules.

The origin of such compounds can be traced back to studies in which inhibition of HIV proteinases cloned and expressed in *Escherichia coli* was found to correlate with antiviral activity and blockage of formation of *gag* polyprotein products in the virus replication cycle (see, for example, Meek *et al.*, 1990; McQuade *et al.*, 1990; Roberts *et al.*, 1990).

For non-nucleoside reverse transcriptase inhibitors, such as TIBO compounds (Das *et al.*, 1996; Ding *et al.*, 1995), the initial discovery of the inhibitory molecules was a serendipitous result of random screening of chemical libraries against whole virus particles; but synthesis of the subsequent analogues has been influenced by theoretical chemistry considerations (e.g. Ding, Hughes & Arnold, 1997). The input of rational drug design is less evident for the nucleoside reverse transcriptase inhibitors didanosine (Franssen *et al.*, 1993) and stavudine (Lea & Faulds, 1996). Direct nucleoside analogues that inhibit reverse transcriptase are, of course, putative examples of 'rational chemical design', although design at this relatively simple level of chemistry is not quite the sophisticated approach that is usually implied by the term these days.

There is one philosophical point worthy of consideration about rational design of antiviral agents. The HIV proteinase inhibitors and non-nucleoside reverse transcriptase inhibitors both represent several, diverse chemical classes. If so many different molecular structures can bind with the same protein pockets to effect inhibition of the respective enzymes, then in theory there is no single optimal inhibitor that can be determined *de novo* by computerized molecular design.

For the reverse transcriptase inhibitors and the proteinase inhibitors, the biggest problem faced in the clinic is the rapid development of resistance that results from mutations in the HIV genome (see, for example, Hammer, Kessler & Saag, 1994; De Clercq, 1998a, b; Boudes & Geiger, 1997). Resistance development, most frequently encountered with the non-nucleoside compounds (Sereni, Lascoux & Gomez, 1996), is the reason why anti-HIV drugs must be given in combinations to be fully effective in the clinic. As the number and complexity of genotypes of HIV resistance to various inhibitors increase, the considerations of theoretical chemistry inevitably become extremely complex (e.g. Boyer *et al.*, 1994). The point may already have been reached where screening of novel compounds (and combinations of compounds) from known series of inhibitors against a battery of resistant virus strains in tissue cultures may be economically more effective than screening against isolated virus proteins (e.g. Saag, Hammer & Lange, 1995; St Clair *et al.*, 1996).

Anti-influenza compounds

The critical pathological significance of the surface neuraminidase enzyme in influenza viruses has long been recognized. The crystal structure of the enzyme was determined more than 15 years ago (Varghese, Laver & Colman, 1983) and (unsuccessful) screening for inhibitors of the enzyme dates back beyond 30 years ago (Edmond *et al.*, 1966). Research in the much more recent past has finally led to the rational design of inhibitory compounds that mimic the transition state of release of *N*-acetylneuraminic acid from its neighbouring sugar in an oligosaccharide chain (Colman, 1994; Wade, 1997). The earliest of these inhibitors, zanamivir (4-guanidino-Neu5Ac2en), is in clinical development at the time of writing (Waghorn & Goa, 1998).

The promise of antimicrobial molecular design for future success

Perhaps it is premature to expect the purely molecular approach to antimicrobial discovery already to have led to definable successes. Many of the molecular genetic technologies that underpin the rational unearthing of novel inhibitors by design and experimentation independent of whole-organism screening are of comparatively recent origin. There may already be new antimicrobial molecules about to emerge from the confidential phase of initial discovery into the full glare of scientific publicity. It is therefore worth surveying the potential for innovation in current biotechnologies to gain a better understanding of what the future may bring.

Genomics, proteomics and the identification of novel targets

The major novel approach to discovery of new molecular targets for antimicrobial drugs involves the newly named sciences of genomics and proteomics. ('Transcriptomics' and 'metabolomics' may also have become household names by the time this volume is published.) Genomics is the study of gene sequences and their annotation, and proteomics is the study and annotation of gene products by means of two-dimensional gel electrophoresis. Both areas of research make heavy use of computer databases for comparison of gene sequences and protein positions on electrophoresis gels. With the aid of appropriate software — something that is being refined very rapidly (see, for example, Spaltmann, Blunck & Ziegelbauer, 1999) — it should be possible to determine the associations between individual genes, their products and thus, ultimately, their function and biological significance in the whole-cell phenotype.

For antimicrobial discovery, the combination of genomics and proteomics will facilitate the discovery of molecular targets that were previously unknown but which play a critical role in the survival of the microbe *in vitro* or in an infected host. Against such novel targets, inhibitors can be discovered either by rational molecular design or serendipitously through experimental screening. Genomics as an approach to antimicrobial drug discovery has been reviewed very recently (Moir *et al.*, 1999) and the general role of proteomics in drug discovery has also been summarized by others (Humphery-Smith, 1998; Mullner, Neumann & Lottspeich, 1998).

Comparison of gene sequence databases allows for differentiation between those open reading frames (ORFs) that show high levels of similarity for deduced amino acid sequences between microbes and mammals and are, therefore, regarded as insufficiently specific to be considered further for antimicrobial discovery, and those that show similarities only between different microbial sequences. Several approaches are available to study and validate as potential drug targets those ORFs and gene products that emerge from computer searches as previously unknown, microbe-specific and potentially essential to the function of a pathogenic microbial cell. (It has been estimated that about 40% of bacterial genes at present have no known function; Moir *et al.*, 1999.) One method for rapidly narrowing down the function of an unknown gene product is to determine the nature of protein–protein interactions in a two-hybrid system (e.g. Cho *et al.*, 1998; Firestein & Feuerstein, 1998) with the unknown gene used as the bait. Another is to transform microbial cells with anti-sense mRNA that blocks the transcription of a gene of interest.

A third approach is to construct gene expression micro-arrays (see, for example, Ramsay, 1998, for review) in which appropriately labelled microbial RNA is hybridized to a tiny two-dimensional matrix of cloned genes from the same organism. Such an approach allows visualization and computer analysis of expression of genes: the approach has already been published for two bacteria (De Saizieu *et al.*, 1998) and differential fluorescent staining methods published for DNA–DNA array hybridization in *Saccharomyces cerevisiae* (Shalon, Smith & Brown, 1996) can be adapted to the monitoring of expression arrays as well. The basic technique can be extended to measure expression of genes in any chosen micro-organism, including strains in which the gene encoding a potential drug target has been selectively deleted. Changes in expression in an engineered mutant can be compared with expression in the parent strain to determine expression pathways associated with the potential target. The goal of all such approaches is ultimately to accelerate the rate of acquisition of knowledge of the physiological behaviour of a chosen molecular target: without such detailed knowledge, the potential for experimentation and rational inhibitor design with the novel target is small.

Screening of potential inhibitors against genetically modified micro-organisms

By overexpression or underexpression of a particular gene in an organism, a mutant is created that can be adapted for screening purposes because it will show either hypersensitivity to an inhibitor (when the gene is underexpressed) or resistance (by overexpression), relative to its parent (Moir *et al.*, 1999). There are possible pitfalls associated with this approach. Genes associated, directly or indirectly, with cell permeability functions may allow all kinds of non-specific inhibitors to show enhanced activity. Alterations in levels of expression of genes already selected as critical for cell function may lead to growth defects that render a mutant useless for screening purposes (see Moir *et al.*, 1999). Alteration of levels of expression of one gene may result in occult changes in expression of one or more other genes so that observed differential inhibitory effects between mutant and wild-type may not be related to the engineered gene at all. Despite such potential problems, it is likely that the approach will become the subject of many publications in the future.

Microbial virulence factors as antimicrobial targets

As research continues to elucidate the nature of microbial attributes that determine the virulence of an organism for a mammalian host, interference with those factors represents an approach to antimicrobial chemotherapy that might lead to the discovery of novel agents. Anti-virulence factor research is currently a busy field of publication. Perfect (1996) has elegantly reviewed the approach as applied to antifungal agents. His comments can be readily extended to other types of infectious microbe. The theoretical disadvantage of virulence factor inhibitors as pharmaceuticals is that in most patients, by the time they require antimicrobial chemotherapy, the virulence factors have mostly played their part and the infection is already established. However, this consideration does not rule out the prophylactic use of such agents, nor of therapeutic use in circumstances where a virulence factor is essential for maintenance of a microbe at an infected site.

Antimicrobial discovery without use of molecular targets

The power of the philosophy underpinning drug discovery via molecular targets stems from the application of detailed knowledge of the structure, physiology and expression of a drug target to the rational design of inhibitors. It is theoretically unlikely that such inhibitors can be more than preliminary discoveries, simply because the obstacle course of drug optimization to clinical development involves very many considerations beyond the molecular fit of a (small molecule) inhibitor into a suitable site on a target protein. The compound that exquisitely inhibits a microbial enzyme or gene target in experimental isolation must also reach that target in the intact cell. That means the inhibitor must penetrate to the right place (the subcellular location of the target) and at the right time (when the target is functioning in its active form during the microbial cell cycle). The inhibitor must also be bioavailable in a mammalian host. In practice, that means it must first reach its target microbe, after administration by the chosen route, in the tissues of an experimentally infected animal; then it must penetrate the microbial cell to find its molecular target at the right place and the right time.

Even when a new inhibitor succeeds in inhibiting microbial growth both *in vitro* and *in vivo*, it still needs to satisfy a plethora of further requirements, including a wide range of tests for toxicity, chemical stability (for production of a formulation that can be distributed and stored) and other considerations, before it can be developed through clinical trials for eventual use in the clinic. The inhibitor–target interaction is of little or no relevance to these considerations, so that the molecular approach can, at best, be regarded only as a preliminary starting point for the whole discovery process. Examples of the positive value of molecular target discovery have been discussed. To gain a balanced view of the role played by molecular target screening, it is perhaps illuminating also to consider examples where, despite knowledge of a molecular target, little or no recourse was made to that knowledge to find new inhibitors.

The quinolone class of antibacterial agents is now represented by more than a dozen compounds in clinical use or development. They are analogues of nalidixic acid, whose antibacterial properties were a chance discovery from screening of a distillate obtained in the course of production of the antimalarial drug chloroquine (Lesher *et al.*, 1962). Efforts to determine the bactericidal mode of action of the quinolone compounds facilitated the characterization of DNA topoisomerases — enzymes involved in DNA supercoiling and uncoiling processes. One of these enzymes (DNA gyrase, or more specifically its GyrA subunit) is the molecular target of the quinolones (Hooper & Wolfson, 1993). Models exist for the interaction between quinolone molecules and the binding of DNA to the gyrase protein (Shen, 1993) and there is extensive knowledge of the physiology and molecular genetics of this target. However, in an extensive review of investigations into quinolone structure–activity relationships (Mitscher, Devasthale & Zavod, 1993), only the breadth of the antibacterial spectrum of the compounds *in vitro* determined the 'activity' component of the studies. In 61 tables listing the minimum inhibitory concentrations of various quinolone analogues for three or more bacterial species, data for inhibition of DNA gyrase were shown only in 15 instances and only *en passant*. It is clear that the inhibitor–target interaction at the molecular level is rarely regarded as a property of significance in optimizing novel structural analogues of this particular antibacterial drug class.

Nevertheless, the knowledge that a clinically successful antimicrobial series is based on anti-topoisomerase activity has inevitably led to studies with various microbial topoisomerases as candidate molecular targets for new drugs. Novel antifungal agents are one example: '…the topoisomerase I from *C. albicans* is sufficiently distinct from the human enzyme as to allow differential chemical targeting and will therefore make a good target for antifungal discovery.' (Fostel,

Montgomery & Shen, 1992); '…there are structural differences between the human and fungal type I topoisomerases which can likely be exploited to allow for the development of antifungal agents which act against the fungal topoisomerase and which have minimal activity against the human enzyme.' (Fostel & Montgomery, 1995); '…these data suggest that there are sufficient structural differences between the topoisomerase I from *Candida albicans* and human cells to allow selective targeting of the fungal topoisomerase I over its human counterpart.' (Fostel, Montgomery & Lartey, 1996). Four years of such publications leave little doubt that fungal topoisomerase I is a validated molecular target for antifungals. However, up to the time of writing (June, 1999) there appear to be no topoisomerase I inhibitors in clinical development as antifungal agents.

The imidazole- and triazole-based antifungal agents, by contrast, represent the largest group of such inhibitors in current clinical use. Their history dates back to dedicated synthetic efforts with imidazole derivatives in the 1960s. Most agents of this class are applied locally for the treatment of fungal infections involving skin and mucous membranes. However, the most recent azole antifungal agents are systemically active when given orally. Although the mechanism of action of azole compounds—inhibition of the 14-demethylation of lanosterol in the ergosterol biosynthetic pathway—had been recognized by the late 1970s (Vanden Bossche *et al.*, 1978), the discoveries of ketoconazole (Heeres *et al.*, 1979) and, even more definitively, fluconazole (Richardson *et al.*, 1985) were based *a priori* on tests for the activity of azoles in animal models of fungal disease, not on tests *in vitro* nor with the 14α-demethylase molecular target. Some investigators did adopt a 'rational synthesis' approach to azole antifungals based on theoretical chemical considerations of the azole–demethylase interaction (Boyle, 1990), but rapid metabolism *in vivo* of the compounds that emerged from these studies led to a shift of emphasis to design molecules resistant to metabolism. The clinically successful antifungal azoles therefore exemplify an antimicrobial discovery process where both molecular design and enzyme screening tests have played, at best, only a minor role. The hurdles of oral bioavailability and pharmacodynamic stability *in vivo* were considerations of greater importance than the use of knowledge of the molecular target for the design of new analogues.

Potential pitfalls of molecular-target-based antimicrobial discovery

So far we have seen actual examples that illustrate the use and non-use of molecular targets in the discovery of novel antimicrobial agents. 'Molecular methods', like most approaches to scientific discovery, possess pitfalls for the unwary, as well as advantages. At each step in the process from genomics and proteomics to 'lead' inhibitors and to candidate molecules for clinical development, assumptions are made that are not always confirmed by real-life experimentation. Consideration of these pitfalls and false assumptions will complete our survey of the pros and cons of molecular antimicrobial discovery.

Molecular genetics, protein structures and proteomics may not successfully identify all new molecular targets

The exponential growth of databases containing complete DNA sequences of microbial genomes has possibly provided a false sense of confidence that statistical comparison of derived amino acid sequences can predict novel drug targets. To discover genes that encode molecular targets, statistical similarities are determined between two sequences in a database. Dissimilarity between mammalian and microbial sequences at and above a chosen cut-off level is an indication *a priori* of a gene specific to one or other of the two entities. However, consideration of known, specific microbial inhibitor targets shows that they often show modest to high levels of sequence similarity with their mammalian counterparts, in violation of the assumption underlying the statistical cut-offs. Human and *E. coli* dihydrofolate reductase enzymes share 28% derived amino acid similarity over the full length of the proteins (Schweitzer, Dicker & Bertino, 1990). Human and fungal cytochrome P450 sterol demethylases show about 40% homology in their derived amino acid sequences (Stromstedt, Rozman & Waterman, 1996). Human and *S. cerevisiae* protein elongation factor 2 proteins show 85% amino acid sequence homology (Justice *et al.*, 1998). Which, if any, of these targets would be chosen on the basis of a computerized genomics analysis depends on the chosen level of similarity cut-off. Of the three examples mentioned, elongation factor 2 would certainly appear to be a most unlikely candidate as a microbe-specific molecular target on a genomics basis. Yet in fact all three enzymes are proven targets for specific antimicrobial inhibition: dihydrofolate reductase for trimethoprim, sterol 14-demethylase for azole antifungals and—as recently reported (Dominguez & Martin, 1998; Justice *et al.*, 1998)—elongation factor 2 for the sordarin family of antifungal agents, first discovered serendipitously almost 30 years ago (Hauser & Sigg, 1971). Although approaches to computerized amino acid sequence comparisons are being constantly refined, the fundamental assumption that a target specific to micro-organisms will necessarily be encoded by a microbe-specific DNA sequence is not applicable in every instance.

Protein folding cannot be predicted from amino acid sequences

To determine the detailed tertiary structure of a protein drug target with confidence, the protein must be capable of crystallization for X-ray analysis. In the absence of crystallography data for a protein, a variety of modelling approaches based on energetic considerations can be used in an effort to approximate likely tertiary structures. However, the complexities of the polypeptide folding process are considerable and current modelling methods offer only about 70% accuracy in predictions of tertiary structure in the absence of crystallographic data (Petrella, Lazaridis & Karplus, 1998; Dokholyan *et al.*, 1999). (The existence of the specialist journal, *Folding & Design*, in which these two references appear is itself evidence of a field of investigation rather than a well-established method!) In living bacteria, chaperonin molecules are involved in establishing the correct polypeptide folding of gene products (see, for example, Weber *et al.*, 1999). This all means that an amino acid sequence derived from a genomic DNA sequence is of little or no use in establishing a dependable protein structure to be used as the basis for computer-assisted rational drug design. At present, once a novel target has been characterized through genomics, there is no known short cut for accurately determining its structure.

Protein structures do not necessarily predict protein functions

Knowledge of the role played by a putative drug target in the microbial cell is important, albeit not always *absolutely* essential, when assays for inhibitory effects on the target are set up. The annotation of gene sequences has led to a frequent assumption that the function determined experimentally for a particular protein in one microbial species is likely to be the function for a homologous protein found by sequencing in another species. Akers (1996) cites the TIM-barrel fold protein domain as an example of a highly precise and consistent, complex protein structure that turns up in molecules with 19 different biochemical functions. *Candida albicans* possesses homologous genes which encode either membrane transport proteins or elongation factors involved in protein synthesis (Sturtevant, Cihlar & Calderone, 1998). Without direct experimental evidence for the function of a gene product, it is possible to assume incorrectly that homogeneity of structure or sequence implies homogeneity of function.

Rational inhibitor design is like shooting at a moving target

Most antimicrobial agents are inhibitors of the catalytic function of enzymes, by either perturbation of function directly at the catalytic site, or allosteric inhibition of function at an extraneous site in the protein molecule. If a thorough knowledge of the crystal structure of a protein is available, the structure can be depicted by three-dimensional computer modelling and attempts made to design small molecules that bind to the active site or other pockets in the structure. However, a protein structure does not remain fixed when an inhibitor binds to it: the side-chains in structural pockets usually adapt both to the shape of the inhibitory molecule and to solvent molecules, a phenomenon that renders attempts to predict binding interactions on the basis of calculated binding energies extremely complex (e.g. Tantillo *et al.*, 1994; Smith *et al.*, 1998). When analogues of known inhibitors are available, approaches such as Monte Carlo simulations exist, which may be able to take account of the conformational changes that occur in protein–small molecule interactions (Smith *et al.*, 1998). However, the chance that structural simulations can be used for *de novo* design of drug molecules seems to be low at present. According to Koshland (1998): 'Perfecting the molecule requires such precision in very small changes that our theory and experiments are strained to make logical predictions that can be used to improve drug design. However, we can use an alternative approach and make many small changes rapidly and at random and then select the ones that work.' This sounds very close to an argument for serendipitous discovery by experimentation rather than rational drug design—although of course computers and their software continue to become exponentially more powerful!

Activities of inhibitors against isolated molecular targets do not always extrapolate to activities against intact microbial cells

It is not easy to find peer-reviewed literature references to failed experimentation. Informally, scientists who work with target-based discovery will acknowledge that inhibitors active against an isolated molecular target often fail to show that activity when tested as inhibitors of intact microbial cells. This limitation has already been alluded to above (the inhibitor must succeed in reaching its target at the correct location and the correct time when the target is exerting its function *in vivo*). The extent to which it is encountered as an obstacle in practice is unknown. At least one curious example of what almost amounts to the opposite phenomenon has been described. Wright & Brown (1999) attempted to discover new and more potent inhibitors in the 6-anilinouracil class of antibacterials by screening novel analogues against their molecular target, DNA polymerase III (pol III). 'Although N3-(hydroxybutylation) did not improve the inhibitory activity of the [new compounds] on the isolated pol III, several positive effects on their respective antimicrobial properties were observed.' In other words, if the molecular target project had not been set up, the enhanced activity found in the parallel intact-cell screen would not have been discovered! The point has been made elsewhere (Moir *et al.*, 1999) that testing for inhibitors of molecular targets in the absence of any parallel cellular assay for the effects of those inhibitors would limit the optimization and development of any potential drugs thus found.

Discussion

This chapter has tried to describe and exemplify the concepts, the known successes, the known failures and the theoretical assumptions and misassumptions associated with antimicrobial drug discovery by the molecular route. Approaches such as the specific inhibition of microbial drug resistance mechanisms, and the use of combinations of inhibitory drugs with each other or with host-cell defence molecules, have not been explored: at the molecular design and screening level these only amount to variations on the same basic theme. The inescapable reality is that a molecular target represents only one brush stroke in the entire picture of discovery of clinically useful antimicrobial agents. It is important (one may logically even claim 'essential'), but it is not a substitute for the other factors critical for a drug to be given to patients.

Had Fleming found in his early experiments that his penicillin was toxic to mice and rabbits (as are many natural products that inhibit microbial growth), it is unlikely that the discovery would have been followed up very far. As it was, the early development and exploitation of the original penicillin molecule were hampered by microbial resistance to the drug and by its bioavailability only by intramuscular injection. Neither of these limitations would have been solved by knowledge of the structure of the molecular target(s) for penicillin nor by testing analogues of benzyl penicillin against those targets *in vitro*. Our current profound knowledge of the targets of penicillin and cephalosporin antibiotics has contributed nothing to the discovery of new agents in these classes; indeed, it is often overlooked that almost all of the several clinically useful subclasses of the penicillin antibiotics were found by means of classical medicinal chemistry in the 1960s and 1970s (Rolinson, 1998).

Screening for selective inhibitors of the β-lactamase enzyme that leads to penicillin resistance, by contrast, is a legitimate example of a discovery at the molecular level (Bush, 1988). However, in its natural state the enzyme concerned is an externalized protein that therefore already resembles an isolated molecular target. It is hardly a representative of the problems associated with inhibiting what are usually *intracellular* molecular targets to be attacked by antimicrobial agents. In fact, molecular screening for inhibitors of β-lactamases was originally done throughout the 1940s and 1950s, but without successful discovery of any *clinically* useful agent (Rolinson, 1991), so it is hardly a glowing example of the accelerated benefits of molecular-level screening.

So far the success stories of clinically applicable antimicrobial discovery through 'molecular methods' are few. That objective assessment forms the basis of the conclusion stated at the outset that the molecular approach has not yet proved to be the optimal route to new antimicrobial agents. That there will be further future successes is not to be doubted. However, serendipitous discoveries of new agents by whole-cell screening are not yet dead, even with antiviral agents (Wentland *et al.*, 1997), and the real folly would be to assume that traditional approaches are no longer worth pursuing at all. It is the *diversity* of technologies available to tackle scientific problems that enriches the process of enquiry, not the exclusive focus on a single approach, however fashionable, at the expense of the rest. It is the thorough knowledge of the structure, function and biochemistry of targets and inhibitors that inspires medicinal chemistry. The challenge, apart from the possible use of antisense nucleotide sequences as drugs, is to find quick and reliable ways to turn sequence data into structural and functional biochemical information. Until that step has been bridged, neither genomics nor proteomics can contribute substantially to antimicrobial drug discovery.

Screening for inhibitors of intact organisms has the advantage that the entire library of possible molecular targets is being tested in their natural setting. The theoretical disadvantage of the approach is that it can miss weak or non-permeating inhibitors that might otherwise prove to be the leads for novel agents. It has also been used so extensively that its inherent capacity for finding new agents has been extremely thoroughly explored and may have reached the point of exhaustion. Using genomics and proteomics opens a more systematic avenue to discover as yet unknown antimicrobial targets. However, this approach is based on assumptions that are not always strictly valid and it has no impact on the numerous difficulties that lie downstream in drug development (including the hurdle of activity against intact organisms).

This point cannot be made too strongly. Finding inhibitors of microbial growth (or of virulence factors, or of any cell attribute) is not the most difficult obstacle. It is finding an inhibitor that is bioavailable, free of toxicity, readily formulated and so on that represents the bulk of the obstacle in drug discovery. Societal pressures for ever-improved safety and efficacy in the drugs used in humans and animals raises the threshold of those obstacles far more than it complicates the discovery of molecules active directly at the microbial level. The high expense of research in molecular genetics is also a factor of considerable relevance to a pharmaceutical industry that is under ever-growing pressures to reduce the prices of its products, as well as to make 'better' ones. In the new millenium, we can surely expect to see a rationalization of the ways in which antimicrobials are discovered. It is now too early to do this with 'molecular methods', but the pressures are sure to come soon.

When the costs as well as the philosophies of research discoveries are taken into account, the balance sheet needs to compare relative investments in chemistry (creation of novel molecules whose possible pharmacological effects can be tested with screening systems) and molecular biology (discovery of previously unknown microbial targets that can be developed into screening systems). Unless and until there is a real possibility of predicting *de novo* which molecular shapes are likely to be successful inhibitors of a molecular target, the creation of chemical diversity is at least as important as the creation of new screening systems.

It is perhaps easier to be optimistic about technologies that are not yet fully tried and proven, which is why reviews of molecular approaches to antimicrobial discovery often seem to err on the side of hyperbole, with the occasional strongly negative exception (Akers, 1996). But the concept of screening against isolated microbial molecular targets is actually not at all new—unsuccessful examples can be found more than 30 years ago (*vide infra*)—it is the technology that has improved. Still, the burgeoning literature from the 1990s far more often describes the future potential of new targets and novel tricks possible with molecular genetic technology than it ever reports clinically useful discoveries.

One very balanced account of the present topic is the review by Gutteridge (1997). By coincidence this also happens to be the only article related to anti-parasitic agents cited in this chapter. Much of Gutteridge's abstract essentially matches the present author's views, and states its points very concisely. 'Absence of a successful track record is inevitable for any newly emerging technology. It is too early to draw conclusions about the relative costs of rational design versus empirical synthesis, since the former is only now beginning to become reality and the latter is in the middle of a (combinatorial) revolution. Similarly, it is too soon to predict with certainty which of these two approaches will prevail in the long run.'

Scientific excitement and medical needs are not necessarily the same thing. The cutting edge in research physics has to do with combining research in cosmology, subatomic particles and quantum theory. Yet the fundamental technology needed to transport people by air or into space dates back many decades and still finds a lot of application, unrefined by either superstrings, quarks or theories of everything. The cutting edge in biomedical research concerns unravelling mechanisms of gene regulation and expression. The fundamental requirements for eradicating microbial invaders from human tissues are generally less profound and have as much to do with pharmacological considerations in the patient as with effects of small molecules on the submicrobial targets themselves. The more we know of agents and targets, the better. The more ways we have available to obtain such knowledge, the better. But the crucial goal of finding agents that are effective and safe in the tissues of *Homo sapiens* is best served by pragmatic use of all available resources, not by assuming that the research cutting edge is inevitably the best place to search.

Postscript

By coincidence, as I took a break after writing what I thought would be the last sentence of this chapter, I browsed through the May, 1999, issue of *ASM News*. This special issue celebrates the centenary of the American Society for Microbiology. A highlighted quote in an article by Venter, Smith & Fraser (1999) reads: 'The great challenges for the future are to define the functions of the proteins encoded within the genome and to decipher the functional networks within these genomes'. This confirms my assessment of the biomedical 'cutting edge' above: but stated this way it sounds to me a long way from giving a life-saving drug, whether pill or potion, to a desperately infected patient. To come up with new antimicrobial agents we need, of course, to utilize the spin-off from fundamental research. However, when we justify doing that research not because of our intrinsic curiosity and excitement but by repeatedly claiming it will lead to medical advances, we are adding a social conclusion to the scientific hypothesis being explored, even before we have started to do the experiments! Society nowadays expects us to justify expenditure on research projects, but predicting applications that are realistically a long way away from the project is not in the best tradition of scientific discovery. The fullness of time will ultimately reveal how much use we can truly make of 'molecular' methods to discover antimicrobial drugs. In the meantime, we need all the approaches we can find, whether old and unfashionable or new and trendy.

Acknowledgements

I am very grateful to Koen Andries, Alison Bird, Walter Luyten, Patrick Marichal and Hugo Vanden Bossche for their kind help with this chapter. Their time, generously given, in discussions and comments on the manuscript has hugely facilitated my writing of this chapter. The opinions expressed, however, are entirely the author's responsibility.

References

Akers, A. (1996). Molecular biology as virtual biology: limitations of molecular biology in pesticide discovery. *Pesticide Science* **46**, 85–91.

Bold, G., Fassler, A., Capraro, H. G. & 18 other authors (1998). New aza-dipeptide analogues as potent and orally absorbed HIV-1 protease inhibitors: candidates for clinical development. *Journal of Medicinal Chemistry* **41**, 3387–3401.

Boudes, P. & Geiger, J. M. (1997). Les inhibiteurs de la protease du VIH: revue generale. *Therapie* **51**, 319–325.

Boyer, P. L., Ding, J., Arnold, E. & Hughes, S. H. (1994). Subunit specificity of mutations that confer resistance to nonnucleoside inhibitors in human immunodeficiency virus type 1 reverse transcriptase. *Antimicrobial Agents and Chemotherapy* **38**, 1909–1914.

Boyle, F. T. (1990). Drug discovery: a chemist's approach. In *Chemotherapy of Fungus Diseases*, pp. 3–30. Edited by J. F. Ryley. New York: Springer.

Bush, K. (1988). Beta-lactamase inhibitors from laboratory to clinic. *Clinical Microbiology Reviews* **1**, 109–123.

Cho, R. J., Fromont-Racine, M., Wodicka, L., Feierbach, B., Strearns, T., Legrain, P., Lockhart, D. J. & Davis, R. W. (1998). Parallel analysis of genetic selections using whole genome oligonucleotide arrays. *Proceedings of the National Academy of Sciences, USA* **95**, 3752–3757.

Colman, P. M. (1994). Influenza virus neuraminidase: structure, antibodies, and inhibitors. *Protein Science* **3**, 1687–1696.

Das, K., Ding, J., Hsiou, Y. & 13 other authors (1996). Crystal structures of 8-Cl and 9-Cl TIBO complexed with wild-type HIV-1 RT and 8-Cl TIBO complexed with the Tyr181Cys HIV-1 RT drug-resistant mutant. *Journal of Molecular Biology* **264**, 1085–1100.

De Clercq, E. (1998a). New options, new challenges. *International Antiviral News* **6**, 177–178.

De Clercq, E. (1998b). The role of non-nucleoside reverse transcriptase inhibitors (NNRTIs) in the therapy of HIV-1 infection. *Antiviral Research* **38**, 153–179.

De Saizieu, A., Certa, U., Warrington, J., Gray, C., Keck, W. & Mous, J. (1998). Bacterial transcript imaging by hybridization of total RNA to oligonucleotide arrays. *Nature Biotechnology* **16**, 45–48.

Ding, J., Das, K., Moereels, H., Koymans, L., Andries, K., Janssen, P. A., Hughes, S. H. & Arnold, E. (1995). Structure of HIV-1 RT/TIBO R 86183 complex reveals similarity in the binding of diverse nonnucleoside inhibitors. *Nature Structural Biology* **2**, 407–415.

Ding, J., Hughes, S. H. & Arnold, E. (1997). Protein–nucleic acid interactions and DNA conformation in a complex of human immunodeficiency virus type 1 reverse transcriptase with a double-stranded DNA template-primer. *Biopolymers* **44**, 125–138.

Dokholyan, N. V., Buldryev, S. V., Stanley, H. E. & Shakhnovich, E. I. (1999). Discrete molecular dynamics studies of the folding of a protein-like model. *Folding & Design* **3**, 577–587.

Dominguez, J. M. & Martin, J. J. (1998). Identification of elongation factor 2 as the essential protein targeted by sordarins in *Candida albicans*. *Antimicrobial Agents and Chemotherapy* **42**, 2279–2283.

Dorsey, B. D., Vacca, J. P. & Huff, J. R. (1997). The invention and development of Crixivan®: an HIV protease inhibitor. In *Anti-infectives: Recent Advances in Chemistry and Structure–Activity Relationships*, pp. 238–250. Edited by P. H. Bentley & P. J. O'Hanion. Cambridge: Royal Society of Chemistry.

Edmond, J. D., Johnston, K. G., Kidd, D., Rylance, H. J. & Sommerville, R. G. (1966). The inhibition of neuraminidase and antiviral action. *British Journal of Pharmacological Chemotherapy* **27**, 415–426.

Firestein, R. & Feuerstein, N. (1998). Association of activating transcription factor 2 (ATF2) with the ubiquitin-conjugating enzyme hUBC9. Implication of the ubiquitin/proteasome pathway in regulation of ATF2 in T cells. *Journal of Biological Chemistry* **273**, 5892–5902.

Fostel, J. & Montgomery, D. (1995). Identification of the aminocatechol A-3253 as an *in vitro* poison of DNA topoisomerase I from *Candida albicans*. *Antimicrobial Agents and Chemotherapy* **39**, 586–592.

Fostel, J., Montgomery, D. A. & Shen, L. L.

(1992). Characterization of DNA topoisomerase I from *Candida albicans* as a target for drug discovery. *Antimicrobial Agents and Chemotherapy* **36**, 2131–2138.

Fostel, J., Montgomery, D. & Lartey, P. (1996). Comparison of responses of DNA topoisomerase I from *Candida albicans* and human cells to four new agents which stimulate topoisomerase-dependent DNA nicking. *FEMS Microbiology Letters* **138**, 105–111.

Franssen, R. M., Meenhorst, P. L., Koks, C. H. & Beijnen, J. H. (1993). Didanosine, a new retroviral drug. A review. *Pharmaceutische Weekblad* **14**, 297–304.

Gutteridge, W. E. (1997). Designer drugs: pipe-dreams or realities? *Parasitology* **114**, S145–S151.

Hammer, S. M., Kessler, H. A. & Saag, M. S. (1994). Issues in combination antiretroviral therapy: a review. *Journal of Acquired Immune Deficiency Syndromes* **7** (suppl. 2), S24–S35.

Hauser, D. & Sigg, H. P. (1971). Isolierung und Abbau von Sordarin. *Helvetica Chimica Acta* **54**, 1178–1190.

Heeres, J., Backx, L. J. J., Mostmans, J. H. & Van Cutsem, J. (1979). The synthesis and antifungal activity of ketoconazole, a new potent orally active broad spectrum antifungal agent. *Journal of Medicinal Chemistry* **22**, 1003–1005.

Hooper, D. C. & Wolfson, J. S. (1993). Mechanisms of quinolone action and bacterial killing. In *Quinolone Antimicrobial Agents*, 2nd edn, pp. 53–75. Edited by D. C. Hooper & J. S. Wolson. Washington, DC: American Society for Microbiology.

Humphery-Smith, I. (1998). Proteomics: from small genes to high-throughput robotics. *Journal of Protein Chemistry* **17**, 524–525.

Justice, M. C., Hsu, M. J., Tse, B., Ku, T., Balkovec, J., Schmatz, D. & Nielsen, J. (1998). Elongation factor 2 as a novel target for selective inhibition of fungal protein synthesis. *Journal of Biological Chemistry* **273**, 3148–3151.

Kaldor, S. W., Kalish, V. J., Davies, J. F., 2nd & 17 other authors (1998). Viracept (nelfinavir mesylate, AG1343): a potent, orally bioavailable inhibitor of HIV-1 protease. *Journal of Medicinal Chemistry* **40**, 3979–3985.

Koshland, D. E., Jr (1998). Conformational changes: how small is big enough? *Nature Medicine* 4, 1112–1114.

Lacy, M. K. & Abriola, K. P. (1996). Indinavir: a pharmacologic and clinical review of a new HIV protease inhibitor. *Connecticut Medicine* 60, 723–727.

Lea, A. P. & Faulds, D. (1996). Stavudine: a review of its pharmacodynamic and pharmacokinetic properties and clinical potential in HIV infection. *Drugs* 51, 846–864.

Lesher, G. Y., Forelich, E. D., Gruet, M. D., Bailey, J. H. & Brundage, R. P. (1962). 1,8-Naphthyridine derivatives. A new class of chemotherapeutic agents. *Journal of Medicinal and Pharmaceutical Chemistry* 5, 1063–1068.

McQuade, T. J., Tomasselli, A. G., Liu, L., Karacostas, V., Moss, B., Sawyerm, T. K., Heinrikson, R. L. & Tarpley, W. G. (1990). A synthetic HIV-1 protease inhibitor with antiviral activity arrests HIV-like particle maturation. *Science* 247, 454–456.

Meek, T. D., Lambert, D. M., Dreyer, G. B. & 9 other authors (1990). Inhibition of HIV-1 protease in infected T-lymphocytes by synthetic peptide analogues. *Nature* 343, 90–92.

Mitscher, L. A., Devasthale, P. & Zavod, R. (1993). Structure–activity relationships. In *Quinolone Antimicrobial Agents*, 2nd edn, pp. 3–51. Edited by D. C. Hooper & J. S. Wolson. Washington, DC: American Society for Microbiology.

Moir, D. T., Shaw, K. J., Hare, R. S. & Vovis, G. F. (1999). Genomics and antimicrobial drug discovery. *Antimicrobial Agents and Chemotherapy* 43, 439–446.

Mullner, S., Neumann, T. & Lottspeich, F. (1998). Proteomics—a new way for drug target discovery. *Arzneimittel-Forschung* 48, 93–95.

Noble, S. & Faulds, D. (1996). Saquinavir. A review of its pharmacology and clinical potential in the management of HIV infection. *Drugs* 52, 93–112.

Perfect, J. R. (1996). Fungal virulence genes as targets for antifungal chemotherapy. *Antimicrobial Agents and Chemotherapy* 40, 1577–1583.

Petrella, R. J., Lazaridis, T. & Karplus, M. (1998). Protein sidechain conformer prediction: a test of the energy function. *Folding & Design* 3, 353–377.

Ramsay, G. (1998). DNA chips: state of the art. *Nature Biotechnology* 16, 40–44.

Richardson, K., Brammer, K. W., Marriott, M. S. & Troke, P. F. (1985). Activity of UK-49,858, a bistriazole derivative, against experimental infections with *Candida albicans* and *Trichophyton mentagrophytes*. *Antimicrobial Agents and Chemotherapy* 27, 832–835.

Roberts, N. A., Martin, J. A., Kinchington, D. & 16 other authors (1990). Rational design of peptide-based HIV proteinase inhibitors. *Science* 248, 358–361.

Rolinson, G. N. (1991). Evolution of β-lactamase inhibitors. *Reviews of Infectious Diseases* 13 (suppl. 9), S727–S732.

Rolinson, G. N. (1998). Forty years of β-lactam research. *Journal of Antimicrobial Chemotherapy* 41, 589–603.

Saag, M. S., Hammer, S. M. & Lange, J. M. (1995). Pathogenicity and diversity of HIV and implications for clinical management: a review. *Journal of Acquired Immune Deficiency Syndromes* 7 (suppl. 2), S2–S10.

Schweitzer, B. I., Dicker, A. P. & Bertino, J. R. (1990). Dihydrofolate reductase as a therapeutic target. *FASEB Journal* 4, 2441–2452.

Sereni, D., Lascoux, C. & Gomez, V. (1996). Les antiretroviraux non nucleosidiques. *Presse Médicale* 25, 361–369.

Setti, E. L., Quattrocchio, L. & Micetich, R. G. (1997). Current approaches to overcome bacterial resistance. *Drugs of the Future* 22, 271–284.

Shalon, D., Smith, S. J. & Brown, P. O. (1996). A DNA microarray system for analysing complex DNA samples using two-colour fluorescent probe hybridization. *Genome Research* 6, 639–645.

Shen, L. L. (1993). Quinolone–DNA interaction. In *Quinolone Antimicrobial Agents*, 2nd edn, pp. 77–95. Edited by D. C. Hooper & J. S. Wolson. Washington, DC: American Society for Microbiology.

Smith, R. H., Jr, Jorgensen, W. L., Tirado-Rives, J., Lamb, M. L., Janssen, P. A., Michejda, C. J. & Kroeger Smith, M. B. (1998). Prediction of binding affinities for TIBO inhibitors of HIV-1 reverse transcriptase using Monte Carlo simulations in a linear response model. *Journal of Medicinal Chemistry* **41**, 5272–5286.

Spaltmann, F., Blunck, M. & Ziegelbauer, K. (1999). Computer-aided target selection — prioritizing targets for antifungal drug discovery. *Drug Discovery Today* **4**, 17–26.

St Clair, M. H., Millard, J., Rooney, J., Tisdale, M., Parry, N., Sadler, B. M., Blum, M. R. & Painter, G. (1996). *In vitro* antiviral activity of 141W94 (VX-478) in combination with other antiretroviral agents. *Antiviral Research* **29**, 53–56.

Stromstedt, M., Rozman, D. & Waterman, M. R. (1996). The ubiquitously expressed human CYP51 encodes lanosterol 14 alpha-demethylase, a cytochrome P450 whose expression is regulated by oxysterols. *Archives of Biochemistry and Biophysics* **329**, 73–81.

Sturtevant, J., Cihlar, R. & Calderone, R. (1998). Disruption studies of a *Candida albicans* gene, *ELF1*: a member of the ATP-binding cassette family. *Microbiology* **144**, 2311–2321.

Tantillo, C., Ding, J., Jacobo-Molina, A. & 7 other authors (1994). Locations of anti-AIDS drug binding sites and resistance mutations in the three-dimensional structure of HIV-1 reverse transcriptase. Implications for mechanisms of drug inhibition and resistance. *Journal of Molecular Biology* **243**, 369–387.

Vanden Bossche, H., Willemsens, G., Cools, W., Lauwers, W. F. J. & Lejeune, L. (1978). Biochemical effects of miconazole on fungi. II. Inhibition of ergosterol biosynthesis in *Candida albicans*. *Chemical and Biological Interactions* **21**, 59–78.

Varghese, J. N., Laver, W. G. & Colman, P. M. (1983). Structure of the influenza virus glycoprotein antigen neuraminidase at 2.9 Å resolution. *Nature* **303**, 35–40.

Venter, J. C., Smith, H. O. & Fraser, C. M. (1999). Microbial genomics: in the beginning. *ASM News* **65**, 322–327.

Wade, R. C. (1997). 'Flu' and structure-based drug design. *Structure* **5**, 1139–1145.

Waghorn, S. L. & Goa, K. L. (1998). Zanamivir. *Drugs* **55**, 721–725.

Weber, F., Keppel, F., Georgopoulos, C., Hayer-Hartl, M. K. & Hartl, F. U. (1999). The oligomeric structure of GroEL/GroES is required for biologically significant chaperonin function in protein folding. *Nature Structural Biology* **5**, 977–985.

Wentland, M. P., Perni, R. B., Dorff, P. H. & 15 other authors (1997). Antiviral properties of 3-quinolinecarboxamides: a series of novel non-nucleoside antiherpetic agents. *Drug Design and Discovery* **15**, 25–38.

Wlodawer, A. & Erickson, J. W. (1993). Structure-based inhibitors of HIV-1 protease. *Annual Review of Biochemistry* **62**, 543–585.

Wlodawer, A. & Vondrasek, J. (1998). Inhibitors of HIV-1 protease: a major success of structure-assisted drug design. *Annual Review of Biophysics and Biomolecular Structure* **27**, 249–284.

Wright, G. E. & Brown, N. C. (1999). DNA polymerase III: a new target for antibiotic development. *Current Opinion in Anti-Infective Investigational Drugs* **1**, 45–48.

Index

Note: page numbers in *italics* refer to figures, those in **bold** refer to tables